The Homeowner's Handbook of Plumbing And Repair

The Homeowner's Handbook Of Plumbing And Repair

Kendall Webster Sessions

JOHN WILEY & SONS

New York   Santa Barbara   Chichester   Brisbane   Toronto

Library of Congress Cataloging in Publication Data:

Sessions, Ken W
  The homeowner's handbook of plumbing and repair.

  Includes index.
  1. Plumbing—Amateurs' manuals. I. Mueller, E.
Frank, joint author. II. Title.
TH6124.S47   696'.1   77-21333
ISBN 0-471-02550-X

Printed in the United States of America

10 9 8 7 6 5 4 3 2 1

Introduction

Why don't people do more of their own plumbing repairs or installations? There's little in the way of a plumbing installation or repair that is especially difficult to do. You need some special tools, of course, but not many. Actually, most of the tools you need for plumbing are the kinds you probably already have in your home workshop. The reason more people don't do their own plumbing jobs is because they lack the know-how—and they haven't been able to get it from books. This book is an attempt to change all that.

Plumbing codes—ordinances affecting pipework and fixture installation—do vary from one locality to the next. No single book can have all the right answers for every problem in every community, simply because communities don't all agree on what is sanitary or what is acceptable in the way of materials or practices. But this isn't as big a hurdle as one might think. The U.S. Government publishes recommendations and standards for plumbing; these have served as guidelines in the preparation of this book. If you're in doubt as to whether or not, for example, it's all right to install plastic pipe in a plumbing system where galvanized pipe has been used in the past, all you have to do is call your appropriate local government representative. Local codes will tell you what you can and can't do; this book tells you how to do whatever your community allows.

The approach used in preparing this book is logical: tool requirements, applicable to all sections, are covered in the Appendix, but you can skip this reference if you already have the tools and the know-how to use them. Maybe you'll never have to melt lead to repair a cast-iron soil pipeline, but the techniques are covered in case you do.

Section One, on water-distribution piping, includes piping sizes, materials, and their characteristics—valuable information to have when you face a choice of materials for a given situation. This section also covers water-line installation and repairs, and tells how to install pipeline branches—indeed, how to solve just about any problem that might occur in a fresh-water system, whether a frozen or broken pipe or insufficient water pressure.

Section Two shows you how to install and repair all types of faucets, shower heads, and valves found in home plumbing systems; and how to install piping insulation, with a discussion of the best kinds to use for the various inside and outside sections of your home system.

Then Section Three tells how to install fixtures and repair them. We have anticipated problems of every description from actual case histories involving plumber calls. This section is divided into categories to make it easy for you to find the kind of help you need when you need it most: kitchen, laundry, bath; and since most leaks around fixtures cause other problems when they go unrepaired for any significant length of time, we include all the information you need to repair adjacent wall cracks and how to replace tiles. This section ends with instruction on water heaters, including those for swimming pools.

Section Four is devoted to drains and vents and the problems you are most likely to encounter in them. Included are the troubleshooting tips professional plumbers rely on.

In Section Six we have borrowed a great deal of material from the U.S. Public Health Service's *Manual of Septic*

Tank Practice, which was developed in cooperation with the Joint Committee on Rural Sanitation. Here are the guidelines you can follow to plan or keep up a sewage-disposal system or to determine the suitability of your soil for sewage absorption.

By reading Section Seven, you can learn to assess the quality of your water and determine whether or not it's to your advantage to employ one of the many available methods of conditioning it. As this section shows, you can drink water from just about any source so long as you process the water properly before you do. If your home is already outfitted with water-conditioning equipment, use the material in this section to learn how it works and how to keep it working right all the time, whether it's a simple chlorinator or iodinator for a private water source, or a fully automated softener.

Section Eight is a summary of ways you can test the plumbing of an older home before you buy; included are pointed questions to ask and specific techniques you can adopt to answer them for yourself—before you move in and learn, too late, that new pipelines have to be installed or repairs initiated that might make the house purchase much less of a bargain than you thought.

In all, we've tried to make this book as comprehensive as possible so that you can be truly independent when it comes to sizing up any kind of plumbing task, large or small. For assistance and valuable contributions, we gratefully acknowledge the willing cooperation of companies like Ruud Manufacturing Company, Harvel Plastics, Inc., Owens-Corning Fiberglas, American Standard, as well as several agencies of the U.S. Government. But we'd like to express special thanks to the F. E. Myers Company of Ashland, Ohio, who provided a wealth of otherwise unavailable material on pumps and water systems.

Kendall Webster Sessions

TABLE OF CONTENTS

Section Seven   Water Conditioning **304**

The Homeowner's Handbook of Plumbing And Repair

Section I

Almost any time you add a fixture to an existing plumbing system you'll have to install new piping. But adding new piping is more complicated than simply inserting a tee in a line and affixing the new pipework. The new pipeline must be properly sized to be sure of getting the right amount of water flow at the new fixture; and the line has to be supported at regular intervals so that it won't develop leaks or be subjected to undue stresses. Hot water pipes have to be insulated so that you won't be paying to heat the air around the pipes rather than the water that flows through them. If the line is outdoors, it must be buried deep enough to prevent freezing in cold weather and carefully enough to avoid damage to the line.

Water-Distribution Piping

Section 1

Selecting and Connecting Piping

Piping Sizes

There are three important dimensions of pipes and tubing: outside diameter, inside diameter, and wall thickness. Generally (it isn't really a rule), a pipe is called *tubing* if its size is identified by actual outside diameter and wall thickness. *Pipe* is usually identified by a "nominal" dimension called *IPS,* for "iron pipe size," and by a wall thickness *schedule* designation. (More on this in the discussion of plastic piping.)

A nominal—this word means *in name only,* not average—dimension such as the standard IPS is close to the actual measured dimension, but it isn't exactly the same. For example, a pipe with a nominal size of 3 inches will have an actual outside diameter of 3.5 inches. And a pipe with a nominal size of 2 inches has an actual outside diameter of 2.375 inches. These nominal dimensions are specified to simplify the standardization of fittings, pipes, and the taps and dies used to thread them.

The wall thickness of pipe is identified by reference to established wall thickness schedules. A reference to 3-inch schedule 40 pipe indicates that the nominal inside diameter is 3 inches but the wall thickness is 0.216 inch. The same pipe in schedule 80 has a wall thickness of 0.300 inch. It is important to understand that a specified schedule does not indicate a set wall thickness unless a nominal diameter is given as well. A schedule 40 pipe in $\frac{3}{4}$-inch diameter has a different wall thickness than a schedule 40 pipe in a 1-inch diameter.

Piping Materials

As the years go by, more and more materials find applications in water supply systems. If you're installing a new plumbing network or portion of a network, you might have to choose between only a couple of candidates, depending on materials currently employed most frequently in your locale; but if you're going to be working on a system that someone else has installed, you're apt to find pipes of cast iron, galvanized wrought iron or steel, copper, or plastics of various types. Each has its own peculiar requirements for installation, cutting, threading, and joining. And should you have to add piping of one material to a water supply system that was built using another material, you should know what is required to make connections that will be secure and trouble-free.

Plastic Piping

The principal types of plastic piping in current use are chlorinated polyvinyl chloride or CPVC, ordinary PVC, and polyethylene. Polyethylene piping is flexible (somewhat similar to a garden hose, but thicker walled generally and of varying diameter) and may be purchased in coils of 100 to 1000 feet. This material, however, is not capable of withstanding great pressures and it cannot be solvent-welded. PVC and CPVC are similar; both are corrosion resistant, solvent-weldable, and comparatively rigid. The chief difference between the two types is the higher temperature rating of CPVC—it's good for temperatures as high as 180°F.

In this section, our discussion centers on the vinyls. This is because connecting polyethylene piping is as simple as connecting the radiator hose in your car: joints are made by slipping the pipe over a male fitting and fastening it with a stainless steel clamp. (The fittings are usually made of nylon, PVC, brass, styrene, or steel and have bulging rings to insure a good fit.)

Cutting and Joining

You can cut plastic pipe and tubing with a hacksaw, saber saw, or power saw; the most important consideration is the number of teeth per inch. You'll get best results with powered saws fitted with fine-toothed saw blades of 12 or more teeth per inch and little or no *set*. (A set of 0.025 inch or less is ideal.) Make all cuts as square and smooth as you can—this is even more important if you plan to thread the pipe rather than solvent-weld it. If you make the cuts by hand, use a miter box to insure a square cut; then use a knife or similar tool to deburr the cuts and bevel the edges with a file.

While you are cutting the pipe, support it well and protect it from nicks and scratches. You can do this easily by wrapping the pipe at strategic spots with a coarse cloth.

Even though the outside diameters of vinyl pipes follow the standard established many years ago by the metal pipe industry, vinyl piping comes in a variety of wall thicknesses—which means that the inside diameter of the pipe you use will depend on the grade or *schedule* of the material. Those of most interest in home water supply systems are schedule 40, schedule 80, and schedule 120. For a 1-inch diameter pipe, the wall thicknesses are as follows:

- 0.133 inch for schedule 40
- 0.180 inch for schedule 80
- 0.200 inch for schedule 120

Threading. Since the diameters of plastic pipes match those of their rigid metal counterparts, standard taps and dies intended for metal pipes can be used; however, threading should be avoided with schedule 40; the thread grooves would be a bit too deep for the relatively thin walls.

For plastic pipe, power threading machines should be fitted with dies having a 5-degree front rake that is ground especially for vinyl cutting. For hand stocks the dies should have a front rake of between 5 and 10 degrees. Harvel Plastics, Inc., a manufacturer of plastic pipes, states that dies designed for brass or copper pipes may be used without special considerations as long as the cutting threads are immaculate. Die chasers should have a 33-degree chamfer on the lead and a 10-degree front rake (with a 5-degree back rake).

There's definitely an art to threading, and if you're inexperienced with plastic pipes it will pay you to exercise a little extra care. Take it slow and easy, and don't exert heavy pressure on the die.

Bear in mind all the time that you're working with plastic, not metal. Unless you plug the end of the pipe with a firm tapered ''cork,'' for example, you're very apt to distort the roundness of the pipe and cause the die to dig into the pipe well.

Hold the pipe in a good vise, and take pains to avoid marking from the jaws. Canvas, emery paper, shim stock and sleeving—these are all effective means of protection. You don't need to use a lubricant because one of the important properties of plastic is its self-lubricating quality. If you feel it necessary to use a lubricant to avoid frictional heating, use water, a soluble oil, or even a vegetable oil from the kitchen.

Two more important points: (1) clean the cuttings from taps and dies immediately after making threads—this will prevent the possibility of transferring gouges into the next pipe section you thread; (2) don't overthread the pipe—if you *do,* you may not be able to obtain a tight, leakproof joint because fittings cannot be run on far enough to make a good seal. Table 1–1 shows the thread dimensions recommended for various pipe diameters.

One of the nice things about plastic pipe is that you don't need anything other than the pipe and the tools to work it with. You don't need a compound, for example, to get a good tight fit that will last. It doesn't hurt to use a standard thread sealing compound, though—and it might give an added measure of security when you thread too far, when you suspect a dull tap or die, or if you anticipate an occasional disassembly. If you do use a sealing compound, use one based on a tetrafluoroethylene resin—Teflon, for example, or Fluoroseal.

Never use a Stillson wrench on plastic pipe. You can get by nicely with hand-tightening most of the time. If you doubt the security of a joint you have hand-tightened, then use a strap wrench. But *always* start a screwed fitting carefully, and snug it up by hand before using the strap wrench. And

Angle between sides of thread is 60°. Taper of thread, on diameter, is $\frac{3}{4}$ inch per foot. The basic thread depth is 0.8 × pitch of thread and the crest and root are truncated an amount equal to 0.033 × pitch, excepting 8 threads per inch which have a basic depth of 0.788 × pitch at the crest and 0.033 × pitch at the root.

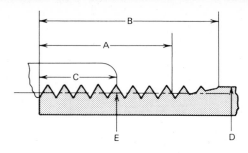

TABLE 1–1. RECOMMENDED THREAD DIMENSIONS FOR VARIOUS PIPE SIZES. (Courtesy Harvel Plastics, Inc.)

PIPE		THREADS					
NOMINAL SIZE	OUTSIDE DIAMETER (D)	NUMBER OF THREADS PER INCH	NORMAL ENGAGEMENT BY HAND (C)	LENGTH OF EFFECTIVE THREAD (A)	TOTAL LENGTH: END OF PIPE TO VANISH POINT (B)	PITCH DIAMETER AT END OF INTERNAL THREAD (E)	DEPTH OF THREAD (MAX)
INCHES	INCHES		INCHES	INCHES	INCHES	INCHES	INCHES
$\frac{1}{8}$	0.045	27	0.180	0.2639	0.3924	0.37476	0.02963
$\frac{1}{4}$	0.540	18	0.200	0.4018	0.5946	0.48989	0.04444
$\frac{3}{8}$	0.675	18	0.240	0.4078	0.6006	0.62701	0.04444
$\frac{1}{2}$	0.840	14	0.320	0.5337	0.7815	0.77843	0.05714
$\frac{3}{4}$	1.050	14	0.339	0.5457	0.7935	0.98887	0.05714
1	1.315	$11\frac{1}{2}$	0.400	0.6828	0.9845	1.23863	0.06957
$1\frac{1}{4}$	1.660	$11\frac{1}{2}$	0.420	0.7068	1.0085	1.58338	0.06957
$1\frac{1}{2}$	1.900	$11\frac{1}{2}$	0.420	0.7235	1.0252	1.82234	0.06957
2	2.375	$11\frac{1}{2}$	0.436	0.7565	1.0582	2.29627	0.06957
$2\frac{1}{2}$	2.875	8	0.682	1.1375	1.5712	2.76216	0.10000
3	3.500	8	0.766	1.2000	1.6337	3.38850	0.10000
$3\frac{1}{2}$	4.000	8	0.821	1.2500	1.6837	3.88881	0.10000
4	4.500	8	0.844	1.3000	1.7337	4.38713	0.10000
5	5.563	8	0.937	1.4063	1.8400	5.44929	0.10000
6	6.625	8	0.958	1.5125	1.9462	6.50597	0.10000
8	8.625	8	1.063	1.7125	2.1462	8.50003	0.10000
10	10.750	8	1.210	1.9250	2.3587	10.62094	0.10000
12	12.750	8	1.360	2.1250	2.5587	12.61781	0.10000

when you do use the wrench, don't tighten beyond an additional half turn for PVC or 1½ turns for CPVC. Understandably, the larger pipe sizes require slightly more wrench makeup.

Solvent Welding. Solvent welding is a very convenient method for joining pipe sections and fittings. You don't need much equipment and you can make joints quickly and simply right on the spot. There are a few guidelines to follow, though, to be absolutely sure you get good joints every time.

As with threading a pipe, you'll have to work where there's no grease or dirt. Cut the pipe square using a miter box, then bevel the edge all the way around the rim. When solvent welding, it is quite important to have a close clearance between the surfaces to be joined; if the outside diameter of the pipe is too large to insert into the socket fitting, dress it down a bit with emery cloth. Then wipe the surfaces clean with methylethyl ketone or carbon tetrachloride.

You can use CPVC cement with PVC pipe, but you can't use PVC cement with CPVC pipe. If you're working with both pipe types, as when running a cold water line above

ground with PVC and a high-pressure hot water line (above or below ground) with CPVC, make all your solvent welds with the material designed for CPVC bonding.

Standard PVC pipe is inherently resistant to most chemicals, so it stands to reason that only a limited number of solvents can be used in bonding operations. To make a secure joint, apply the solvent to both members to be joined with a small paintbrush or the brush that comes with the solvent. As soon as the socket fitting and the male pipe mating section are coated, push them together snugly (so the pipe section bottoms in the socket) and twist the pipe about ⅓ turn. Bear in mind that without the solvent, the rigid plastic pieces won't fit perfectly regardless of how well they are machined by the manufacturer; but when the solvent is applied and the surfaces begin to soften, 100% contact is achieved with the twisting motion, which expels all the air and distributes the solvent-cement completely. In a while the solvent dissipates and the material hardens, forming a strong bond. You must make the connection after the softening process has begun and hold

the joint immobile for several minutes. This is a lot tougher than it sounds, because the pipe sections have a tendency to push themselves apart as the material begins to swell.

Some PVC solvents are very slow to dry; they don't leave the joint completely even after weeks. You can determine whether or not the solvent you are using falls into this category by reading the label on the container. If it does, you should avoid using too much of it, particularly when you work with the thinner-walled PVC.

Even though initial set takes place after several minutes, no handling strength at all is developed following a solvent bond until about 30 minutes has passed. Don't apply more than 10% of the rated stress for at least four hours. The bond will reach full strength only after 48 hours or so; therefore, you should avoid using the pipe for this entire period if your water pressure and temperature are close to the upper rating of the material you have used.

If you opened a can of solvent months ago, then saved it for another repair, you'll probably find it unusable. Don't use the solvent if the material has turned to a jelly-like consistency in

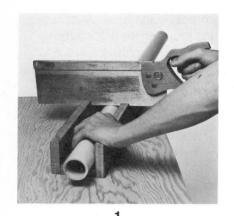

1

2

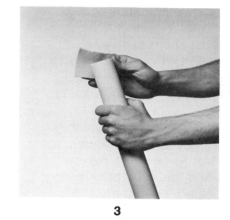

3

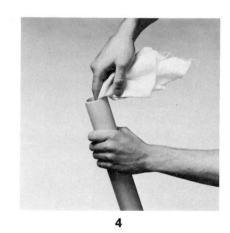

4

5

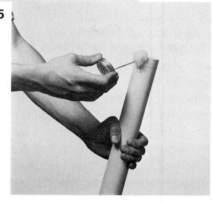

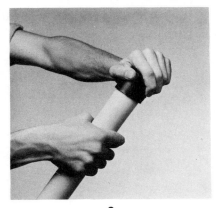

6

the can or if any solidification has taken place. Solvent doesn't cost much, and if it doesn't have its original bonding qualities you are risking not only your time but valuable materials by attempting to save pennies in reclaiming outdated cement.

   Probably the most important element to consider in the installation of a plastic piping system is "dry-fitting." Position all the pipes and fittings exactly as you want them in the finished installation. Then carefully mark the pipes and fittings with a crayon or laundry marker where they go together so that final alignment is simplified in the bonding operation. The installation goes easier when you can apply the solvent, push and twist pipe and fitting to previously marked alignment points, then move along to the next joint.

Hot-Gas Welding. Regular PVC pipe can be heat-welded in a manner quite similar in principle to the oxyacetylene method normally used for metal piping. It's not a good idea to weld CPVC, however, even though you can buy welding rod specifically for this purpose. According to Harvel Plastics,

1 **Using miter box, cut ends of pipe square**

2 **File edge of pipe to get a 10–15° chamfer**

3 **Sandpaper area to be cemented until gloss is removed**

4 **Wipe out fitting socket and pipe surface with soft cloth to remove dirt, moisture, and sand "dust"**

5 **Apply solvent to both surfaces to be joined, coating both surfaces completely**

6 **Push fitting socket and pipe length together firmly so that pipe bottoms in socket; as pipe bottoms, twist pipe about a third of a turn. Hold joint immobile for several minutes**

the weldability of CPVC is much inferior to that of standard PVC, and considerable experience is required to get consistently good joints that will hold up under all conditions of rated stress. The temperature range between a cold weld that is unsatisfactory and a burned and decomposed weld is narrow when you are working with CPVC. Even when you have installed a fitting that leaks, you'll probably be much better off replacing the fitting than trying to repair it with a hot-gas weld.

With PVC piping, hot-gas welding is a versatile and convenient method for making permanent joints and repairs to joints made by other means. But if you have no experience with welding in general, you would be well advised to steer clear of this technique because it does require a considerably higher degree of skill than the other joining methods.

For hot-gas welding, commercially available hot-gas or air welding equipment is used with a tip temperature of around 600°F and a welding temperature of around 500°F. Welding rods are employed just as in metal work, but they are made of polyvinyl chloride.

Flanged Joints. When you anticipate making a connection that might have to be dismantled periodically, you should use flanged joints. They are more expensive than threaded or solvent-welded joints, but you are sure to find them more economical in the long run; plastic piping isn't intended for applications requiring repeated disassembly, and premature leakage and below-spec performance will inevitably result from any type of joint other than a flanged fitting in such cases.

In making a flanged connection, you have to use a gasket of the same shape as the flange surface. You tighten the flange bolts enough to compress the gasket but not so tight as to distort the flange itself. Table 1–2 gives the torque recommended for flanges of $\frac{1}{2}$ to 12 inches. When you use this table, consider the torque specifications as being applicable to bolts that are well lubricated; a dry, unlubricated bolt can't reliably be torqued to a closely specified value.

TABLE 1–2. PLASTIC FLANGE BOLT TORQUE RECOMMENDATIONS. (Courtesy Harvel Plastics, Inc.)

FLANGE SIZE (INCHES)	BOLT DIAMETER (INCHES)	TORQUE (ft-lb psi *)
$\frac{1}{2}$	$\frac{1}{2}$	10–15
$\frac{3}{4}$	$\frac{1}{2}$	10–15
1	$\frac{1}{2}$	10–15
$1\frac{1}{4}$	$\frac{1}{2}$	10–15
$1\frac{1}{2}$	$\frac{1}{2}$	10–15
2	$\frac{5}{8}$	20–30
$2\frac{1}{2}$	$\frac{5}{8}$	20–30
3	$\frac{5}{8}$	20–30
4	$\frac{5}{8}$	20–30
6	$\frac{3}{4}$	33–50
8	$\frac{3}{4}$	33–50
10	$\frac{7}{8}$	53–75
12	1	80–110

To give bolt stress of 10,000 to 15,000 psi. Bolt torque refers to a well lubricated bolt. Use soft rubber gaskets between flanges. Use metal bolts and nuts.

Installation Considerations

Plastic pipe for water service comes in the standard pipe length of 20 feet. It's considered rigid, but you'll have to use more support than you would for metal pipe. You won't experience much difficulty in making pipe installations once you have mastered the techniques of making good connections. Both PVC and CPVC can be used underground and will interface nicely with pipes of other materials. There are a few special considerations that apply specifically to plastic pipe, however. Vinyl piping increases in length as its temperature goes up, which must be taken into account when you use it for high-temperature applications. And you must employ some special techniques if you want to incorporate permanent bends.

Thermal Expansion Factor. Vinyl piping has an unusually high thermal expansion factor, which must be taken into account any time you make an installation using this material. You can expect a 1½-inch change in length per 20-foot section for a temperature change of about 170°F. For this reason, manufacturers of plastic piping produce special joints, called *expansion joints,* for long runs. For most home installations, though, where pipe runs are usually less than 100 feet, you don't have to use these special joints. But you do have to install the piping so that it may contract and expand normally. This is achieved by affixing the support hangers so that the pipe can slide back and forth, or by using suspension hangers that have a built-in flex capability. Some professional plumbers put 360° bends in very long runs and allow the loop to absorb the expansion and contraction forces; but this is not considered good practice unless the pipe is to be used at less than its rated pressure and temperature, because the pipe tends to weaken at the bend.

Bending. When bending rigid plastic pipe, you have to apply heat of 250 to 275°F. You can do this with a flameless hot gas torch or a hot-air oven, or by immersion in hot oil. Uniform heat distribution is required throughout the bend area and over the entire circumference of the pipe. Take care not to overheat the pipe or keep the pipe at bending temperature too long; otherwise the pipe may lose its form and you won't be able to use it at all.

Bend the pipe around a regular pipe bender grooved to the proper diameter, with a radius at the bend that is at least five times the outside diameter of the pipe itself. Any greater curvature risks flattening the plastic pipe and consequently changing the flow characteristics as well as weakening the pipe's structure at the point where strength is needed most.

Other proved bending techniques include filling the plastic pipe with sand for shape retention during the bending operation and inserting a coiled pipe spring before bending.

Plastic pipe has a tendency to recover its original shape, so when you bend it, go slightly beyond the desired radius to allow for the springback. As soon as the bend is formed, cool the pipe as quickly as possible with water or air.

Compatibility with Other Materials. When you want to add plumbing to an existing fresh-water distribution system, as when installing a dishwasher or adding a bathroom, you might find it convenient or economical to employ plastic piping for the modification, even though the existing pipework might be of copper or galvanized steel. But there's more to this than making simple connections.

Generally, you can insert a male plastic thread into a metal socket fitting or coupling if heat is not involved, so long as you anchor both fittings well immediately adjacent to the joint. What you *can't* do is connect a metal pipe to a plastic socket fitting or coupling. The best practice is to avoid *any* direct connection that does not employ flanges. The rates of expansion are considerably different for metal and plastic, and direct connections without flanges (and compressible gaskets) are risky even under the best of conditions. The flange bolt torque recommendations given in Table 1–2 should be used as a guide when mating metal to plastic pipe.

Copper Pipe and Tubing

Copper pipe and tubing with soldered joints or flared-tube connectors are quite common in modern water supply systems. Copper is highly regarded for its corrosion resistance and its ability to convey large volumes of water with comparatively small pipe diameters. Copper pipe is easy to handle, relatively easy to install, and flexible enough to allow *shallow* bending without the use of special tools.

The copper pipes most commonly used are designated as types K, L, and M. Type K is well suited for underground service, type L is good for general plumbing use, and type M is intended for use in applications where soldered fittings are to be employed exclusively. Types K and L are available in 20-foot lengths of hard-tempered copper or in 50- or 100-foot reels of soft temper. Type M is available in 20-foot lengths of hard temper only.

The word *temper* refers to the brittleness of a pipe. When you repeatedly bend steel, it becomes soft and weak at the bend and will

ultimately break there. With copper, repeated bending or stressing by excessive pressure or vibration will make the piping brittle and stiff, but, like steel, it will break at the stressed point eventually. Steel is softened by heating it until it becomes cherry red, then allowing it to cool slowly; the slower the cooling process, the softer the steel. But just the opposite is true for copper; you heat it the same way, but you try to cool it as quickly as possible, usually by quenching in cold water. The *faster* the cooling process with copper, the softer the copper becomes. This method of softening copper is called *annealing*.

Cutting

Without question, a tube cutter is best for cutting copper piping. First, mark the copper where it is to be cut, then position the cutter so that the cutting wheel is directly over the mark, as shown in example 1–1. Turn the adjustment wheel knob clockwise until the cutter wheel touches the pipe securely. Revolve the tube cutter around the pipe so that the cutter wheel rolls around on the mark, then tighten the knob one turn or so and

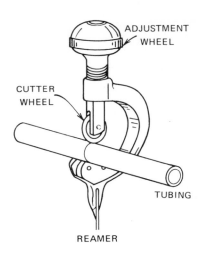

example 1–1. Combination tubing cutter and reamer. Tighten the adjustment wheel gradually as you roll the cutter wheel around the pipe on the "cut" mark. With each rotation the cutter bites deeper into the pipe wall.

repeat the process. Each time you roll the cutter around the pipe, you have to tighten the adjustment wheel a bit. Eventually, after about six turns, the pipe will separate on the line.

Once you have severed a copper pipe with a tube cutter, you'll notice a burr around the inner rim of the pipe; this must be removed before the pipe is put into service. Most tube cutters have a built-in reamer designed for this purpose. Just put the reamer into the end of the copper pipe as shown in example 1–2 and turn the tube cutter clockwise so that the cutting edge scrapes away the burr.

Joining

In water service, the copper piping of an existing installation might contain joints of *sweated* solder or *flared* fittings, and sometimes both. *Flaring* is a method of forming the end of the copper pipe into a funnel shape so that it can be held in a threaded fitting when making a line joint; *sweating* involves applying solder to a slip fitting so that an immobile joint results.

Flaring. Probably the most common plumbing error when working with

REAMER ON END OF TUBE CUTTER

example 1–2. The reamer portion of the combination cutter and reamer has a sharp edge that removes burrs easily and allows you to put a slight bevel on the cut pipe end.

copper pipes and tubes is making the flare in the end of a pipe section before slipping the fitting into place. The correct procedure is to cut the pipe, ream it, clean it so that all filings are removed, then slip the fitting over the end of the tube; the flare will prevent the fitting from coming off. The Appendix shows how to make secure flared joints with copper tubing.

Example 1–3 shows some typical fittings used for copper; note that some of these are strictly flared fittings, others are for sweat-solder applications, and a few are designed for crossover from one to the other type.

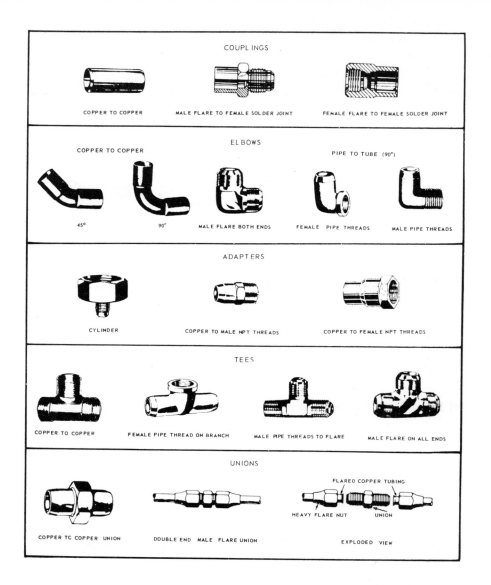

example 1–3. Representative copper fittings; shown are couplings, elbows, adapters, tees, and unions, both flare and solder-type.

Sweat Soldering. When making a soldered joint using a sweat fitting, clean an inch or more of the end of the copper tube inside and outside with steel wool or triple-aught (000) sandpaper until the metal shines. Then spread a thin film of paste flux on the tube end with a clean brush (a basting brush works nicely for this) or other applicator.

When the end of the pipe has been prepared, insert the pipe into the fitting as snugly as possible. Apply heat directly to the metal fitting with a flame torch until the melting point of your solder is reached. Feed the solder at the edge of the fitting. Capillary action must spread the solder evenly and completely over the surfaces. (The solder won't do its sealing job if the fit is sloppy, though.) When a continuous ring of solder appears around the end of the fitting (at the junction of the pipe and fitting seam), the joint should be complete.

If you have to disconnect a soldered fitting, the process is the same. Disconnection is often required when you remove old faucets that have developed leaks or are broken, or when you need to replace a kinked pipe section, especially one of the leader tubes to a kitchen faucet or water closet intake valve. Apply heat directly to the fitting or the thickest part of the existing weld. When the solder begins to sweat or melt, you can pull the fitting free of the pipe. When working with hot piping, don't forget to use gloves—fittings get much too hot to handle without them.

It's very important that you do not consider a joint *really* complete until you have cleaned it with a stiff brush, soap and water, or emery cloth. If you forget this operation and leave a flux residue on the joint, you are inviting corrosion—flux is an extremely corrosive compound and is helpful only during the soldering operation.

Silver Soldering. Silver soldering is more common in refrigeration work than in working with fresh water pipes, but many plumbers install fittings with silver solder simply because they are accustomed to working with it more often, so you are apt to find fittings bonded using this somewhat expensive method. The process, also called *silver brazing,* achieves fusion with a silver alloyed soldering material that has a melting temperature of more than 800°F.

As with sweat soldering, capillary attraction is important in the silver brazing process. Not quite as important in silver soldering is the initial fit, since the solder used in this process tends to serve as a filler. Still, there's a practical limit to the allowable clearance between surfaces; the greater the clearance the more difficult it is to make a permanent, watertight bond.

Two methods are used to make joints between pipes and fittings with silver solder: the *insert* method and the *feed-in* method. The feed-in method, described below, is the same as that described for sweat soldering. Whichever you use, make sure both parts are adequately supported during the brazing process. And you must keep the joint immobilized until the filler metal has completely solidified.

To use the insert method, insert a strip of the filler metal into the joint just before assembly. Clean the joint exactly as you would with sweat solder joining, and apply flux with a brush. Then fit the two parts together and align them, and apply high-temperature flame on the thinner portion of the tube at the joint as shown in example 1–4, which also shows the recommended path of the torch while heating the tube.

As the pipe heats, the metal expands until the surface of the tube is enlarged to the same diameter as the inside of the fitting, which seals the clearance area and forces the flux out of the joint. The flux flow—which should be even over the complete circumference of the joint—indicates that the expansion process is complete. At this point, direct the flame to that portion of the fitting hub most distant from the junction of the tube and fitting. Keep rotating the flame over the joint segment until the solder appears at the newly formed seam of the two metal parts.

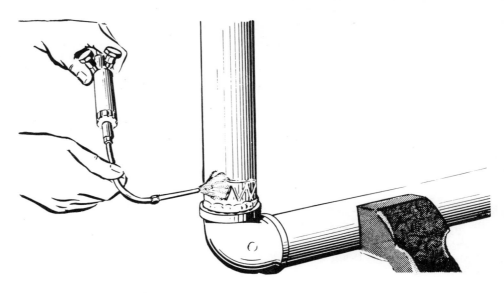

example 1–4. How to silver-solder a copper fitting. Lines indicate path of torch during the heating process.

Bending

Soft-drawn copper bends nicely without requiring special tools if the bends aren't too severe. But if the bend you need is 45 degrees or greater, shaping the tubing without the benefit of a bending device is risky; either way, you should employ some precautions to prevent partial collapse or flattening at the bend.

If you don't plan to be doing much work with copper pipe and don't own a bender, you need not invest in one for a single job. The photo sequence shows a procedure for hand bending that will minimize the risk of ruining a section of tubing:

1 Cut the copper tubing to the required length using a tubing cutter or hacksaw.
2 Plug the end nearest the point where the bend is to be positioned.
3 Using a funnel, fill the pipe section with sifted dirt or sand. You don't need to completely fill the pipe; just make sure sand fills the pipe where the bend is to be so that the sand can absorb some of the stress and prevent kinking.
4 Without allowing any of the sand to escape to the other end, grasp the pipe tightly and begin the bending process carefully, using your knee as a fulcrum. Don't hurry!
5 When the pipe is bent to the desired angle of curvature, empty the filler material from the pipe section.
6 Clean the inside of the pipe thoroughly with soap and water using an appropriate-sized swab tied to a strong length of twine or wire. Make sure the wire is longer than the pipe section, and push it through the pipe first; then pull the string so the swab wipes the entire inner surface as it comes through the pipe. Repeat the swabbing operation until the cloth comes out clean.

1

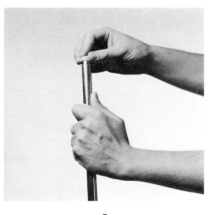

2

3

4

5

6

7

1 **Use tubing cutter to cut copper pipe to required length**

2 **Use wood, putty, or other material to plug end nearest bend point**

3 **Pour sifted dirt or sand into pipe so that portion to be bent is filled**

4 **Without allowing sand to escape bend area, carefully pull pipe over knee until desired bend is achieved**

5 **Remove plug and pour sand from pipe. Tap pipe to remove all debris**

6 **Tie long string or wire around a soft cloth of about 4 inches square**

7 **Dip swab in warm soapy water and pull through pipe until pipe is clean**

Cast-Iron Piping

Cast-iron pipe that is intended for water service comes in lengths of 12 to 20 feet and has hub-and-spigot joints or gland-type joints. Gland and mechanical joints are made with rubber sealing rings held in place by metal follower rings bolted to the pipe. This arrangement allows the pipe to contract and expand with temperature variations without damaging the joint. Cast-iron pipe is commonly used for underground pressure service in older homes and where codes require it.

Cast-iron pressure pipe is measured by the *inside diameter;* most other types are measured according to the nominal outside diameter.

Cutting

You can use either a hand-operated chain cutter or a powered hacksaw to cut cast-iron pipe. After cutting, you don't need to ream the inner rim as you would with softer metals, but you might occasionally have to use a file to dress down a rough cut.

In water service lines, hub-and-spigot cast iron pipe is joined with lead, lead wool, or a sulfur compound. The packing material might be specially treated paper or asbestos rope.

Before going to work on joining cast-iron pipe lengths, check every piece you intend to use to be sure none of them has cracks. You can't always spot a cracked cast-iron pipe with the eye, but there's no mistaking the *sound* of one. Tap each length *lightly* with a hammer; a good section has a pleasant ring that resounds for a while following the tap; a cracked length will give off a dull thud or a short buzz not unlike a torn loudspeaker cone.

Joining

Once you've determined that the pipe sections are all right, wrap the yarn (this technique is covered in detail in the Appendix) around the spigot end (the end opposite the hub or bell), and place this in the hub of the previously laid length. Straighten the pipe and adjust the yarn with a yarning iron. Use enough yarn to fill the joint within about 2 inches of the face of the hub. Then clamp a joint runner in place so that it fits tightly against the outer edge of the hub.

For horizontal pipe runs, it's a good idea to use asbestos to make a tight seal between the runner and the pipe so the hot lead won't run out of the joint. Pour molten lead into the V-shaped opening left at the top by the clamped joint runner. The lead fills the space between the yarn and the runner.

You have to pour the joint in one continuous operation if you want good, repeatable results. After the joint hardens, remove the runner. Then calk the lead to expand it fully as it cools. But *be careful*—if you tap the calking iron too heavily, you'll break the hub on the pipe section laid earlier, and if you don't tap heavily enough, you'll leave a joint that will leak when water pressure is applied.

If you encounter water in the trench, you can avoid the lead-pouring operation by using lead wool rather than molten lead. You have to use more yarn when you do it this way, and the calking operation is more time-consuming.

A sulfur compound is melted on the job, like lead, but at a lower temperature. It is then poured into a joint prepared as for a cast-lead joint. The primary advantages of the sulfur compound are that it is light in weight, requires no calking, and gives strong joints that are not likely to blow out. Joints of sulfur sweat a bit at first but they tighten up in a short time. Since the joints are rigid, they should not be used to connect a newly laid line to an old one; otherwise the settlement of the new line may cause a broken place. The connections between the new line and the old one should be of lead.

Galvanized Piping

Galvanizing is a process of coating iron or steel electrically with zinc to make the metal resistant to corrosion. Wrought iron and steel are both galvanized in the same manner, but iron piping is considerably more expensive than steel, so the latter is more common in fresh water pressure-pipe service. Galvanized pipe comes in lengths of 20 feet (some lengths may be as much as two feet shorter; others that much longer, with 20 feet being the average).

Joining

Detailed information on pipe cutting and threading is included in the Appendix. You should be thoroughly familiar with the processes of cutting and threading as well as the tools required before attempting serious pipe work in actual plumbing systems. If you gain some experience with galvanized pipe, you'll be able to adapt to other materials; galvanized pipe is the mainstay of plumbing and it is likely to be so for years to come.

To obtain a tight threaded joint, the threads must be clean and in good condition, without nicks or dulled spots. If the pipe has been banged around much, check the threads visually; if necessary, run a die over the threads to sharpen and straighten those that have been damaged.

Unless you have just finished threading the pipe section you'll be using, clamp the piece in a vise and clean both ends with a stiff wire brush. Then apply pipe dope to all exposed male ends, exercising care not to let the lubricant get into the conduit itself. (The compound you use will depend on what the pipe is to be used for; steam lines require something more substantial than water pipes.)

Start the joint by hand and tighten it as much as you can without a tool. If you are working with a fitting, screw it on by hand; then cinch up the next pipe section and tighten the whole assembly with a pipe wrench. With galvanized pipe, you don't want to cover all exposed threads; this would result in a wedging action that might split the fitting (the threads on galvanized pipe are tapered). Don't use an oversize wrench, because this would prevent your getting a good "feel" of the joint. If you have properly threaded the pipe initially, there should

be two or three unused threads left on the pipe section joined to the fitting.

Fittings

For iron and steel galvanized pipe, the fittings you will use are usually cast iron; these fittings include elbows, crosses, tees, and unions. But small pipe sections, called *nipples,* and couplings should also be in your pipe fitting inventory.

Types of Fittings

Elbows. The four basic elbows are the 90-degree, 45-degree, street, and reducing. Of these, all but the street elbow are conventional curves with a female socket on each end, as shown in example 1–5. The *street elbow,* shown in outline form with other fittings in example 1–6, has one female and a male end, and is used most frequently in close quarters where there is insufficient space for a conventional elbow and nipple.

The *reducing elbow,* as the name implies, allows you to make a pipe bend while going from a pipe diameter of one size to that of another. You can buy reducing elbows for almost all standard pipe diameters, but the most common are those with openings one

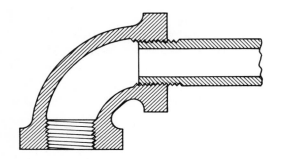

example 1-5. The conventional elbow in cross section; note that elbows have a socket at each end. (A street elbow, with a socket at one end and a threaded male fitting at the other, is an exception.)

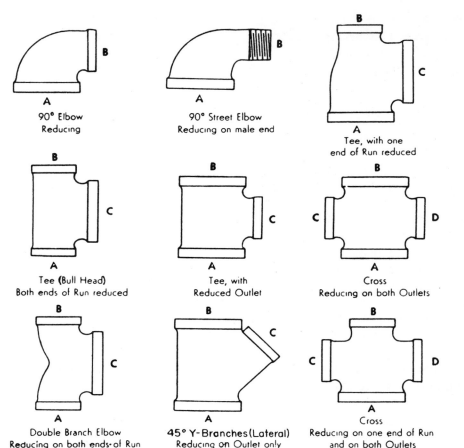

A
90° Elbow
Reducing

B

A
90° Street Elbow
Reducing on male end

A
Tee, with one end of Run reduced

A
Tee, with one end of R
and Outlet redu

A
Tee (Bull Head)
Both ends of Run reduced

A
Tee, with Reduced Outlet

A
Cross
Reducing on both Outlets

A
Cross
Reducing on one end
and on one Outle

A
Double Branch Elbow
Reducing on both ends of Run

A
45° Y-Branches (Lateral)
Reducing on Outlet only

A
Cross
Reducing on one end of Run
and on both Outlets

A
Cross
Reducing on one Outle

example 1-6. Various galvanized fittings—elbows, tees, a wye, and crosses.

size apart—a ¾-inch opening on one end, for example, and a ½-inch opening on the other.

Crosses. Crosses are fittings that have four branch openings (see example 1–6). You can buy cross fittings with almost any combination of opening sizes; a few of those available are shown in the example.

Tees. The conventional or *straight tee* has a straight-through portion and a 90-degree takeoff on one side only; all openings are the same size. A *reducing tee* has one opening of a different size than the other two. As with other reducing fittings, tees are available for most standard-size pipe diameters.

Unions. Pipe unions are used for joining two pipe sections already fixed in place; unions are designed for making easy connections and disconnects. They are not one-piece assemblies like other fittings, but typically contain two or three discrete elements.

Couplings. The three most common types of couplings are the *straight coupling,* the *reducer coupling,* and the *eccentric coupling.* Couplings are short lengths of cast metal pipe threaded inside on each end; they are the female counterpart of the nipple, and are used in pipework installations where pipe lengths are placed sequentially. The straight coupling has female threads of the same size on both ends and is used for joining two pieces of pipe in a straight run that requires no other fittings. A reducing coupling, or *reducer,* is used to join pipe lengths of two different sizes. The eccentric reducer is a version of an *offset* fitting, a reducer where the openings are not quite concentric.

Nipples. Nipples are pieces of pipe that are less than a foot in length and threaded on the outside on both ends; they are used to make an extension on a fitting. Nipples are available in a wide variety of precut lengths and diameters, and are generally made of the same material as the pipe (rather than cast).

Plugs, Caps, and Bushings. *Plugs* are solid male pieces that can be screwed into a female fitting to seal off an opening; they are typically fitted with a slot (for screwdriver tightening) or a raised square section (for wrench tightening). A *cap,* used for the same purpose as a plug, has female threads and mates with the male threads of a pipe end that is to be sealed off. It often has recessed socket, slotted, or raised square portions to allow tightening with a hand tool. The *bushing* is a thick-walled length of pipe threaded inside and out; it is used to change the effective diameter of a pipe opening so that it can be mated with a fitting of another size.

Dielectric Fittings. Not all fittings are conductive metal, even when they are used for connecting metallic pipes. A fitting that is nonconductive is called a *dielectric* or insulating fitting. They are commonly used to connect underground tanks or hot water tanks. But one very important use is connecting pipes that are constructed of different metals. Dissimilar metals, copper and iron, for example, act as a rudimentary cell (battery element) in the presence of certain chemicals. Without a dielectric separating the two types, a constant current might be generated, resulting in the destructive corrosion of the joint as one metal loses its electrons to the other.

How To "Read" a Fitting

As you examine example 1–6, you'll note that each opening of every fitting is given an alphabetic designation; these letters identify the openings and give you the sequence by which a fitting is specified. For example, elbows are specified by naming the largest opening first. The first one shown in the figure might be a $\frac{3}{4}$ x $\frac{1}{2}$ or a 1 x $\frac{3}{4}$. Tees are specified by the straight-through opening first; and if one is larger than the other, that is the one given first. The same is true for crosses; the largest opening is given first, then the one opposite.

Selecting the Right Pipe

"Fast" and "Slow" Pipe

When you install a plumbing network or a new branch to an existing one, you can't just "buy a bunch of pipe." If you want predictable water pressure at the faucets and fixtures, you have to know something about your existing water supply system and something about the materials you choose. Unfortunately, all pipe has resistance to water flow. The older the pipe, the greater the resistance is going to be. The surface tension of water is such that it doesn't seem to "want" to flow; it "grabs" the inside surface of the piping through which it's being pushed. That tenacity has to be overcome before satisfactory water flow can be achieved. The rougher the inside surface of the piping, the greater the resistance of the pipe to water flow. The drop in pressure as a result of a pipe's resistance to the flow of water is called *head loss.* Piping that offers a very low resistance to water flow is termed *fast;* old piping and piping that has a naturally high

resistance to flow are considered *slow.* If you use fast pipe, you can get by with smaller diameters; conversely, if you use slow pipe, you must use larger diameters for a given water pressure. It naturally follows that longer runs mean higher resistances and shorter runs, lower.

Pressure vs Flow

Pressure isn't much more than a word to most of us; we don't relate a specific value to the water flow we'd like at the outlet of the faucet. What we *can* relate to our everyday experiences is water flow itself—the number of one-gallon jugs a given tap will permit us to fill in one minute. The higher the water pressure in a pipe of a given diameter, the more gallon jugs we can fill in that one-minute period. It takes considerably more pressure to force 10 gallons of water per minute through a $\frac{1}{2}$-inch pipe than to force the same amount in the same time through a larger-diameter pipe. This should seem entirely logical: the water at the input end of the pipe can't all go through at the same time, so some of it has "to wait in line." If the pressure of the line is increased in some way, there would be less "waiting" because the water would be pushed through faster. If the $\frac{1}{2}$-inch line were very short, the wait would be negligible, as would be the loss at the output end. To understand why, think of what happens at the end of a play when the theater is full. Those who are standing near the exit can whiz out in short order, but those who are sitting down front have to wait for all the people in the aisles to leave. To the people down front, the aisle is like a long section of narrow-diameter pipe; to those standing at the door when the exit is opened, the aisle is like a short length of narrow-diameter pipe.

Now suppose we have a water source capable of delivering 10 gallons per minute. What size pipe would we need to deliver that same flow rate 100 feet away? If you've been paying attention, you know it can't be done regardless of the pipe diameter—there has to be *some* loss. The losses can be minimized by using fast pipe that is very large in diameter, but they can't be overcome completely.

It is important to understand the relationship of water pressure to rate of flow, since this is the principal basis for selection of pipe in a water system.

As indicated earlier, it takes a specific water pressure to deliver a flow at the source of 10 gallons per minute. The instant you connect a pipeline of any size to that source, you will see a pressure drop at the far end of the line; and that pressure drop will cost you in gallons of flow. What you have to know is *how much* pressure drop, or head loss, to expect for a given installation.

At a flow rate of 10 gallons per minute, it has been estimated that a new 100-foot length of $\frac{3}{4}$-inch galvanized pipe would offer a head loss of 10 pounds per square inch. This means that the pressure of your water source would have to be increased by 10 psi in order to simply maintain the water pressure at its current level. If the 100-foot pipe were 25 years old, the head loss would be twice that amount. If the far end of the pipe were higher in elevation than the source, the head loss would similarly be increased. Since most of us don't have the capability of increasing our water source pressure, we try to get by with minimizing our losses.

It happens that vinyl pipes are faster than metal. Copper is faster than galvanized steel, but plastic is faster

than copper. But the differences in the inside surfaces of these conduit types aren't significant in fresh-water (pressure) systems when *new pipe* only is being compared. Plastic pipe, however, ages better than metal, so the head loss increases with age in metal—whereas it stays essentially the same with PVC and CPVC.

Example 1–7 is a nomogram that you can use to determine the diameter pipe needed for specific applications. The chart is based on plastic pipe, but the values are close enough to those of new metal pipe that the differences can be ignored.

To use the nomogram you have to establish two independent variables, as explained in the example. Then place a straightedge across the page so that those two variables are touched at their precise value points. Wherever the ruler or straightedge crosses a line, you'll see a specific value pinpointed. Some examples of practical usage appear at the lower right corner of the nomogram.

Flow Loss Characteristics of Water Flow Thru Rigid Plastic Pipe

This nomograph provides approximate values for a wide range of plastic pipe sizes. More precise values should be calculated from the Williams & Hazen formula. Experimental test value of C (a constant for inside pipe roughness) ranges from 155 to 165 for various types of plastic pipe. Use of a value of 150 will ensure conservative friction loss values.

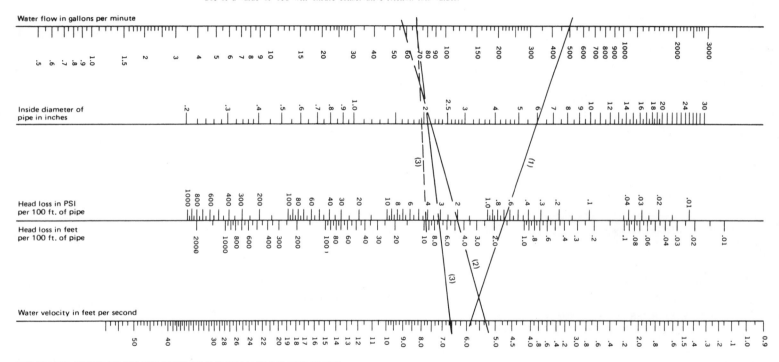

THE VALUES OF THIS GRAPH ARE BASED ON THE WILLIAMS & HAZEN FORMULA

$$f = .2083 \left(\frac{100}{C}\right)^{1.852} \times \frac{g^{1.852}}{d^{4.8655}}$$

WHERE: f = Friction head in feet of water per 100 feet.

d = Inside diameter of pipe in inches.

g = Flowing gallons per minute

C = Constant for inside roughness of the pipe (C = 150 for thermoplastic pipe).

The nomograph is used by lining up values on the scales by means of a ruler or straight edge. Two independent variables must be set to obtain the other values. For example line (1) indicates that 500 gallons per minute may be obtained with a 6 inch inside diameter pipe at a head loss of about 0.65 pounds per square inch at a velocity of 6.0 feet per second. Line (2) indicates that a pipe with a 2.1 inch inside diameter will give a flow of about 60 gallons per minute at a loss in head of 2 pounds per square inch per 100 feet of pipe. Line (3) and dotted line (3) show that in going from a pipe 2.1 inch inside diameter to one of 2 inches inside diameter, the head loss goes from 3 to 4 pounds per square inch in obtaining a flow of 70 gallons per minute.

Nomograph courtesy of Plastics Pipe Institute, a division of The Society of The Plastics Industry.

example 1–7. Flow-loss characteristics of various pipe sizes.

Measurement

Six methods are commonly used to measure fresh water conduit: illustrated in example 1–8, these are end-to-end, center-to-center, end-to-center, end-to-back, center-to-back, and back-to-back. When you read installation instructions for any fixture or when you make pipe-run measurements, it is important to understand the convention and stay with it *consistently;* if you don't, you'll find some pipes too long and some too short for the task at hand.

An *end-to-end* measurement indicates that a pipe is threaded on both ends; the measurement is made from one end of the pipe to the other and includes both threads. A *center-to-center* measurement means that there is a fitting on each end of the pipe; the measurement is made from the center of the fitting at one end to the center of the fitting on the other. When a pipe is to have a fitting on one end only, the *end-to-center* measurement is usually stipulated; in this case the measurement is made from the far end of the pipe to the center of the fitting. *End-to-back* refers to pipe with a fitting on one end only;

Pipeline Installation

The discussion on the preceding pages of this section helps you to select the pipe material that best suits your needs and shows you how to make watertight connections using a variety of joining techniques; it also proves helpful in determining the proper diameter of pipe to install for any application. Once these initial hurdles are overcome and you have decided how the new pipes are to be routed, you can begin in earnest the task of pipeline installation.

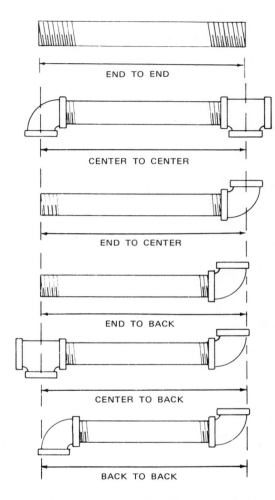

example 1–8. The six methods for measuring piping.

END TO END

CENTER TO CENTER

END TO CENTER

END TO BACK

CENTER TO BACK

BACK TO BACK

in this case the measurement is made from the back of the fitting to the far end of the pipe. *Center-to-back* measurements are made on some pipes that have fittings on both ends; as implied, the extremes are the center of the fitting at one end and the back of the fitting at the other. *Back-to-back* measurements, made on pipes with fittings on both ends, are taken from the backs of the two end fittings.

Trenching and Bedding

Where your water system is already installed and you wish merely to add a fixture or group of fixtures within one of the existing rooms, you don't have to worry about trenching. But if you plan to run piping to a pool in the backyard, to a sprinkler system, or to an annexed part of your home, you have to bury the pipe.

Proper trench depth is determined by intended service and local conditions, but regardless of locale, the trench must be deep enough to allow burial of the service pipes below frost level to prevent freezing. The ideal depth is one foot below the frost line. Permanent lines that will be subject to heavy traffic should be buried at least two feet deep; otherwise a depth between 12 and 18 inches should be sufficient, particularly where the pipe to be installed has a relatively small diameter. In addition, it is imperative to check your local codes for specific requirements regarding pipe depth.

If you plan to make the pipe joints right in the trench, then you will have

to make the trench as wide as possible, as much as 2 feet in some cases. But you can minimize trench width by making all the joints outside the trench and then lowering the line into position with leveling supports. If you plan to use vinyl piping and the run is relatively long, 100 feet or so, a wide trench is also desirable; this allows you to snake the pipe from side to side to accommodate a great deal of expansion and contraction without stressing the joints, pipes, or fittings. Of course, special *expansion joints* are available. These allow the pipe sections to extend or compress with temperature changes of the ground or the water being carried.

The trench bottom should be continuous, relatively smooth, and as free as possible from rocks and debris. If the trench is dug in hardpan, rocky soil, or where large subsurface boulders jut into the diggings, the trench bottom should be padded with sand or compacted fine-grained soil. This especially applies where the pipe to be laid is plastic.

Sharp bends and dead ends should be anchored by rodding or concrete anchors. (Use of more than one concrete anchor should be avoided where plastic pipe is being used to convey water at a high temperature. A larger-diameter plastic nipple should be used to pass through a concrete pit wall; this precludes scratching and excessive abrasion with pipe contraction and expansion.)

Generally, the pipe should be supported over its entire trench run on stable material. Blocking should not be used to change the pipe grade or to provide intermittent support over low sections in the trench. A better alternative is to let the trench bottom do the grading if any is required.

Testing

Don't fill the trench with dirt until you have tested the installation to be absolutely certain there are no leaks or bad joints in the piping. There are several methods for making pressure tests in newly installed water lines, but these are the most common: the *air test* and the *water test*.

Air Test

To make an air test, plug all openings in the system with regular plug fittings. Then connect a gaged source of compressed air to the inlet. Slowly bring the pipe up to its rated pressure while you watch the pressure gage; hold it there. Now apply a soapy water solution to every joint in the run and examine each for evidence of leakage. If leaks are present, the location of each will be revealed by bubbles created by escaping air. Check all joints at the same time, and mark any leaks with chalk or soapstone. Once testing has been accomplished, relieve the air pressure, *then* repair the leaks and retest using the same procedure.

Water Test

If you opt for the water test, you will have to wait several hours before the filling-in operation. Fill the pipe with potable water; apply and maintain the pressure with a hand pump. You don't have to apply a water solution to the piping in this case—it wouldn't help. Instead, let the water pipe get up to its rated pressure, then wait for 4 to 8 hours while making periodic inspections for any evidence of leakage. If no leaks are spotted, relieve the pressure and connect the pipe to the water line.

Backfilling

The pipe should be surrounded with relatively fine-grained soil or backfill materials (with a particle size not exceeding $\frac{1}{2}$ inch). Backfilling should be done in layers so that even distribution of dirt pressure around the pipes can be assured. You can tamp the layers manually or with water. Be very careful when backfilling to keep the pipe straight and minimize settlement. Never throw the fill material directly onto the pipe, since it would very likely throw the pipeline out of alignment and possibly place extra stress on the joints. This is no time to develop a pipe leak. Drop the fill material on either side of the pipe. When compacting sand or gravel, vibratory methods are recommended to assure complete settling.

For best results, the soil should be nearly saturated. When you have water available, and where the pipe run is not especially long, you can use water in the backfilling operation. But you have to be careful that you use enough fill material to insure coverage of the trench even after the compacting soil dries. Don't use additional backfill until the drying initial backfill is hard enough to walk on.

Here's one effective method of backfilling a short water-pipe run: fill the ditch completely with loose soil, but avoid letting dirt fall directly onto the laid pipe. Attach a piece of pipe to a water hose and push it through the loosely placed soil until it touches the water pipe. Then turn on the water and let it run until it reaches the surface. This technique allows all the earth to be replaced except for the portion equal to the volume of the laid pipe.

In all instances the trench should be filled completely. All backfill should be placed and spread in uniform layers to eliminate voids. Large rocks, frozen clods, and all debris greater in diameter than 3 inches must be removed when it shows up in the backfilling operation. Rolling equipment or heavy tampers should be used only to consolidate the final backfill.

Above-Ground Pipe Runs

When a pipe carries liquids it gets heavy; unless you have taken precautions to support an above-ground pipeline adequately, you are running a considerable risk of an eventual leak or joint rupture. A sagging water pipe may not seem very serious, but when it strains to the extent that valves and joints begin to leak, you'll gain some new respect for the process of installing regularly spaced pipe supports.

The main supply pipe, regardless of whether it's horizontal or vertical, has to be supported adequately to take its weight off the fitting and to prevent future leaks. Example 1–9 shows several support hangers that can serve satisfactorily for this. There is no way a single support type or style can be recommended for all installations because installations vary. For long overhead runs and where the supply pipe is particularly heavy, it's best to use supports that can handle a lot of weight. And, needless to say, the anchor for the support should be at least as solid as the support itself.

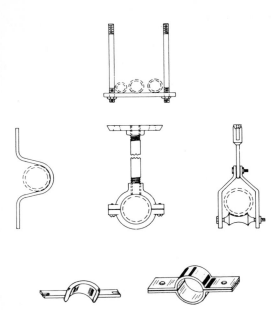

example 1–9. Pipe support-hanger variations. This is not all-inclusive; there are many variations of the basic hangers shown here.

It's generally wise to avoid using a pipe support as an anchor. Pipes that are to carry high-temperature fluids will tend to expand, and the expansion will be even more pronounced with vinyl pipes than with metal. Make sure the support does not prevent the pipe's regular expansion and contraction. With plastic pipe, keep sharp support edges from gouging the pipe; pipes should be free to slide a bit back and forth in their supports. If there is any chance that the sharp edges of a support will dig into the surface of a plastic pipe, use a shim or soft strip of material between the support and the pipe.

Fixture supply risers are pipes taken off the main water distribution run to service fixtures. These risers may be in the wall or left exposed. If they're to be enclosed in a wall, they must be made secure (no expansion consideration) and tested first. All vertical fixture risers should be supported at each floor level or at each change of direction. Remember this, above all: vertical fixture risers should *never* depend on the horizontal fixture supply branch for support.

Pipe Repairs

Even in the best plumbing system, you can expect a failure from time to time. When it happens, do your best to find the trouble and repair it fast. Any delays can cost you more than you might think in dollars and time. A break in a pipe can release water under pressure, which can in turn weaken foundations and other supporting structures. A neglected frozen pipe can lead to the rupture of other lines depending on that one for supply. The key to successful maintenance of any plumbing system is the speed with which repairs are made.

Pipe Freezing

Probably the single most common plumbing problem in water pipes is cold-weather freezing. When water freezes, it expands. The expansion takes place regardless of how strong the pipe might be, and when the process of expansion is complete, something has to give; either the pipe expands with the water or it breaks. Unfortunately, since metal contracts in cold weather, the more common result of freezing is pipe rupturing. It is absolutely essential to begin a thawing operation immediately to preclude pipe breaks—the longer you wait to start the thawing procedure, the greater your chances of having to replace entire pipe runs.

Rapid Thawing

Before you begin to thaw a frozen pipe, open all faucets and valves in the affected line. Then apply heat locally to the lowest open end of the frozen section. This is extremely important. If you were to start thawing at the center of a pipe, the locally applied heat could turn the water into steam, and with no escape route the steam thus generated would very likely cause an explosion right in the center of the pipe. By starting at the faucet, the steam that is generated always has an escape route—through the faucet and into the air.

As to the source of heat, many have successfully employed candles, a bunsen burner, or a wad of lit papers. But the quickest and most convenient source of a continuous hot flame is a small propane torch, (available at most hardware stores). Be especially careful when you use a torch, though; and don't use one at all where the pipe is situated next to a wall or wood framing joist. Outdoors and in clear areas, light the torch and allow the flame tip to touch the pipe, but keep the flame moving constantly so that the heat will be applied evenly rather than in occasional hot spots. You can apply heat to polyethylene and vinyl piping too, but you must be particularly careful with these plastic conduits to avoid concentrating the heat into one place.

Pipe-Heating Cable

One very common method for preventing the freezing of piping works equally well for thawing pipes once they have been hit by a hard frost, but the defrosting process takes a while. What you do is wrap pipe-heating cable—available at hardward stores and drug stores during the freezing season in many areas of the country—around conduit most likely to freeze. These cables don't draw much current, averaging around 50 watts per cable, and so don't cost any more to operate than a small light bulb; they can save you a great deal of time and trouble if you employ them wisely.

Example 1–10 shows how the cable is wound around the exposed pipe. Usually, the two wires of the cable are bound together in a single ribbon that has much the same appearance as TV antenna twinlead, so the winding process is even simpler than the sketch indicates. For full protection, buy a cable that is three times as long as the pipe area to be protected. Then wind the cable around the pipe so that the turns are spaced at about 2 feet. The colder the weather the closer, of course, you should space the turns.

If you wind cable about the expected trouble spots before the freeze hits, all you have to do when

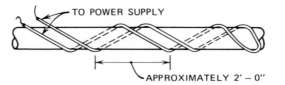

example 1–10. A ribbon-type heating cable wound about a pipe to thaw or to prevent freezing.

bad weather comes is plug the end of the cable into a handy electrical outlet. It's also a good idea to apply at least a temporary insulating covering to the heating tape; this will prevent heat loss and give you more freeze protection per watt of electricity used.

Don't forget a cable-wound pipe once you've made the installation. Most of the cables are inexpensive and won't stand up long to the rigors of severe weather changes outdoors. Every two weeks during the cold season examine the cable, the pipe, and the plug-in connection at the convenience outlet.

Thawing Inaccessible Pipe

Where the frozen pipe extends from a wall and is thus inaccessible, you can use the thawing approach of example 1–11. Remove the fittings or faucet from the pipe end so the pipe will be open as close to the wall as possible. Use a pipe or tube that is smaller in diameter than the pipe you're trying to thaw; attach an elbow and a vertical member as shown, then pour hot water into the top of the vertical member as you push the assembly back as far as possible into the frozen

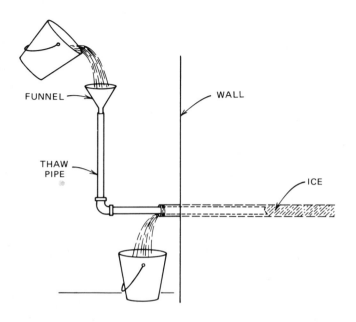

FUNNEL

WALL

THAW PIPE

ICE

example 1–11. Thawing an inaccessible pipe. The large vertical section on the insert tube is used to increase the head of the hot water being poured into the existing water pipe. (This operation is performed after removing the faucet and all other piping pieces as close to the wall as possible.)

pipe section. Place a bucket under the frozen pipe section's opening to catch the flowback and pour hot water into the pipe using a funnel. If you don't clear the pipe by the time the elbow reaches the opening, screw an extension onto the smaller-diameter piece and try again.

It isn't always necessary to use the vertical section, but an advantage is that it gives added head to the water, assuring a flowback of the water pouring into the pipe. Withdraw the pipe quickly after the flow begins, but don't stop pouring in water until the thaw pipe is completely withdrawn and the fixed pipe is cleared of ice.

Draining A Plumbing System

Plan to close down your house during part of the winter season? Be sure to drain your plumbing system. You'll probably find a drain valve located at the lowest point of your water distribution piping system. Drain the water closet, pumps (except submersibles, of course), water storage tanks, hot water heater, and any other appliance or fixture you can think of that might contain water that could freeze. When you have drained everything, pour a pint of antifreeze into the water closet trap, and about a cup into each other fixture trap.

Water Hammer

When you shut off the water at any faucet, is there a resounding *bam* that sounds like someone is striking the water pipe with a hammer? A moving column of water has inertia that is proportional to its weight and velocity. When you stop the flow suddenly (by turning off a faucet or a valve) you are converting this momentum into a high-pressure surge that manifests itself as a very loud shock load. The longer the line and the faster the liquid's velocity, the greater that shock load will be. In extreme cases, and where repeated hammering has occurred, the shock loads can be sufficient to break pipes and rupture fittings and valves.

Don't confuse water hammer with faucet chattering. Chattering occurs during the period water is flowing and may be attributable to flaws in the faucet, whereas water hammer always occurs as a result of sudden cessation of water flow.

There are several ways to eliminate water hammer, but most of them involve rerouting water pipes to avoid a long straight run. Usually, a couple of 90-degree turns in the line are all that is required.

One method will reduce water hammer even if it doesn't stop it altogether. Install a short (18 to 24 inches) length of pipe just ahead of the faucet, extending it vertically from the supply pipe. The pipe has to be capped to keep it watertight.

To provide reduced water hammer, quieter overall operation of faucets, and higher residual water pressures, the following water velocity limits or specific pipe sizes are recommended:

PIPE	DIAMETER	FLOW VELOCITY
Service mains	$1\frac{1}{4}$ inches	4 feet per second
Service branches	$\frac{3}{4}$ to 1 inch, depending on number of fixtures per branch	6 feet per second
Automatic washer, hose bibbs, wall hydrants	$\frac{3}{4}$ inch	6 feet per second
Bath, dishwasher, sinks	$\frac{1}{2}$ inch	6 feet per second

If you are planning a long run of pipe, you can predict the water hammer (in terms of surge pressure) with the aid of the nomogram in example 1–12. Once this is known, you can add the calculated surge pressure to the anticipated regular water pressure, which will give you the minimum pressure rating of the pipe you use.

At the outset you have to know the velocity of the water that will be going through the pipe (in feet per second), the total straight-run length of the pipe, and the closing time of the valve or faucet (in seconds). Place a straightedge from the appropriate value of *liquid velocity* (line A) to the corresponding pipe *line length* (line D).

example 1–12. Nomogram for predicting pressure surges from water hammer. Instructions for using nomogram are provided in accompanying text.

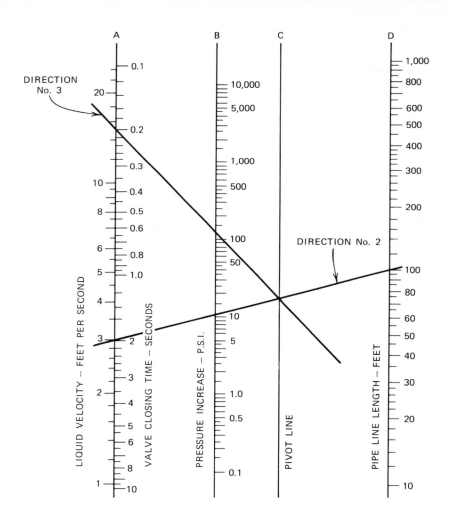

Make a small pencil mark on pivot line C where the straightedge crosses it. Now place the straightedge from the pivot line mark to your valve closing time (line A). Line B will now indicate the liquid momentum surge pressure that should be added to the expected pipe pressure to determine the pressure rating of the pipe for a given application.

As you can see by studying the nomogram, very short straight pipe runs don't contribute much to water hammer; and it is often possible to eliminate the problem completely by breaking up a long long straight run with a couple of short bends.

Water-Pressure Variations

When a variation in water pressure (and constancy of flow) occurs frequently, the most obvious place to start looking for the source of the trouble is in the piping. Check to see if the pipes are of the proper diameter for their length. Very often a family's needs will increase to the point where an original installation is outgrown.

If the pipes look all right for the job, there's a chance that mineral deposits are building up inside the supply piping. A large enough accumulation of limelike deposits can create a stricture in a pipe, effectively reducing its diameter. You can check this by disconnecting the pipes at a union and checking inside at that point. The only long-term correction for this problem is to replace the piping with something larger in diameter.

If the pressure variations are comparatively recent, try to pinpoint the cause by relating the time of origin to some change you have made in the plumbing. Have you added a new branch to an existing one? If so, you may find that the piping is overloaded as a result of the additional fixtures. A variation in water pressure may also occur when there is excessive friction in the pipe owing to too many fittings and changes in direction. If you can reroute the piping so as to cut down the number of turns, couplings, unions, and other fittings between the source and the faucets, you can sometimes increase the pressure substantially.

Electric water pumps are a notorious source of pressure problems. If your house is being served by a private water system (spring, pond, cistern, well) that requires its own water pump, check the pump installation first. When the pressure variations are quite rapid, there's a good chance that the tank associated with the pump is waterlogged or leaking. A leaking tank isn't always easy to repair, but waterlogging usually can be corrected by recharging the tank with air. (There is more on this in Section Five.)

Pipe Leaks

When a leak develops at the threaded joint of a pipe, one of the most likely suspects is a ruptured pipe. A casualty of water hammer is fracturing at the end of a pipe, because the end of the pipe is thinner than the body as a result of threading, which decreases the wall thickness and weakens the pipe accordingly. The risk of fracture as a result of water hammer increases when the threads of the pipe and fitting don't mate perfectly, that is, when the pitch is not the same on male and female ends and there is a small space adjacent to each thread.

In cold climates, freezing often causes pipes to rupture or crack. Sometimes the break isn't noticeable immediately because the crack may have begun as a hairline, enlarging itself with water hammer, extreme temperature changes, or time—or all three. Replacement is the only solution to a problem involving a broken or cracked metal pipe or tube. Some have experienced some success at solvent-welding a repair on a plastic pipe, but this is definitely not recommended for several reasons. First, there is nothing secure about a repair effected to the *surface* of a pipe—a new break or loosening at the repair is only a matter of time. Second, a plastic pipe is rated for certain pressures and temperatures. Making a repair using solvent may permit the pipe to be used but it will not permit it to be used at its full rated pressure or temperature.

Loose or cracked fittings are a fairly common cause of leaks in pipes, and they are considerably easier to replace than the pipes themselves. If you can't definitely see the leak at a fitting or joint—if you can't determine whether the fault is in the pipe or the fitting—then consider the fitting at fault and replace it first. And if you have to replace a pipe, first examine the piece you are replacing to see if it's corroded or limed up inside; if so, replace the section with a larger-diameter length.

To replace a leaky pipe, the first step is to cut the water supply to the line containing the leaky pipe. If you can control the water routing with a valve adjacent to the pipe, do it—that way the water can stay turned on throughout the rest of the house. If the leak is in a small section of pipe, remove that entire section. If there is no union on either end of the pipe, sever the pipe where the leak exists; the two remaining pipe sections can now be removed easily because they can be moved independently. You can reuse the fitting on one of the ends, but the fitting on the other end should be replaced with a union. When the leak exists at one small spot in a long length of pipe, you may repair it by replacing the leaky portion only. Cut the piece out and replace it with a pipe or nipple of the correct length; use a standard fitting at one end of the section and a union at the other.

Water-Pressure Surging

The problem of constantly surging water pressure is unique to water systems served by individual water pumps. The tank associated with a water pump is supposed to release a certain amount of water before the pump is turned on automatically—usually the amount is 8 gallons, but this is an average figure. When there seems to be no measurable quantity of water from the tap between surges, this indicates a tank that's become waterlogged or leaky. Or the pressure control associated with the tank and pump has been set incorrectly.

Look at the pressure gage adjacent to the tank while the water is running; you should be able to hear a distinct relay click when the water pressure drops to some specific value—usually 20 pounds per square inch—and another click when the pressure reaches maximum—normally 40 pounds. If the pressure goes from 20 to 40 to 20 rapidly, this signals waterlogging; there is too much water in the tank and not enough air. An air recharge should correct the problem.

Leaks in Water Tanks

Water corrosion is a common cause of leaks in tanks; but sometimes valves may fail to open when they should, causing a pressure increase that can rupture a tank. Generally, when a leak develops in a tank, it's an indication that the walls are thin and ready to break in other places. If the leaks are occurring along the seam only, replacement may not be necessary; a soldering or brazing operation may do the job nicely.

A relatively small leak can be repaired temporarily with a toggle bolt, a rubber gasket, and a brass washer, as shown in example 1–13. If the hole is too small for the toggle bolt, ream or drill it until it can accommodate the bolt. Push the bolt into the hole so the gasket is compressed between the tank and the brass washer, and tighten the bolt securely.

As an effective tank patch for a large hole, you will need both a temporary and a permanent patch. The temporary patch is a tapered soft wood plug. Insert it into the hole and tap it lightly with a hammer until the seal is watertight. Then saw off the portion of the plug that protrudes to the outside of the tank. This will hold until you have the chance to drain the tank for the permanent repair.

The permanent repair is made over the top of the temporary one. First, drain the tank. Then scrub the patch area thoroughly with steel wool, emery cloth, or a wire brush. Try for a large patch—about 6 by 6 inches—of shiny new metal, in the center of which is the soft wood plug installed earlier. Now cut a patch of rubber—the thicker the better—about 5 inches square from an old innertube. Cement this in place over the plug so that about an inch of shiny metal is exposed all around the rubber patch.

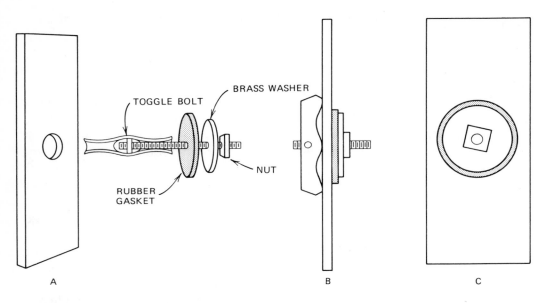

The last step is applying a metal patch that is larger than the rubber gasket but smaller than the periphery of the cleaned portion of the tank. The metal you use has to be the same material as the tank; otherwise, the dissimilar metals of the tank and the patch will cause rapid corrosion because of galvanic action, as electrons from one metal escape and join the other, weakening one of the metals in the process. Put the metal patch in place and solder or braze the outer circumference of the patch in one continuous seam.

TOGGLE BOLT

BRASS WASHER

NUT

RUBBER GASKET

A

B

C

example 1–13. The toggle bolt provides an effective tank repair. Use a rubber gasket and a brass washer as shown. If necessary, drill the existing leak hole larger to accommodate the toggle bolt.

Insulation Repairs

Insulation that is properly installed should last a long time. But occasionally insulation will get waterlogged and become ineffective, or it has to be removed in order to effect a repair in a valve or in the pipeline where it's installed. In some cases insulation can be repaired by merely covering the "bad" part with waterproof canvas or tar paper. But before attempting a repair, it would be wise to determine the cause of the problem. Otherwise, the repair may be a waste of time—the problem will recur.

If the insulation is of the rigid variety—in two half-sections—and does not appear to be waterlogged, you can remove the old worn-out covering and replace it with fresh waterproof paper or plastic. Example 1–14 shows how this is done. If you use tar paper, allow the paper end to overlap as shown, and arrange the covering so that the end of the paper winding points down. This keeps moisture from getting under the covering. Tape or band the insulation at the center, then do the ends.

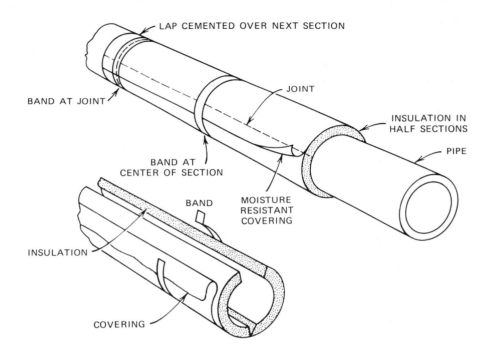

example 1–14. Insulation can be re-covered so long as it hasn't become waterlogged. Use a tar paper or plastic adhesive-backed wall covering. The end of a section of insulation should be about in the center of the applied covering, and the covering should be well overlapped to keep moisture out. Note that the edge, or end, of the covering is pointed down to protect against water entry.

For repairs to small sections of pipe, you can use one of the many available adhesive-backed plastic wall coverings for a temporary repair. Install it as you would tar paper—band it in place even though the adhesive can hold the plastic without the additional assist.

Sometimes a pipe hanger will come unfastened or break, causing the pipe to sag. Don't repair the insulation in such cases without repairing the hanger first. Make sure the pipe does not leak and that it is securely in position before rewrapping the pipe with insulation material.

Section 2

Most of the repairs required in home plumbing systems are simple enough to do without special training or experience. Most householders call in a plumber when a faucet's leak turns to a torrent or a drain clogs "hopelessly" because of a lack of knowledge of how a faucet is put together or where to gain access to the blockage in a drain pipe, not because of the complexity of the job. This section should give you all the information you need to repair water faucets, water-line pipes and valves, and the insulation that goes over in-house hot water pipes.

Faucets, Valves, and Piping Insulation

Section 2

Faucets
and
Valves

In the context of this section, a faucet and a valve are similar but not the same. A *faucet* allows water to be drawn from a pipe or a main. A *valve* controls the flow of water *through* a pipe or main; it may be automatic or manually operable, depending on its purpose.

Faucet
Repairs

Example 2–1 illustrates a simplified fresh water distribution system for a tub, shower, kitchen sink, and other utilities. Briefly, water flows from the domestic water supply through the main and is diverted or split at the water heater. The cold water piping continues its routing throughout the house accompanied by a second water line that carries hot water. On the heater, a pressure relief valve vents steam or water excesses for safety. In all but very old installations the cold and hot water lines are brought together in sinks, tubs, lavatories, and showers, where a mixing faucet (inset, example 2–1) allows a flow of water at any desired temperature. In earlier houses, hot and cold water are accessed with individual taps or bibbs or with two-handle combination faucets.

In principle all faucets are alike, though they differ considerably in construction and in outward features. Unfortunately, you can't learn one repair procedure that will apply to all faucet types. The following repair procedures include instructions for both mixing and bibb-type faucets. Do read the *general* information under each "trouble" heading before starting the repair.

Dripping Faucets

One of the most costly wastes in the average household is water, the result of faucets in need of nothing more than a simple washer. One leaky faucet that *drips*—not *flows* —continuously can waste as much as 8000 gallons of water a year! Imagine what several leaky faucets can do to your pocketbook.

It may not be possible for you to know in advance the size of washer your faucet uses, but most hardware stores sell "washer kits" that contain a wide variety of sizes. The cost is negligible, rarely more than a dollar; they will pay for themselves easily by the water they conserve.

Begin your repair job by shutting off the water flow to the faucet. The water supply line in most kitchens and bathroom sinks is immediately beneath the fixture. Tubs and showers present special problems because their water supply lines are rarely easily

example 2-1. Simplified diagram of typical home water distribution system. Arrows indicate direction of water flow. Faucets and valves are represented by enclosed Xs. The inset shows a modern single-lever mixing faucet.

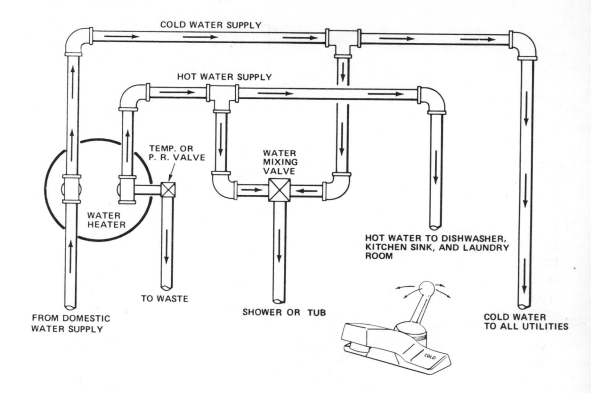

COLD WATER SUPPLY

HOT WATER SUPPLY

TEMP. OR P. R. VALVE

WATER MIXING VALVE

WATER HEATER

TO WASTE

FROM DOMESTIC WATER SUPPLY

SHOWER OR TUB

HOT WATER TO DISHWASHER, KITCHEN SINK, AND LAUNDRY ROOM

COLD WATER TO ALL UTILITIES

COLD

accessible. If you can't find a control valve in the vicinity of the plumbing fixture, your only recourse is to shut off the master water supply line at the main. With the proper tools at hand, however, and a handful of washers of various sizes, you can be reasonably sure the repair won't take long, unless you run into such complications as a burred surface under the old washer, or worn-out packing.

Individual Faucets. Example 2–2 shows various faucet types. The following repair procedure applies to both bibb-type units and combination types that have individual handles for cold and hot water; it does not apply to mixing faucets that have a single lever or knob controlling both hot and cold water flow.

Disassemble the portion of the faucet that requires the repair; there is no need, for example, to dismantle the cold-water side of a combination faucet that has a leak in the hot-water side. Start by removing the handle, packing nut, and stem. Example 2–3 shows all faucet parts and their names, as they apply to the basic individual and combination assemblies. The stem itself may have either

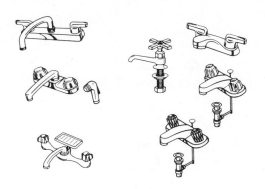

example 2–2. Traditional faucets. Those with individually controllable hot or cold water taps are similar in design. (Courtesy Melard Manufacturing Corporation.)

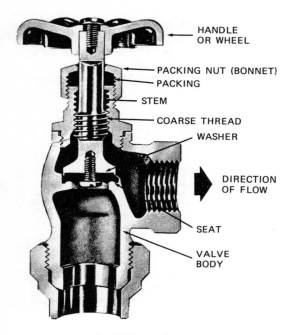

HANDLE OR WHEEL

PACKING NUT (BONNET)

PACKING

STEM

COARSE THREAD

WASHER

DIRECTION OF FLOW

SEAT

VALVE BODY

A   BIBB—TYPE VALVE

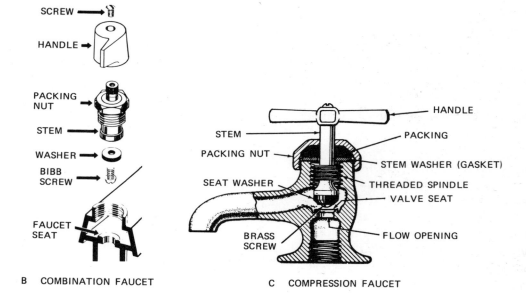

SCREW

HANDLE

PACKING NUT

STEM

WASHER

BIBB SCREW

FAUCET SEAT

B   COMBINATION FAUCET

STEM

PACKING NUT

SEAT WASHER

BRASS SCREW

HANDLE

PACKING

STEM WASHER (GASKET)

THREADED SPINDLE

VALVE SEAT

FLOW OPENING

C   COMPRESSION FAUCET

example 2–3. Standard globe valve (A); it is similar in construction to the conventional faucet or bibb cock. B shows one side of a combination faucet. The traditional compression faucet is pictured in C. (Part B courtesy Melard Manufacturing Corporation.)

left-hand or right-hand threads, depending on whether the tap is for hot or cold water.

Once the packing nut has been loosened completely, you can unscrew the stem by placing the faucet handle in place temporarily and using it as a wrench.

The replaceable washer is attached to the bottom of the stem with a small machine screw, usually a standard 8–32 (that is, gage 8 with 32 threads to the inch). Use a conventional flat-blade screwdriver to remove the washer-holding screw.

When you remove the old washer, examine the cup area where the washer fits. If the cup area is corroded or coated with mineral deposits, it must be cleaned before reassembly. You may scrape the residue out with a small screwdriver, but be sure to avoid nicking the surface or sides; small nicks and scratches could create channels that wouldn't be sealed with the new washer even when it's compressed fully.

You may find ball-bearing washers in some older faucets. These ball bearings, between the stem and washer holder, permit movement of the washer independent of the stem.

This allows the washer to stop its rotation on the slightest contact with the seat, thereby reducing the frictional wear on the washer.

Occasionally you'll come across a faucet that has managed to escape attention for many years, and find the washer holding screw can't be removed regardless of the force applied to it. The screw head will become distorted and the screwdriver-blade channel will be destroyed. When this happens, try to lock a pair of plier jaws onto the protruding part of the head (Vise-Grip pliers are ideal for this) and hold the body of the stem with a vise or another pair of pliers. Be sure, however, to hold the stem in such a manner as to avoid damage to the stem threads from the plier or vise jaws.

If the holding screw breaks during this process—it's not uncommon—you must either replace the stem or retap the existing one. If you have a hardware store in your neighborhood, you can take the old stem with you and buy another that matches it in design. Larger hardware stores carry "blister-packed" stems and other faucet replacement parts. If you can't

find one and don't feel it worthwhile to replace the faucet itself, you must drill a new hole through the brass screw and tap new threads.

It's a good idea to use a brass screw instead of one of nickel or steel for holding the washer in place. Brass does not corrode under normal conditions, and it generally matches the material of the stem itself. When two dissimilar metals come into contact with each other and maintain contact for long periods, a rudimentary galvanic cell is formed if the pH of the water is very low. (An actual battery may be formed from cells containing dissimilar metals immersed in acidic solution.)

If a washer requires frequent replacement, it may be the wrong type or the seat may be rough or nicked, scoring the washer as soon as it's installed. Use a flat washer on seats that have a crown or a rounded ridge. Tapered or rounded washers should be used with tapered seats. Note that the compression faucet of example 2–3C has a tapered washer rather than a flat one. The repair procedure isn't different for this type; the important thing is to properly orient the washer. Remove the faucet handle, packing nut, and stem; then remove the brass screw holding the washer to the bottom of the spindle. Replace the washer with a new one that is flat on one side and crowned on the other so it can get both horizontal and vertical compression to provide a good, tight seat.

Use a good-quality hard composition washer because leather (common on older, larger faucets) and other soft-material washers do not give good longevity, particularly in hot-water lines.

Examine the valve seat and repair or replace as necessary before replacing the spindle and stem. If you don't do this, you'll find the new washer giving adequate service for only a short time.

Reface or ream solid-type seats with a faucet seat dresser consisting of a cutter, stem, and handle. (You may also use a fine-grit globe grindstone—an abrasive ball mounted on the end of a 5- or 6-inch shaft that has a cross handle on its opposite end.) Rotate the tool with the cutter centered and held firmly on the worn or scored seat. But be very careful not to overream. Remove all grinding debris before reassembling the faucet.

Solid seats can be replaced with renewable types by tapping a standard thread into the old solid seat and simply inserting the renewable one. Remove renewable seats with a regular seat-removing tool or an Allen wrench. (Replaceable seats have square or hex-shaped water passages, into which the wrench fits.) If the seat seems frozen to the body, you can normally loosen it by applying a little kerosene to the spot where they're joined. Tapping, reseating, or replacement of faucet seats can be accomplished without removing the faucet from its fixture.

Mixing Faucets. Single-lever faucets aren't nearly as standardized as compression faucets, but the general construction characteristics are similar enough to make repairs simple if you have the materials. Some firms specialize in the manufacture and sale of specialized repair kits for these faucets, so your most serious problem will be to determine which of the kits contain parts that fit your particular faucet.

Some manufacturers, such as Melard Manufacturing Corporation (153 Linden Street, Passaic, New Jersey 07055), include a special tool with their kits that can be used to loosen and retighten setscrews, remove and replace "spanner" type nuts, and so on. Usually, this tool is all you'll need other than conventional household tools. Typically, the tool is small and simple—not an expensive accessory at all but adequate for the one-time repair task at hand.

Faucet repair kits typically cost less than two dollars and contain not only the special wrench to dismantle the faucet but the O-rings, metal springs, rubber seats, bushings, washers, and so on needed for any mixing-faucet leak, whether from the spout or around the neck or lever. It's generally a good idea to use all kit components rather then just those components that correct a specific problem; this will give your faucet a whole new lease on life.

Example 2–4 shows typical mixing faucets in exploded view; the nomenclature matches those names you'll see identifying the kit parts.

The wrench that comes with the repair kit (example 2–5) will come in handy right from the start; use the Allen end to loosen the setscrew so that you can pull off the handle (see

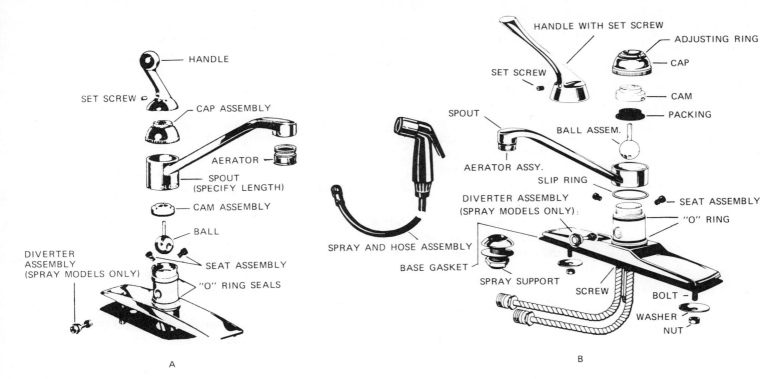

HANDLE

SET SCREW

CAP ASSEMBLY

AERATOR

SPOUT
(SPECIFY LENGTH)

CAM ASSEMBLY

BALL

DIVERTER
ASSEMBLY
(SPRAY MODELS ONLY)

SEAT ASSEMBLY

"O" RING SEALS

A

HANDLE WITH SET SCREW

ADJUSTING RING

SET SCREW

CAP

CAM

PACKING

SPOUT

BALL ASSEM.

AERATOR ASSY.

SLIP RING

DIVERTER ASSEMBLY
(SPRAY MODELS ONLY)

SEAT ASSEMBLY

"O" RING

SPRAY AND HOSE ASSEMBLY

BASE GASKET

SPRAY SUPPORT

SCREW

BOLT

WASHER

NUT

B

example 2–4. Mixing faucets in exploded view,
with nomenclature of integral
components. Pictured: Delta (A)
and Peerless (B) faucets.

example 2–5. A simple wrench such as this typically accompanies a mixing-faucet repair kit. On one end is an Allen wrench; on the other, a spanner. (Courtesy Peerless Faucet Company.)

example 2–6). Then unscrew and remove the cap assembly (you may need the special tool again). Pull up on ball stem to remove the cam and ball assembly. Now all you have to do is slide the rubber boots over the two springs (seat assemblies) and replace those in the dripping faucet. Also replace the cam assembly, which consists of a nylon bushing and a mating bell-shaped washer. Check the drawings of example 2–4 once more to make sure you have put everything in its proper place, then start reassembling the faucet. Be sure to tighten the adjusting ring or cap assembly (example 2–6B) until no water leaks around the stem when the faucet is on and pressure is exerted on the handle, which forces the ball into its socket.

Leaky Faucets

Faucets don't always leak at the water outlet; one particularly troublesome leak source is the point where the stem enters the faucet body. When this happens, the packing is loose or the bibb gasket or O-ring is bad.

Individual Faucets. With the main water supply to the faucet shut off, remove the faucet handle or wheel and then remove the packing nut or bonnet. Examine the bibb gasket, O-ring, or packing inside the nut. The packing should be a compressible material superficially resembling a washer; if there is no packing, you'll see an O-ring around the body of the stem as shown in example 2–7A. Remove and replace the O-ring (kits are available from hardware stores for virtually every faucet type); if packing material is required, pry out the old material with a small screwdriver and replace with new packing (example 2–7B). Now put the faucet back together again, turn on the water supply, and you're all set for another two or three years.

Mixing Faucets. There are several potential sources of leaks in mixing faucets, the most likely being a loose adjusting ring. For this repair, leave the main water supply turned on. Use the kit wrench to remove the faucet handle (example 2–8), then turn the

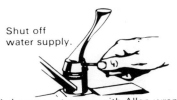

Shut off
water supply.

A. Loosen set screw with Allen wrench end of Peerless wrench and lift off handle.

B. Unscrew cap assembly and lift off.

C. Lift out ball and cam assembly by pulling up on stem.

D. Remove old seats and springs and insert new ones. Reassemble faucet in reverse order.

example 2–6. Mixing faucet repair procedure. (Courtesy Peerless Faucet Company.)

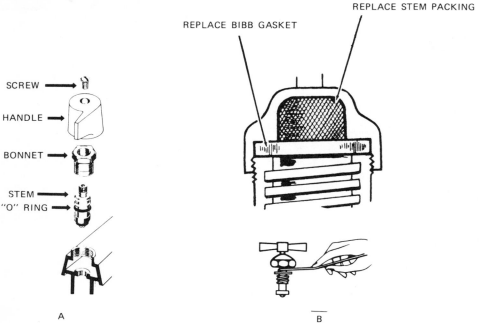

REPLACE BIBB GASKET

REPLACE STEM PACKING

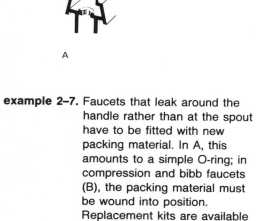

SCREW →

HANDLE →

BONNET →

STEM →
"O" RING →

A

B

example 2–8. Around-the-handle leaks on mixing faucets can often be remedied by removing the handle and tightening the adjusting ring.

example 2–7. Faucets that leak around the handle rather than at the spout have to be fitted with new packing material. In A, this amounts to a simple O-ring; in compression and bibb faucets (B), the packing material must be wound into position. Replacement kits are available for both. (Part A courtesy Melard Manufacturing Corporation.)

wrench end for end and tighten the
adjusting ring. Replace the handle and
check the faucet for leaks again. If the
leak persists, you have two
possibilities remaining: faulty O-ring
seals or a deteriorated plastic ball.

Start with the most likely problem:
O-rings. Shut off the water supply
valve and dismantle the faucet
according to the procedure of example
2–6A, B, and C. Then gently rotate the
spout and lift it off to expose the
O-rings (example 2–9A). Replace
existing O-rings with the new ones
contained in the repair kit. Make sure
that you match the O-rings size for
size; even when these O-rings have
the same overall diameter, they're
often of several varying thicknesses. If
you replace a thick O-ring with a thin
one, or vice versa, the faucet won't go
together easily, and when it does, it
will leak.

As with all mixing-faucet repairs,
remember the advisability of replacing
all replaceable parts at the same time,
whether the existing part appears to
need replacement or not.

If, when you reassemble the
faucet, the leak is back, the trouble
lies with the little plastic ball (or its
equivalent). This part comes in a

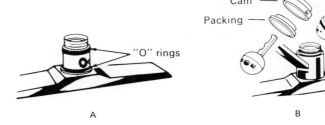

A

B

example 2–9. To replace O-ring (A), cut the
old ones off, then stretch new
ones and snap them into place,
being sure to get different sizes
into proper grooves. Sketch B
shows ball replacement: slip
packing and cam assembly
onto new ball and insert into
socket.

separate kit, which you'll have to buy or send for. Once you have the piece, the repair is easy: just drop it into position (example 2–9B) and put the faucet back together.

Sink-Spray Repair

Most sink-spray problems occur at the nozzle or at the place where the spray hose connects to the faucet. Both these conditions almost always require replacement of the complete spray and hose assembly. When the problem is nothing more serious than a leak that is situated away from the spray nozzle and connector, the repair is simple; all you'll need are these items: (1) 3-inch length of $\frac{3}{8}$-inch-diameter threaded brass tubing (the type used as the threaded shaft in household lamps is ideal); (2) a sharp knife; (3) contact cement.

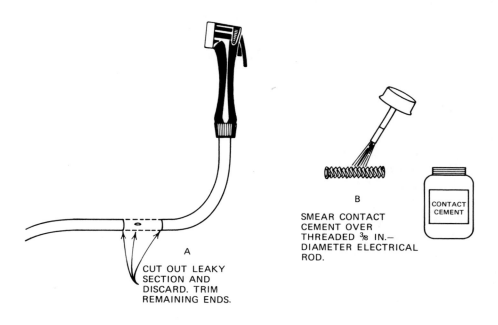

A

CUT OUT LEAKY SECTION AND DISCARD. TRIM REMAINING ENDS.

B

SMEAR CONTACT CEMENT OVER THREADED $\frac{3}{8}$ IN.— DIAMETER ELECTRICAL ROD.

example 2–10. Repairing a sink-spray hose is simple enough if you have a length of $\frac{3}{8}$-inch-diameter threaded rod such as the type used in lamp making. Sequence shows repair procedure.

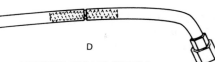

C

WHILE CEMENT IS STILL TACKY, SCREW ROD INTO HOSE ENDS.

D

FINISHED REPAIR SHOULD BE FREE OF SURFACE BUMPS AND SHOULD SLIDE READILY INTO SINK RECEPTACLE.

The repair procedure is shown in example 2–10. Use a sharp knife to cut out the worn section and trim the remaining ends so that they'll butt together with a flush fit. Smear a generous amount of contact cement over the threaded lamp rod and allow it time to dry to a tacky-but-not-wet feel. Screw the rod first into one hose end and then into the other, but exercise care not to let the inside get jammed with cement. Smear a little additional cement around the joint and let it dry thoroughly. Wipe off excess cement, turn the water back on, and the repair is finished.

If your existing sink spray is leaking at the spray head itself, you might be able to solve the problem with a simple washer. At the lower end of your hand-held spray, unscrew the knurled collar to expose the washer between the collar and spray head. The washer is a standard part available from most hardware stores—indeed, the washer is often the same type as that found in the faucet. Examine the retaining clip, too, to make sure it hasn't loosened or become bent out of shape. Replace it if necessary.

If the existing sink spray leaks at a point within the spray head, or if it leaks at the connectors, replacement is necessary. Most replacement sink sprays, including the Melard unit pictured in the sketches that follow, are *universal types;* that is, they are designed to fit virtually every faucet type you're likely to come across. The Melard spray, for example, comes with a special connector adapter that permits installation in either male or female faucet mountings.

First, remove the spray head from the hose (we're talking about the parts you get when you buy a replacement kit, not the parts that are to be replaced). Remove the retaining clip (example 2–11) so that the entire collar assembly can slide off. When you've done this, push the new hose piece up through the existing sink spray next from the bottom, then replace the collar, retaining clip, and spray head.

Unscrew the existing hose connector where it taps into the faucet (example 2–12A), then discard the old hose portion. If the faucet tap point is a female-thread fitting, prepare the male end according to the

SPRAY HEAD

RUBBER WASHER

CLIP

COLLAR

example 2–11. Unscrew the knurled collar from the lower portion of spray head to replace leaky washer. Replace clip if necessary, then reassemble spray head.

example 2–12. To install new sink spray with a universal kit, remove old spray hose (A), then insert hose end from new unit. The Melard kit's hose connector screws directly into a female receptacle (B); if the existing connector is a male part, an adapter included with the kit can be used as shown in C.

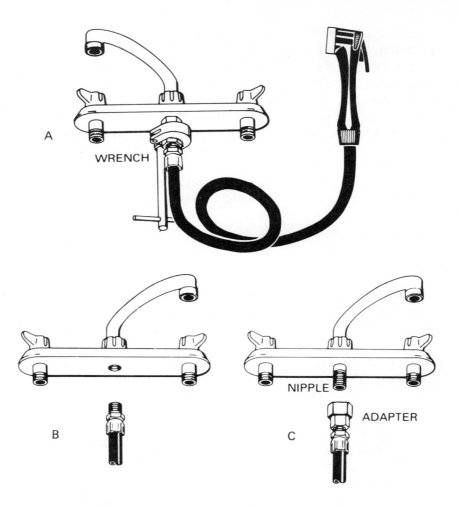

A

WRENCH

B

C

NIPPLE

ADAPTER

manufacturer's instructions and screw it directly into place (example 2–12B). If the existing tap point is a male end (example 2–12C), use the adapter furnished with the universal replacement kit. Again, prepare the male ends as required and screw the hose end onto the adapter. Tighten securely.

Shower-Head Problems

About the only trouble you're apt to have with a shower head is uneven flow or insufficient water pressure. You can't do much at the shower head to change the total water pressure applied to the head, so repairs are in general confined to fixing flow problems.

Erratic flow is almost always attributable to the presence of mineral deposits on the back of the perforated faceplate or in the tiny water channels of the head. Remove the perforated faceplate and clean off the back of it with a fine-grit sandpaper or steel wool. You can almost always free stopped-up shower-head holes with an appropriate-size needle.

Repairs such as the one described are generally only stop gaps. If the water in your area contains a heavy concentration of minerals, you'll find the trouble recurring.

Noisy Faucets

Occasionally a faucet will be noisy when water is flowing through it. This may be attributable to a loose washer, but it's equally likely that the stem's threads are so worn that the stem and handle vibrate according to the water pressure. To determine the source of faucet chattering, turn the water on and press down hard on the faucet handle (toward the pipe). If the vibration stops when you apply fairly heavy pressure to the handle, the faucet's stem doesn't fit properly (because of worn threads or because someone has mistakenly replaced the stem with a part from another faucet). If the vibration continues when you apply pressure, the problem is most likely a worn-out washer.

Before you buy another stem to replace one in a chattering faucet, you should determine whether the worn-out threads are on the stem or the stem receiver. If the receiving threads are worn excessively, a new stem will not eliminate the trouble entirely. Some faucets—the more expensive ones—are built with replaceable stem receivers, but many are not. If yours does not have a replaceable stem receiver, you must replace the complete faucet.

One particularly effective method of stopping water hammering is to install an air chamber in the water supply line to the faucet giving the trouble. Example 2–13 illustrates a typical installation. As shown, the air chamber is nothing more than a capped length of pipe that extends from the most convenient horizontal run to any level higher than the faucet itself. In the pictured installation, the chamber is positioned behind the wall; but in cases where the fixture already exists and no access to the behind-the-wall piping is provided, the chamber may be positioned against the wall on the inside, next to the fixture. Ordinarily, putting in the air chamber is no more complex than replacing an elbow with a standard tee fitting.

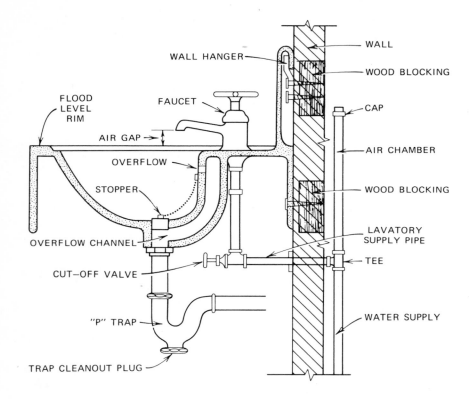

WALL

WALL HANGER

WOOD BLOCKING

FAUCET

CAP

FLOOD
LEVEL
RIM

AIR GAP

AIR CHAMBER

OVERFLOW

STOPPER

WOOD BLOCKING

OVERFLOW CHANNEL

LAVATORY
SUPPLY PIPE

CUT-OFF VALVE

TEE

"P" TRAP

WATER SUPPLY

TRAP CLEANOUT PLUG

example 2–13. Cutaway of lavatory showing supply and drainage piping. Note air gap below faucet discharge point; the air gap should be twice the diameter of the drain opening.

Valve Repairs

After extended use and following repeated repairs, some valves will no longer give tight shutoff; they must be replaced. When this is necessary, it's usually advisable to upgrade the quality of your installation with equipment having better flow characteristics and longer-life design and construction materials.

In some cases, ball valves will deliver more water than globe valves. And some globe valves deliver more water flow than others, even for identical pipe sizes. As a general rule, Y-type globe valves—in straight runs of pipe—have better flow characteristics than straight stop valves. Example 2–14 shows the principal features of the most common controllable valve types.

Globe Valves

The *globe valve* got its name from its globular body. It is used for control of liquids, gases, and vapors by means of throttling. In the home, this type of valve is most frequently found on main supply lines. Example 2–15 is a

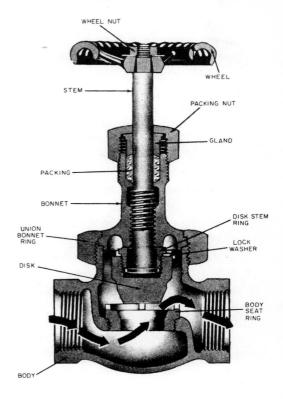

example 2–14. Various valves. In A, note the large passage for water in the globe valve. B shows a Y-pattern globe valve, which allows almost straight-through water flow with little resistance. The ball valve (C) allows straight-through flow because the hemispherical stopper simply rotates to obstruct or allow water flow.

example 2–15. Globe valve cutaway showing valve parts with their nomenclature and the flow of water through the valve.

detailed cutaway of a globe valve showing all the parts and their nomenclature. The closure disk and seats increase resistance to flow and permit close regulation of the amount of water passing through the line. In this type of valve, the fluid flow is normally proportional to the number of turns of the wheel in opening or closing the valve.

Globe valves can be fitted with fiber disks that are suitable for almost any type of service. If your home is heated by radiators, then you'll likely find a valve of this type somewhere in the steam supply line. When the globe valve is used to control the flow of steam, then metal disks instead of fiber must be used.

The globe valve is designed to minimize erosion, in which case lower maintenance costs are likely where a wide range of flow is desired; but operating costs may be greater since globe valves offer more resistance to flow than a gate valve and cut down line pressure.

Gate and Ball Valves

Gate valves and ball valves allow water to flow through in a straight line. The *ball valve,* shown in example 2–14C, is shut off when the hemispherical stop cock is rotated so that it blocks the flow; in the open position, the hemispherical section rotates to a position 90° from the pipe. The gate valve (example 2–15) contains a gatelike disk that is moved by a stem screw attached to the hand wheel. Both these valve types offer little friction to water flow and contribute negligible water-line pressure drop (providing the valve disk or stop cock is kept fully closed or fully opened).

The *gate valve* differs from the globe valve in that it releases a variable amount of water with each turn of the wheel; the second turn, for example, may release three times the water of the first turn, and the third may release five times as much. Gate valves and ball valves should *not* be used for throttling; they should always be operated in either the fully opened or fully closed position. It may be impossible to throttle flow with a ball valve, but it can be managed with a gate valve. A partly closed gate will cause vibration and chattering. If left in the partial-throttle position, the valve will suffer damage to the seating surfaces.

Ball and gate valves will last a long time if they are opened and closed at infrequent intervals. But if they are operated many times a day on a daily basis, they'll wear out quickly. If you find it necessary to operate such a valve frequently, replace it with a globe valve.

When you repair a gate valve, you'll find the most wear on the downstream faces of the seat and disk, because the line pressure forces all the wear on these surfaces. Since the upstream faces are seldom damaged in this way, the valve will have a longer life if worn gate valves are reversed.

There are four types of disks commonly used in gate valves: solid-wedge, flexible-wedge, split-wedge, and double-disk. Perhaps the most common of these is the solid-wedge disk, a single moving part that will not jam because of misalignment of mating parts, whether the stem is up, sideways, or down. This type is ideal for supplying steam to a radiator distribution line, and it's well suited for water and other fluids because it allows good flow of turbulent streams without chattering or vibration.

Double-disk valves and valves with split-wedge disks have wedges (the cylindrical section under the disk in example 2–16) that come in several parts. As the valve is tightened after closure, the wedge or spreader forces the disk outward and hard against the body seat. The first opening turn releases the disk sections, and continued turns raise them clear of the seat openings. Such parallel seats can be repaired or replaced easier than those accommodating a tapered-wedge disk (the type pictured).

It is generally advisable not to use a double-disk valve in a radiator line when the stem is to protrude downward at any angle, even slightly. Such use often results in vibration and causes the disk and seats to wear inordinately fast.

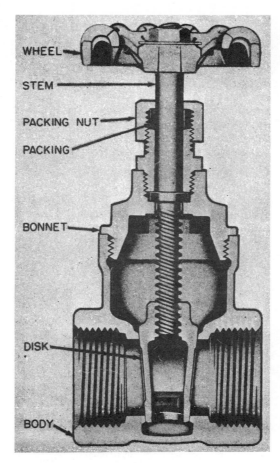

WHEEL

STEM

PACKING NUT

PACKING

BONNET

DISK

BODY

example 2–16. Gate valve cutaway.

Repair Procedures

For the most part, valve repair will be limited to globe and gate valves; other types of valves may be repaired in much the same manner as globe valves, with slight modifications as necessary or appropriate. Valve repair other than routine packing is generally limited to overhaul of the seat and disk assemblies. However, all other parts of the valve should be inspected carefully and replaced if necessary.

Spotting-In Valves. The method used to determine visually whether or not the seat and the disk make good contact with one another is called *spotting-in.* To spot-in a valve seat, first apply a thin, even coat of prussian blue (a common lubricating oil and dye available at hardware stores, sheet-metal shops, and so forth) over the entire machine-face surface of the disk. Then insert the disk into the valve and rotate it a quarter turn using a light downward pressure. The prussian blue will adhere to the valve seat at points where the disk makes contact.

After you have noted the condition of the seat surface, wipe all the dye off the disk face surface. Apply a thin, even coat of prussian blue over the contact face of the seat and again place the disk on the valve seat and rotate the disk one quarter turn. Examine the resulting blue ring on the valve disk. The ring should be unbroken and of uniform width. If the blue ring is broken in any way, the disk is not making a proper fit.

Grinding Valves. To grind-in a valve, apply a small amount of grinding compound to the face of the disk. Insert the disk into the valve and rotate the disk back and forth about a quarter turn, shifting the disk/seat relation through several rotations. During the grinding, the compound will gradually be displaced from between the seat and disk surfaces; therefore, it is necessary to stop every minute or so to replenish the compound. When doing this, you should wipe both the seat and the disk clean before applying the new compound to the disk face.

Once the irregularities have been removed, spot-in the disk to the seat as described in the preceding subsection. Grinding is also used to follow up all machining work on valve seats or disks. When the valve seat and disk are first spotted-in after any machining operation, the seat contact will be very narrow and located close to the bore. Grinding-in, using finer and finer compounds as the work progresses, causes the seat contact to become broader. The contact area should be maintained as a perfect ring, covering approximately one-third of the seating surface.

Don't overgrind a valve seat or disk. Overgrinding tends to produce a groove in the seating surface of the disk. It also rounds off the straight angular surfaces of the disk. Machining is the only means by which overgrinding should be corrected.

Lapping Valves. It is important to avoid the temptation of using the metal valve disk as a lapping tool. A lap resembles a reseating tool; it has a circular surface of exactly the same size as the disk and is used to true the valve seat surface following a grinding operation. With the proper compound applied, the lapping tool allows you to remove slightly larger irregularities from the valve seat than can be taken out with the grinding operation.

The compounds used for grinding and lapping come in a variety of abrasive grades, from very coarse to superfine. When the seat is extensively corroded or out of round, use the coarse grade. The medium grade should be satisfactory for smoothing over gouges that aren't too deep and for following up after working with the coarse grade. The fine grade is used when the reconditioning operation nears completion. Finish lapping and grinding-in should be accomplished with the superfine grade. Always use clean compound for lapping, and replace the compound often. Try to spread the compound evenly and not too heavily, then lap no more then necessary to produce a seat that is even and smooth.

In the lapping operation, occasionally lift the lap and reseat it again in another position on the seat; this changes the relationship periodically between the lap and the valve seat so that the lap will gradually and slowly rotate around the entire seat circle. You have to keep a more or less constant check on the working surface of the lap to be sure it's flat and smooth; if a groove develops, the lap is being affected by your operation more than the valve seat—and you will have to reface the surface of the lapping tool.

Two more points to remember in any valve reseating type of lapping operation: (1) always use a fine grinding compound to finish the lapping job, and (2) don't forget to spot-in and grind-in the valve to the seat following the lapping operation.

Faucet and Valve Replacement

Faucets are generally a bit more complicated than valves to install. A valve may be inserted in any existing pipe line exactly as if it were a pipe fitting; that is, the valve screws into position and remains there. A faucet, however, has a finish that must not be marred. It must be installed so that it is functional without sacrificing appearance, mounted solidly enough to insure that it will become as one with the fixture it is associated with, and stay that way through many thousands of uses.

But this doesn't mean that it is particularly difficult to install a faucet properly. On the contrary, there are replacement assemblies for virtually every type of faucet, and they come from the package with instructions written with the do-it-yourselfer in mind.

The first step in any faucet replacement operation is to examine the fittings that connect your existing faucet to the water supply lines. Because the new faucet will have to connect to these same lines, you must

be sure that the replacement faucet you buy will fit. It's also a good idea to measure the distance from the sink's faucet-mounting holes to the supply line connection point; if this distance is greater than 8 or 9 inches, you may need a length of extension tubing. Before buying any extension, however, see if there is any way to get the supply line end closer to the sink bottom. (Sometimes this can be achieved by merely straightening a gently curved water supply line or removing the subfloor slack in one.)

Another important point to remember: the faucet's fittings are immobile; the fittings that will be wrench-tightened must be on the supply lines.

The faucet most likely to need replacement is the one serving the kitchen sink; this typically gets 10 to 20 times more use than lavatory or bath or shower faucets. So let's install a typical modestly priced mixing faucet in a kitchen sink to replace a worn-out combination faucet. Our new faucet is the Peerless unit pictured in example 2–17, which contains the faucet assembly, rubber gasket, mounting hardware, and sink spray.

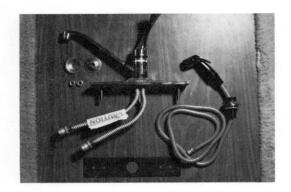

example 2-17. This replacement mixing faucet (from Peerless) comes with everything but a basin wrench, and you might not even need one to do the job.

The manner in which the existing faucet mates to the supply lines will tell us whether or not we need an adapter for the replacement. If the old faucet's water inlets are $\frac{1}{2}$-inch IPS (iron-pipe size) male fittings that are within 8 or 9 inches from the sink mounting spot, we're in business. If the existing inlets are nonstandard, say, female fittings or male fittings other than $\frac{1}{2}$-inch IPS, we'll need some form of adapter.

So our first step is to examine carefully the existing arrangement under the sink. Don't be discouraged if you see a sweat-soldered connection between copper tubing and iron or plastic pipe under the sink; it may well be that someone else has already installed an extension.

The first thought on seeing installations such as the two pipes shown in example 2-18 might be that considerable work lies ahead. At the left, a galvanized pipe terminates in a valve that is sweat-soldered to a copper tubing; at the right a polyvinyl supply line terminates in a standard valve that is itself nonmovable (how can we attach a connector when neither of the two fittings can be rotated?). Also, note that both lines

example 2-18. Connections like these might look like a plumber's nightmare. But chances are you won't have to break the lines at these valves. When faced with hopeless looking supply lines like these, make a closer inspection; you'll probably find conventional fittings at the faucet (under the sink).

are well below the required proximity to the sink.

More often than not, such hard-to-manage connections as these have already been accommodated by the plumber who installed the faucet in the first place. Instead of selecting this point for the connection, trace the tubing all the way to the faucet. Nine times out of ten you'll find one of the three connections shown in example 2–19, and the connection point might be right at the base of the faucet itself.

A word about these connectors: the slip-joint connection shown in example 2–19A might be entirely adequate for mating to a rigid tube, but it is anything but that for mating to the corrugated lines of the typical replacement faucet. The slightest movement of the corrugated line with respect to the existing supply line will result in a leak. By playing around with the lines, you can probably stop the leak temporarily, but it will recur over the lifetime of the faucet. If your supply line has this slip-joint type of fitting, be sure to buy a replacement faucet that has rigid inlets. If you have to purchase an extension line, make sure you get the type with either ball-nose

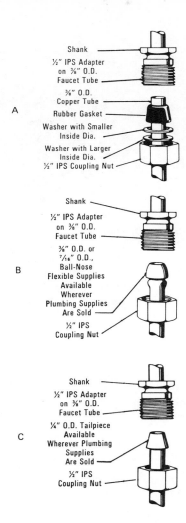

A

Shank
½″ IPS Adapter
on ⅜″ O.D.
Faucet Tube
⅜″ O.D.
Copper Tube
Rubber Gasket
Washer with Smaller
Inside Dia.
Washer with Larger
Inside Dia.
½″ IPS Coupling Nut

B

Shank
½″ IPS Adapter
on ⅜″ O.D.
Faucet Tube
⅜″ O.D. or
⁷⁄₁₆″ O.D.,
Ball-Nose
Flexible Supplies
Available
Wherever
Plumbing Supplies
Are Sold
½″ IPS
Coupling Nut

C

Shank
½″ IPS Adapter
on ⅜″ O.D.
Faucet Tube
¼″ O.D. Tailpiece
Available
Wherever Plumbing
Supplies
Are Sold
½″ IPS
Coupling Nut

example 2–19. The three most common water-line connections. A shows a slip-joint connection that can't be used with corrugated faucet lines; B and C are "ideal" fitting types—ball-nose (B) and tailpiece(C).

(example 2–19B) or tailpiece (example 2–19C) ends.

The ideal tool for removing the existing faucet is a *basin wrench,* a swiveling tool that lets you get plenty of turning torque in tight places (see drawing in example 2–12A). But since the installation is a one-time job and won't require occasional disassembly, you can probably get by all right with a *good-quality* pair of locking pliers. The quality is stressed because the pliers will have to stand up under side-twisting torque—the kind that cheap import wrenches will buckle under. The photo in example 2–20 shows how to use the locking pliers for disconnecting the supply line when it's very close to the faucet.

Adjust the pliers so they snap shut without excessive pressure when the jaws line up with the flats of the connector. If any play exists in the pliers, tighten the handle adjustment slightly and snap into place again. Start loosening the connector; as soon as the connection is loose enough, complete the loosening operation with your fingers.

The next step is removal of the large nuts holding the faucet in place. You can try to do this with the locking

example 2–20. You can try Vise-Grips on the existing fasteners if you like, but don't gamble with more than one set of flats; if you deface one set, give up and get a basin wrench.

pliers, but it might be difficult because the nuts are typically thin and hard to get a good bite into with the plier jaws. Don't ruin more than a single pair of flats with the locking pliers. If the pliers chew the nut flats, don't move the jaws to another set. If you do, you may wind up with a tougher job than you bargained for.

Before giving up on an apparently frozen nut, move the pliers to the other faucet holddown fastener; you might have better luck here. If you get one side free, the other side will loosen more easily because you can have someone jiggle the faucet while you work at the remaining nut.

When you can't free the existing faucet with locking pliers or with other tools you have on hand, get a basin wrench. (Now you're going to be glad you refrained from chewing up the flats on those existing holddown fasteners.)

Don't bother to disconnect the sink-spray hose from the existing faucet while it's still in position; this can be removed easily once the faucet is free.

If the faucet you're going to install has an integral sink spray with hose, the next step is to remove and discard the old sink-spray "nest." This is done easily from the top of the sink if someone is available to hold the nut on the sink underside. But be careful when you do this operation; the sink hole is quite apt to have a sharp edge, as is the collar or nest that you are removing.

Remove the nut and washer from the new nest, push the threaded portion down through the hole, and secure it from underneath with washer and nut (example 2–21).

Now slip the rectangular faucet gasket into position on the bottom of the faucet as shown in example 2–22; this gasket is a must for a solid installation, because it seals the faucet to the sink. Be careful to avoid unnecessary bending of the corrugated water lines affixed to the faucet; they aren't built to take much abuse.

Feed the spray hose end up through the center hole remaining, and attach the hose to the connector on the bottom of the faucet as shown in example 2–23. When the connection is secure, you are ready for the tricky part—getting the three lines from the faucet down through the opening in the sink without forcing the corrugated

example 2–21. Installation of sink spray nest. If a helper is available, have him spin the nest from the top while you hold the nut from the bottom.

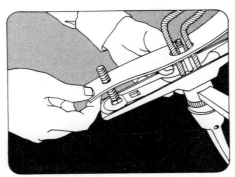

example 2–22. Attach rubber base gasket to bottom of faucet by slipping it over protruding pieces.

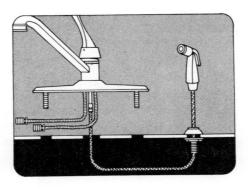

example 2–23. Bring connector end of spray hose up through center hole and attach to faucet without disturbing existing bends in faucet lines.

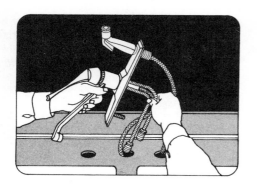

example 2–24. Without straightening flexible corrugated copper tubes, feed entire "under" assembly of faucet unit down through the three sink holes. Center the faucet over the holes on top of sink and fasten in place with mounting washers and nuts provided. (Courtesy Peerless Faucet Company.)

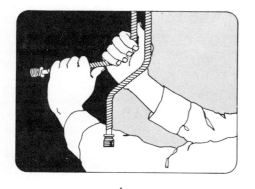

A

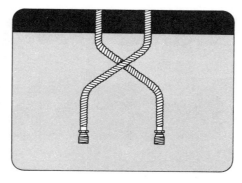

B

example 2–25. Gently cross left (cold) copper tube to align with right water supply line (cold) from floor or wall (A). Properly crossed tubes will resemble B. (Courtesy Peerless Faucet Company.)

lines and without nicking or slicing the flexible plastic or rubber spray line. This isn't difficult if you remember to keep the corrugated lines bent as shown in example 2–23. This bend staggers the two end fittings so that they will feed down through one at a time.

Be sure to orient the faucet (example 2–24) so that when it is seated in position the handle points *toward* you. (It's all too easy to install mixing faucets without seeing until after "completion" of the job that it is backwards.) When you're satisfied as to the position of the faucet, fasten it in place using the washers and nuts in the new faucet package.

Caution is the most important word at this point. The flex lines coming from the faucet are extremely susceptible to kinking. You probably won't get a refund or a new faucet if you return one that has a kinked line, so be very careful. Under the sink, *gently* cross the two lines as shown in example 2–25. The cold side—it should be marked—must cross to the right in order to mate with the proper supply pipe; the hot side crosses to the left.

Before connecting the lines from the faucet to the supply, very carefully shape each faucet line so that the termination point and fitting are on precisely the same axis as the supply line. The fittings should touch and be capable of being finger-threaded without force. If the axes of the supply line and the faucet line are not precisely the same, you risk a cross-threading and possible leakage at the connection point. When the fittings are lined up properly, smear pipe dope on the threads and connect them, tightening as much as possible without wrenches.

When the connections are cinched as much as they can be by hand, use a wrench to hold the fitting on the corrugated line immobilized while you tighten the mating connector with another wrench (example 2–26). *Don't turn the corrugated-line fitting*—it is soldered in place and must be held absolutely immobile during the tightening process.

The last step is removing the aerator from the faucet spout (example 2–27A) and flushing the line. This is important in every repair job involving a break in the water line. Small particles enter the water line during

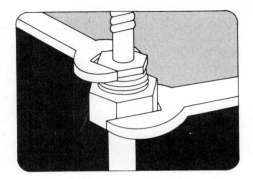

the repair; they tend to plug the aerator holes if left in the line. Flush the lines for a full minute with the aerator removed. While the water is running, check for leaks in the supply lines at every connection point under the sink. Install the aerator again and the job is complete.

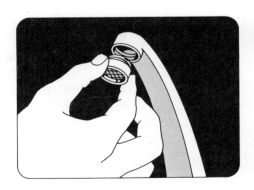

A

example 2–26. Connect both hot and cold corrugated tubes to the supply lines from the floor or wall. Be careful when you connect tubes. Use one wrench to hold hex of adapter on end of corrugated tubing and another to tighten nut on water supply line from floor or wall.

example 2–27. After installation is complete, remove spout aerator. Turn faucet on full-mix position, then turn on water-supply valves under sink and flush both hot- and cold-water lines for one minute. While water is running, check for leaks below sink. Do not turn faucet on and off repeatedly during initial flushing operation. (Courtesy Peerless Faucet Company.)

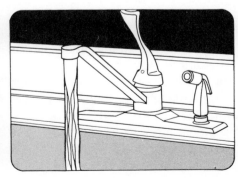

B

Insulation

Before you read anything about insulation or how it's applied, learn two words well—they make a phrase that can be your guide to an extremely cost effective piping installation: DON'T SKIMP! Use the best insulation material you can get, and use it whenever there's a chance that a pipe can radiate its heat to the surrounding air; the money you invest in insulation will be returned to you many times over in the years to come. Insulation isn't a paramount consideration with cold water pipes, it serves to stop condensation of moisture on the pipes and protects furniture underneath from the water produced by the constant sweating. But with hot water pipes, insulation will prevent you from wasting dearly bought heat. Your hot-water heater won't come on quite so often when you use well-insulated pipes; when the heater does its job, the water will stay hot longer and be delivered to the taps at a higher temperature. But you have to remember the rule: *Don't skimp.*

The purpose of insulation is twofold: (1) to retard the flow of heat from piping that is hotter than the surrounding atmosphere, and (2) to prevent heat flow to water lines that are cooler than the ambient air. Most of us get by without insulating our cold water pipes, but neglecting insulation of hot water lines can be extremely costly—and dangerous, as anyone who's ever come into body contact with a steam pipe or hot water line can attest.

Insulation is a general term referring to the composite pipe covering, which consists of insulation material, lagging, and fastening. The insulating material resists heat flow through it; *lagging* is the protective (and confining) covering placed over the insulating material; fastening is used to attach the insulating material with its protective lagging to the pipe.

Temperature Range

Insulating material covers a wide range of temperatures, from the extremely low temperature of air conditioner ductwork in homes with central cooling to the extremely high temperatures of radiator and hot-water piping. Contrary to popular belief, there is not a single material that can meet the requirements of all temperatures with equal efficiency. Cork and rockwool are low-temperature insulators; asbestos, magnesia, diatomaceous earth, aluminum foil, clay limestone, mica, fiberglass, and special silica compounds are excellent high-temperature insulators.

Material Requirements

New insulative materials are being developed all the time. When you evaluate a material for your own needs, there is a great deal more to consider than ease of application. For example, the "right" material for pipe covering should be able to withstand the highest or lowest temperature to which it might be subjected without its insulating quality being impaired even slightly. It should have sufficient structural strength to withstand (1) the normal handling required to install it, and (2) vibration during service—without disintegration, settling, deformation, or deterioration in any way. It should be predictably stable in its chemical and insulation characteristics. Ideally, it should be repairable if torn or degraded. Of course, insulation should be fireproof, waterproof to the greatest extent possible, and not attractive to vermin.

Materials Overview

Cork pipe covering is made from granulated cork; it is the bark taken from cork trees and pressed into molds of various pipe sizes or blocks. Cork pipe covering is furnished in half sections so that installed piping can be covered with minimum effort. Cork insulation is ideally suited for cold water and refrigeration lines. However, it should be coated with a vapor barrier to prevent moisture from entering the insulation material, causing it to lose insulative value or to come apart.

Rockwool is supplied in wire-reinforced pads. This material is suitable for high-temperature use and is particularly useful for insulating large areas.

Asbestos fibers, available in sheets, pads, and tapes, are suitable for insulating temperatures up to 850°F. This insulation material is cheaper and lighter than the diatomaceous earth type, and is durable and rugged. The pads or blankets are used for insulating flanges or valves that must be taken down fairly often. The pads are molded to fit any shape, and the outer surface is fitted with metal hooks to facilitate installation and removal. The blankets are made in various thicknesses and widths, and are also fitted with hooks. Tapes are used for covering small piping with curves and bends. Tapes can be used for temperatures up to 750°F, and tend to reduce fire hazards, but they have only fair insulating quality.

Diatomaceous earth (formed from skeletons of certain microscopic plants) materials are combinations of earth and magnesium or calcium carbonates, bonded together with small amounts of asbestos fibers. These materials are heavier, more expensive, and less insulating than others, but their high heat resistance allows their use for temperatures up to 1500°F. When practical, pipe coverings are made up with this material as an inner layer, and with an outer layer of magnesia–asbestos material. This lightens the overall weight.

Fiberglass slabs and batts are used widely for insulating. The fibrous glass has a low moisture-absorbing quality and offers no attraction to insects, vermin, fungus growth, or fire. The slabs are first cut to shape, then secured in place by mechanical fasteners (as quilting pins), and covered with glass cloth facing and stripping tape (held in place by fire-resistant adhesive cement).

Insulating cements are composed of many varied materials, differing widely among themselves as to heat conductivity, weight, and other physical characteristics. Typical of these variations are the asbestos cements, diatomaceous cements, and mineral and slag wool cements. These cements are less efficient than other high-temperature insulating materials, but they are valuable for patchwork, emergency repairs, and for covering small irregular surfaces (valves, flanges, joints, and so on). The cements are also used for a surface finish over block or sheet insulation, to seal joints between blocks, and to provide a smooth finish over which asbestos or glass cloth lagging may be applied (example 2-28).

Magnesia (85%) is one of the most-used types of insulation; it is particularly suitable for steam and hot-water lines and can be used on other pipes where the temperature does not exceed 600°F. Magnesia comes in three different forms:

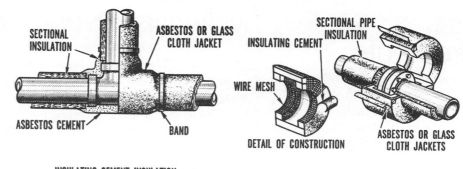

INSULATING CEMENT INSULATION FOR SMALL FITTINGS AND VALVES

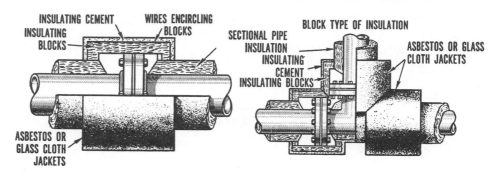

example 2-28. Permanent insulation for fittings, flanges, and valves.

molded, powdered, and blocked. Each is designed for a different purpose. The molded type is used on pipe of various sizes—it comes in three-foot lengths. Available pipe sizes range from $\frac{1}{2}$ to 12 inches in diameter. The *block* type is suitable for boilers or other places where a large flat surface needs cover. The *powdered* type, when mixed with water to form a paste, is used as a covering for fittings, valves, and the like.

Molded insulation is split in half and covered with a layer of cloth that may be opened somewhat like a hinge. After the molded insulation is placed on the pipe, an adhesive insulation or wheat paste cement is applied to the loose edge of the cloth, which is then pulled tightly around the insulation. After drying, the paste holds the insulation in place. For added strength, metal bands are placed at frequent intervals around the insulation. If the insulated pipe is outside, or where it will be exposed to the weather, the insulation should be covered with tar paper. Give the tar paper a coating of melted tar to waterproof and hold it in place.

The procedure for insulating with the powdered form of magnesia involves mixing the powder with water to form a mud-like compound. Apply this mixture with a trowel to the fitting or valve to be insulated. Unions should *not* be insulated.

The following method is commonly used to insulate fittings:

1 Apply an even coat of the mud-like mixture.
2 Build up recesses and low spots with chunks of asbestos.
3 Apply a second coat of the mixture, building up the surface area to be even and smooth, with no recesses.
4 Add a covering of canvas.

Hair felt and tar paper provide an excellent insulation for cold-water lines. In applying this type of insulation, first wrap a layer of tar paper around the pipe, then follow with a layer of hair felt, tying each layer separately with any type of string or cord. Repeat this operation until you have three layers of felt and four layers of tar paper. Complete the operation by adding a layer of waterproof paper.

Where insulation is required for below-ground piping, it needs more protection from the elements. Even if you use a poured-concrete trench, you can still insulate effectively. Just be sure to protect against rain and ground water. This is accomplished by using coal tar as a sealer or by wrapping with tar paper or aluminum foil.

Applying Insulation

Do not allow the insulating materials to become moist. Moisture impairs the insulating value of the material and may cause eventual disintegration. Large air pockets in the insulation cause large heat losses, so be sure to fill and seal all cavities or cracks. Hangers or other supports should be insulated to prevent loss of heat by conduction.

All sections or segments of the pipe covering should be tightly butted at joints and secured with wire loops, metal bands, or lacing. Block insulation should be secured with steel wire and galvanized mesh wire or expanded metal lattice. Insulating cement should be used to fill all crevices, to smooth all surfaces, and to coat wire netting before final lagging is applied.

Moisture-proofing is just as important in high-temperature insulation as it is in low-temperature insulation. In the former case, heat is lost because of evaporation, while in the latter case condensed moisture may freeze. In either case, insulating efficiency is impaired and eventually the insulating material disintegrates.

Although the same insulating material employed on the piping may be used, the insulation of pipe fittings, flanges, and valves requires additional consideration. Example 2–28 illustrates several types of insulation for flanged pipe joints.

For piping components any one of the following methods of fabrication is acceptable:

1 Covers may be made in two halves of thermal insulating felt enclosed in asbestos cloth. Each half-cover should be sewn and quilted with wire-inserted asbestos yarn, or fastened with mechanical stapling to provide uniform thickness, strength, and rigidity.
2 Covers for use at temperature of 850°F and below should be filled with asbestos felt. Wire-inserted asbestos cloth should be used on the inside surface of the covers.
3 Covers for use at temperatures above 850°F should have filling consisting of inner layers of fiberglass felt, outer layers of asbestos felt, and should be covered on the inside surface and on the ends with nickel–chromium alloy wire mesh and on the outside surface with asbestos cloth. Asbestos roll felt, $\frac{1}{8}$ inch thick, should be inserted between the asbestos felt and the asbestos cloth to retain the cylindrical shape of the cover.

Precautions

The following general precautions should be observed with regard to the application and maintenance of insulation:

1 Fill and seal all air pockets and cracks. Failure to do this will cause large losses in the effectiveness of the insulation.

2 Seal the ends of the insulation and taper off to a smooth, airtight joint. At joint ends or other points where insulation is liable to damage, use sheet-metal lagging. Flanges and joints should be cuffed with 6-inch lagging.

3 Asbestos cloth covering fitted over insulation should be tight and smooth. It may be sewed with asbestos yarn or cemented on.

4 Keep moisture out of all insulation work. Moisture is an enemy of heat insulation fully as much as it is of electrical insulation. Any dampness increases the conductivity of all heat-insulation materials.

5 Insulate all hangers and other supports at their point of contact from the pipe or other unit they are supporting; otherwise a considerable quantity of heat will be lost via conduction through the support.

6 Sheet-metal covering should be kept bright and not painted unless the protecting surface has been damaged or has worn off. The radiation from bright-bodied and light-colored objects is considerably less than from rough and dark-colored objects.

7 Insulate all flanges with easily removable forms that can be made up as pads of insulating material wired or bound in place, and cover the whole thing with sheet-metal casings that are in halves and easily removable.

8 Once installed, heat insulation requires careful inspection, upkeep, and repair. Lagging and insulation removed to make repairs should be replaced just as carefully as when originally installed. When replacing insulation, make certain that the replacement material is of the same type as used originally. Old magnesia blocks and sections broken in removal can be mixed with water and reused in the plastic form for temporary repairs.

Section 3

Plumbing-Fixture Installation and Repair

Section 3

Most kitchen and bathroom fixtures can't be repaired when they break; removal and replacement are required. The material in this section should be sufficient to help you clear clogged fixture drains, install a new sink or water closet, or create a new shower facility even when there is no rough-in plumbing available for one. Also, there are tips for fixing up the enclosure areas around tubs and shower stalls when they develop cracks, leaks, or broken tiles—the typical results of a plumbing problem that's gone too long before repair.

Throughout these pages we'll assume that your rough-in need not be changed in any way. (The term *rough-in* refers to piping, fittings, drains, and so on, and their positions of concealment under the floor or behind a wall serving a fixture.)

When you are contemplating an extensive plumbing job such as installation of a whole new bathroom, be sure to check with local authorities for special requirements governed by regional codes or ordinances; often major alterations require formal authorization from local authorities and possibly inspection of the completed work.

80

Kitchen and Laundry Plumbing

Most kitchen sinks these days are double-bowl types, available in a wide range of materials: brushed or stainless steel, enameled steel, plastic, fiberglass, and porcelain. Laundry tubs are also dual units, but they are typically very deep (to accept the entire output of a modern clothes washer without overflowing even when the drain is plugged) and generally fabricated from cast metal.

Kitchen sinks should be built into a cabinet or hung from a wall bracket screwed to a mounting board; example 3–1 shows the rough-in measurements for a kitchen sink suspended by a wall bracket. The bracket should be positioned so that the mounted sink will be at a convenient height for normal use. The distance between the finished floor and the top of the drainboard shouldn't be less than 36 inches; remember that this is one sink where you and other members of your family will be standing often.

After screwing the bracket for a wall-hung sink into place, lower the

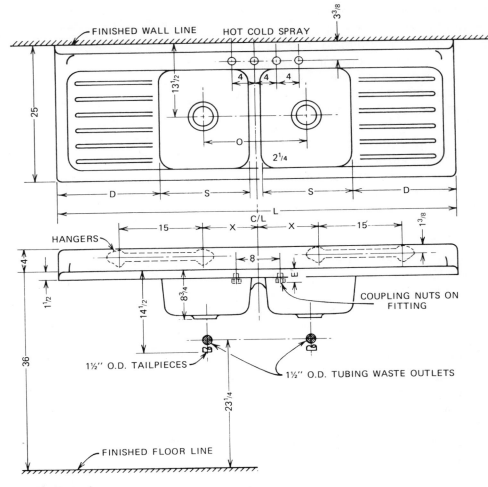

example 3–1. Rough-in measurements and plan view of one type of kitchen sink.

sink into position so that the lugs cast into the sink's back fit right down into the corresponding notches in the bracket. Screw the strainer and tailpiece into the sink bowl (if they're not integral parts of the assembly), and connect the P-trap to the existing waste outlet.

Most modern kitchen sinks are designed to drop into the top of a counter cutout, with the cutout positioned at the central point of a work triangle (shown with crosshatching in example 3–2). In this case, all you have to do is drop the sink into position and secure it with the fasteners integral to the sink. The rough-in plumbing will accommodate the fixture's inlet and outlet connections.

Even though plumbing fixtures may vary a great deal in appearance and features, they do have some common characteristics. In the typical plumbing network shown in example 3–3 for a two-story house with basement, note the similarity between the installations of tub, kitchen sink, and laundry fixture. (Note also the economy of this vertical layout, which allows the service of all floors with a minimum amount of piping.) As

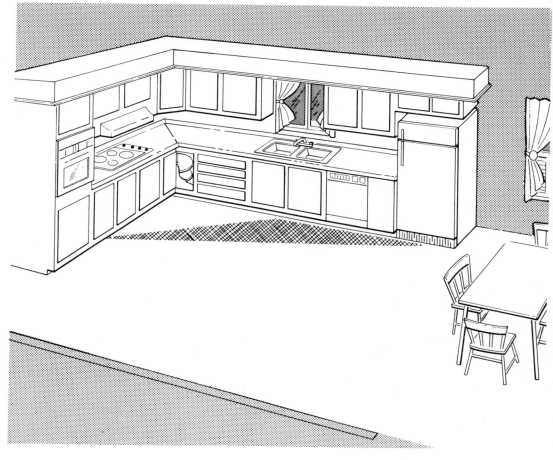

example 3–2. Typical well-planned kitchen layout. Note position of sink relative to stove and refrigerator.

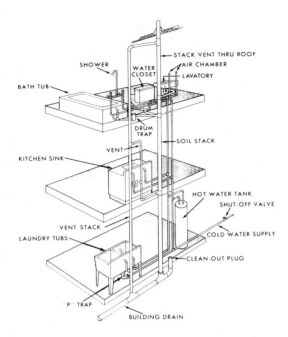

example 3-3. Typical kitchen, laundry, and bathroom plumbing arrangement; note economical use of piping.

The labels in the figure read:

SHOWER
STACK VENT THRU ROOF
WATER CLOSET
AIR CHAMBER
LAVATORY
BATH TUB
DRUM TRAP
SOIL STACK
VENT
KITCHEN SINK
HOT WATER TANK
SHUT-OFF VALVE
VENT STACK
LAUNDRY TUBS
COLD WATER SUPPLY
CLEAN-OUT PLUG
P-TRAP
BUILDING DRAIN

shown, kitchen sinks and laundry tubs are connected to the waste-drain system using P-traps, which allow clogged up sinks to be unstopped with very little effort and serve to prevent the backflow of wastes and gases into the fixture (and room).

Most people probably don't realize that sewer-gas backflow can result in serious illness or even death to the occupants of a house. That's why it's absolutely essential to install a P-trap properly and provide an effective waste-line vent. The *P-trap* —so named because of its resemblance to the letter P—comes in a variety of sizes between $1\frac{1}{4}$ and 6 inches; those you'll encounter in household use are typically 2 inches or less. There are two basic design variations, both pictured in example 3-4. In A, the trap has a cleanout plug that allows easy draining of the trap to reclaim a lost article such as a small jewelry item or other obstruction.

The water seal is the amount of waste water held in the bent portion; this should be a minimum of 2 inches to be effective, but never more than 4 inches. The trap should be the same size as the drain and as close as possible to the fixture outlet. (It's bad form to use a long vertical run between fixture outlet and trap, which gets messy when you have to remove the trap to clear a drain of an obstruction.)

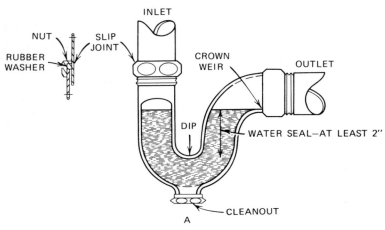

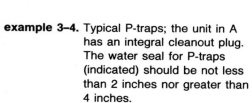

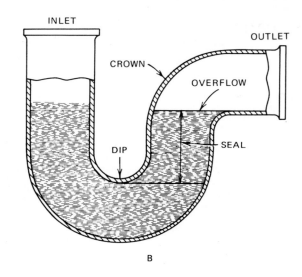

example 3–4. Typical P-traps; the unit in A has an integral cleanout plug. The water seal for P-traps (indicated) should be not less than 2 inches nor greater than 4 inches.

Food-Waste Disposer

Installation of a food-waste disposer is a great way to eliminate the mess and fuss of handling garbage. Local codes usually govern the installation requirements, but some generalizations apply.

A disposer unit mounts immediately below the sink, and the mouth of the disposer typically replaces the sink drain; all plumbing drain connections are then made to the body of the disposer unit. Installation of one of these units does not defer the need for a sink P-trap—the trap is installed immediately downstream from the disposer.

These are two types of food waste disposers in current use: the *batch-feed* type and the *continuous-feed* type. Batch-feed disposers are loaded to capacity and are activated when the cover is positioned over the opening; a failsafe feature prevents operation when the cover is not in place. Continuous-feed disposers permit continuous loading during operation. Operation is activated by a remote switch nearby.

Both types have inherent safety hazards, since the heart of the disposer function is a grinding action. It is beyond the scope of this book to cover electrical requirements and installation of electrical switches and wiring, but there are a few considerations that you'd do well to remember: (1) Locate the operating switch so that a good reach is required to touch it but it is within easy eyeshot of the sink, as shown in example 3–5. This prevents accidental turn-on by someone who has an arm down inside the fixture. (2) Use a momentary-contact operating switch rather than one that stays in the on position. This is assurance that the disposer won't be operating unless someone is physically actuating the switch.

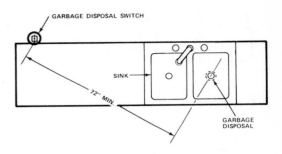

example 3–5. Ideal layout for garbage disposal switch relative to the sink.

Automatic Dishwasher

Plumbing is no problem for the portable automatic dishwasher; its use involves nothing more than wheeling the appliance to the kitchen sink and making one simple plug-in connection to the faucet. But if you plan to add a built-in dishwasher to a finished kitchen, then you're in for some work—not much, but it has to be done right.

Built-in dishwashers can be placed under the kitchen counter near the center or at the end; they can also be free-standing. They should be located near the kitchen sink or other plumbing so that the water inlet and drain can be connected easily. (Be sure to allow adequate space for loading through the front-opening door.)

Before connecting a built-in dishwasher to the electrical, water supply, and drain terminals, check local codes; many have specific requirements as to plumbing air gaps, electrical wire sizes, and current-handling capacity of breakers or supply switches.

An automatic dishwasher has only one water supply connection—hot water (between 140 and 160 degrees). Connection requirements differ from one model to the next, but generally all have a water-supply line that connects easily to any available outlet. The ideal connection point for the washer supply is between the hot-water supply valve and the kitchen sink. All you have to do is break the line and install a tee fitting that mates with the supply line and the dishwasher input line. If the lines are dissimilar, you can buy an adapter that effectively changes the dishwasher's connector to the type that matches your water-supply pipe.

Automatic dishwashers typically have a fill air gap integral to the unit itself, which acts as a shock absorber of sorts. It eliminates water hammering and spitback, smooths the delivery of hot water to the washing machine, and prevents dishwater from flowing back into the supply plumbing. Don't confuse this with the *drain air gap,* which most local codes require in the plumbing connection for the drain. There are two more or less universal rules regarding dishwasher drain air gaps: (1) The air gap should be above the flow level of the fixture, and (2) the waste-water discharge from the dishwasher must be connected to the inlet side of the sink's P-trap (see drain system pictorial, example 3–6).

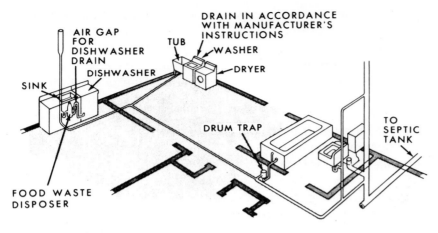

AIR GAP FOR DISHWASHER DRAIN

DRAIN IN ACCORDANCE WITH MANUFACTURER'S INSTRUCTIONS

TUB

WASHER

DRYER

DISHWASHER

SINK

DRUM TRAP

TO SEPTIC TANK

FOOD WASTE DISPOSER

example 3–6. Drain arrangement when kitchen includes dishwasher and food-waste disposer.

Clothes Washer and Dryer

The electrical requirements for home laundry systems are typically stringent and tightly controlled by local codes. But once an electrical outlet is provided (washers usually require 120-volt service; dryers, 240-volt), you need only plug in the electric cords and make simple plumbing connections.

If no facilities for supply and discharge of water exist, there are a few guidelines to follow, but these won't absolve you of the responsibility to check for local codes or ordinances affecting home laundry installations.

Plan to situate the washer, at least, as close to the laundry tub as possible; this will simplify the water-supply problem. Don't connect the washer hoses to the existing laundry-tub faucets; this will either render your laundry tub useless for any function other than clothes washing or will cause you to repeatedly connect and disconnect the washer's supply hoses (which tends to cause premature hose wearout).

You need two hose bibbs (faucets with a hose fitting, as pictured in example 2–3), one for hot and one for cold water. If possible, see to it that these are installed within 2 feet of the washer (washer hoses are typically only 5 feet in length, and they're connected to the washer at nearly floor level). To install these in a central position of an existing pipe run, you need a tee to replace an existing coupling; the bibb will connect to the tee of each pipe. If you install them at some point beyond the supply pipe run, you should install additional pipes of the same diameter as those of your water-supply network (covered in previous section).

Most clothes washers have an integral discharge hose that terminates at a cane-shaped copper or aluminum tube. This tube may be inserted into any handy drain vent that is at least 2 inches in diameter. The drain vent opening should be as high as possible and situated near the washer. Keep the drain vent clear and unobstructed to assure a good air gap between the washer discharge nozzle and drain-vent mouth.

The dryer does not need to be located adjacent to the washer, although this does permit easy transfer of wet clothing from the washer to the dryer. If the washer and dryer are situated next to each other, connect the two chassis with a braided copper grounding wire. And be sure that the chassis of both units also connect via a grounding wire to the water supply line. (Your electrician will do this for you when he makes the electrical installation.)

The dryer discharges hot humid air through a 4-inch-diameter opening. In most climates it is absolutely essential that the output air from the dryer be ducted to the exterior of your home. The most common duct is collapsible plastic over a wire framework. A standard 4-inch hose clamp fastens the duct to the dryer's discharge port, and a similar clamp is used to fasten the opposite end to a through-wall, window, or roof vent. (These vents are quite easy to install and are available at hardware and most larger mail-order department stores.)

If you live in a climate where the winters are cold and dry, however, you can save a great deal on basement heating costs by allowing the dryer to vent into the basement area during the cold season. If you do allow the dryer to vent into your basement or other large room, however, you must install a filter screen at the outlet to avoid discharging breathable particles into the air. And you shouldn't let the dryer vent into a room unless someone is in attendance to circumvent the vent should moisture begin to precipitate on glass or metal surfaces.

A dryer is a natural humidifier and air heater. If you use it selectively during winter months, you can cut your room heating costs drastically and at the same time improve your breathing air considerably. But it is extremely important not to overdo it; electronic equipment can be destroyed by excessive moisture in the air, and wood surfaces can warp and peel. If you see moisture buildup anywhere in the room, it's time to vent the dryer to the outside again. And when you do vent into the room, make sure the discharge point is well away from the dryer itself, which has electrical parts that are not moisture-proof.

Bathroom Plumbing

Even though you'll not likely be able to repair a broken lavatory or resurface a worn out bathtub, you can replace older units with modern types, and you can repair wall cracks, set new tile, and make other modifications that result in a more livable and convenient bathroom facility.

Lavatory

Until comparatively recently, lavatories were usually wall-hung units; today this type (example 3–7) is common but in the process of being displaced by the in-counter or vanity-cabinet bowl. (The in-counter lavatory bowl is designed to drop into a prepared countertop cutout; it comes in an unending variety of shapes (example 3–8 shows a few representative lavatory fixtures from American Standard). Regardless of the type of lavatory you buy or install, the rough-in measurements are the same.

Wall-hung and pedestal lavatories are suspended from a bracket screwed to the wall. Installation of a countertop lavatory should proceed in accordance with instructions supplied by the manufacturer, but the photo sequences (pages 91–95) picture the installation steps. The installation procedure for wall-hung units is fairly universal:

1 Mark the wall at the correct height for the lavatory, and secure the hanger to the wall. (For wall-hanger mounting details, see example 2–13.)

2 Position the lavatory on a hanger.
3 Install lavatory faucets (or mixing faucet assembly).
4 Install permanent overflow (PO) plug and tailpiece assembly (example 3–9). (If your fixture's faucet assembly contains an integral PO plug, make the connections as shown in example 3–10, then readjust the movable arm as necessary to permit full closure and drain opening.)
5 Connect tailpiece of PO plug to P-trap for drainage.
6 Connect water-supply lines to faucets.
7 Fill the basin with water. During this procedure, examine the water-supply line downstream from the shutoff valve—particularly at the supply-line connection points—for evidence of leakage.
8 Allow the water to drain from the lavatory while you examine the tailpiece area and P-trap connections to check for water leakage.

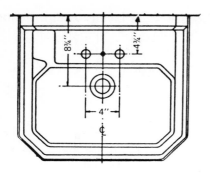

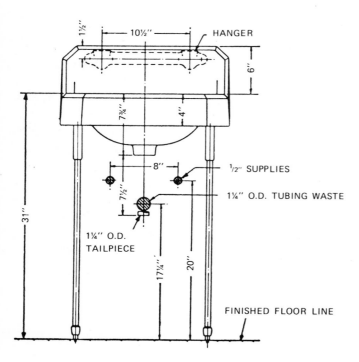

8¾"

4¾"

4"

℄

1½"

10½"

HANGER

6"

7¾"

4"

8"

½" SUPPLIES

7½"

1¼" O.D. TUBING WASTE

1¼" O.D.
TAILPIECE

31"

17¼"

20"

FINISHED FLOOR LINE

example 3–7. Standard lavatory rough-in
measurements (wall-hung or
pedestal mount).

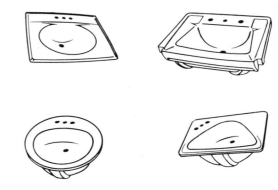

example 3–8. Several of the available
in-counter lavatory bowl shapes.
(Courtesy American Standard.)

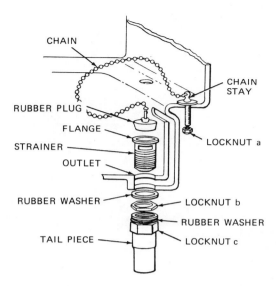

CHAIN

CHAIN STAY

RUBBER PLUG

FLANGE

STRAINER

OUTLET

LOCKNUT a

RUBBER WASHER

LOCKNUT b

RUBBER WASHER

TAIL PIECE

LOCKNUT c

example 3–9. Detail of permanent-overflow plug that uses rubber stopper and chain stay.

example 3–10. Lavatory fixtures of recent vintage incorporate permanent-overflow plugs like that shown in example 3–9, but the stopper mechanism is controlled by an undersink lever throttled by a top-mounted button or pull bar. (Courtesy American Standard.)

1

3

4

2

5

6

7

VANITY LAVATORY INSTALLATION

1 Connect faucet to lavatory following maker's instructions

2 Place cabinet in position against wall over inlet pipes. Allow no space between wall and cabinet back

3 Place lavatory on cabinet and check position with level. Add shims under cabinet base to true-up as necessary

4 Connect water supply to faucet under sink and tighten nuts securely

5 Install tailpiece and soft rubber washer. Make sure stopper lever arm points toward back wall

6 Connect lift rod to drain lever, then operate drain lever and adjust position of setscrew so drain opens adequately and closes completely

7 Connect P-trap and secure with wrench. Fill lavatory and then drain it while you inspect undersink area for leaks

1

2

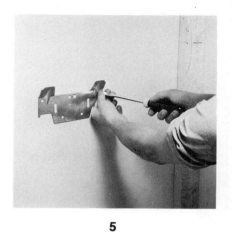

3

4

5

6

WALL-HUNG LAVATORY INSTALLATION

1 **Determine preferred position of lavatory between two adjacent studs, then mark horizontal line on the two studs exactly 30 inches from floor**

2 **Cut a length of wood 6 by 17.5 in. from inch-thick stock**

3 **Cut inch-deep slots in marked studs using chisel or saber saw. Slots should extend 6 inches below 30-inch marks so that wood length can seat flush with wall**

4 **Attach cut board to stud with long wood screws. Double-check flushness so that wall panel can be applied directly over surface without bulging**

5 **Place wall paneling in position and secure it, then position lavatory hanger on wall over inset wood crosspiece. Secure hanger with wood screws**

6 **Attach tailpiece and faucet to lavatory following maker's instructions**

7 **Place lavatory on wall hanger and connect drain and supply lines**

7

Tubs and Showers

When a tub is to be replaced in an area that has already contained such a fixture, the installation is elementary. The only pipe that is directly connected to the tub is the inch-and-a-half drain with its combination overflow. The hot and cold water lines and the shower diverter are installed in the wall area immediately above the tub. In making the installation, set the tub in the desired location and level it. Place a carpenter's level across the tub, then build up the corners as necessary with integral leveling strips, wedges, or shims. When the tub is secured in position and level, connect the drain of the tub to the existing P-trap. The rough-in measurements are shown in example 3–11.

If you plan to install a shower stall where none has been, you'll have to do your own rough-in. But several companies manufacture prefabricated panels and stalls that make this relatively easy to do. Owens-Corning, for example, markets a complete line of fiberglass (Fiberglas, as a matter of

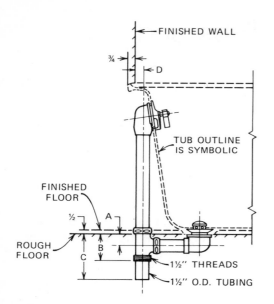

example 3-11. Typical tub drain; alphabetized arrows show important rough-in dimensions, which vary from unit to unit.

fact, is the company's trade name) showers and tub/shower combinations.

The U.S. Government has conducted an engineering analysis of fiberglass and porcelain tub surfaces; the conclusion was that the fiberglass surface is more slip-resistant than the porcelain but not quite as slip-resistant as the surfaces offered in some special "safety" bathtub models. Unlike the safety tubs, however, which offer some cleaning and sanitation problems, the fiberglass models are easy to keep clean and sanitary (because they don't have to depend on special recesses or surface variations to maintain the required slip resistance).

The inherent resilience of fiberglass, as used in the construction of bathtubs and shower stalls, provides another desirable safety feature that is unattainable in the porcelainized cast-metal tub or the tiled floor of a shower stall: fiberglass tends to lessen the severity of many tub and shower falls.

You need four basic pieces (example 3-12) to build your own new shower or shower/tub with Fiberglas prefab pieces. The big advantage of

example 3-12. Prefabricated Fiberglas panels allow dimensions of a shower stall (top) or shower/tub combo to vary. Walls and floor sections are moved into a house easily; they are positioned and fastened in place after the stall area is roughed-in.

this *prefabricated* concept is that you don't have to use a preassembled stall—which might prove difficult to carry through the doors of a completed house.

The shower stall is assembled in place—it can't be preassembled outside the stud pocket (the 2'' x 4'' studs that support the walls)—but it can be installed any time during the construction of a room or house as long as the room wall surface is finished.

Example 3–13 shows the framing. Only the basic dimensions are given here, since stud spacing depends on whether a 3- or 4-foot shower stall is desired. The critical studs—those that must be precisely at the spacings shown—are circled in the drawing. For installation against a finished wall, of course, the dimensions must be increased to accommodate the thickness of the wall: the width dimension of the basic stud pocket must be increased by two times the wall thickness, and the depth must be increased by the wall thickness; the circled studs must be moved accordingly. The blocking must be positioned exactly as illustrated.

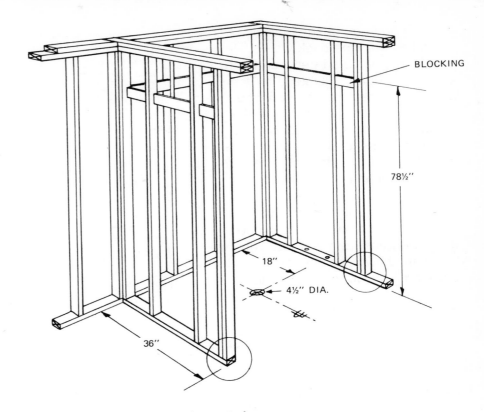

example 3–13. Framing requirements for installing a Fiberglas prefab shower or shower/tub. (Courtesy Owens-Corning Fiberglas.)

The drain rough-in dimensions are 18 inches from the back wall of the pocket and centered between the two side walls. Rough plumbing for the drain has to be located according to these dimensions. Example 3–14 shows the openings in wood and concrete floors to accommodate the molded-in drain. The waste line must extend one-half inch above the subfloor or slab.

The supply faucet should be located at the central point (if one is used) between front and back (that is, between shower-stall opening and the back wall). If two faucets are to be used, they should be centered on this point, and spaced according to the requirements of the faucet assembly. The faucets should not be mounted more than 4 feet above the surface of the finished floor. When you install the shower head, make sure it extends down far enough so that it is within 6 feet of the finished floor surface; if it is above this level, water might spray over the walls of the shower and cause mildewing and rotting.

You can use 16- or 24-inch grab bars with the Owens-Corning Fiberglas shower stalls, but they have to be

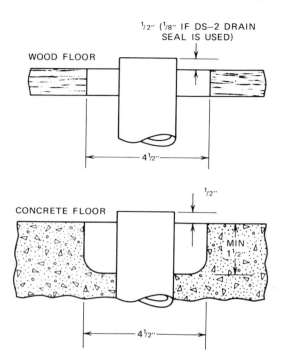

example 3–14. Drain rough-in for wood and concrete floors.

installed at specified locations before the walls go in.

The tub or shower floor portion of the unit sits on the subfloor. To install it, just clear the area of debris, then drop the piece into position. Use a level and shim up the unit as necessary, then nail it into place around the rim. It should be nailed to all studs to assure proper alignment of wall joints (necessary for leakproofing).

There are two ways to locate the faucets accurately. Measure the distance carefully from the end of the finished wall (inset, example 3–15), or position the appropriate end-wall (where the plumbing will protrude) in place temporarily while you mark the faucet locations from the back (example 3–15).

You must seal the drain joint in accordance with local codes if they specify detailed requirements. You may have to use oakum and lead (see Appendix). The oakum and lead installation is pictured in example 3–16A; (B) shows how to make a "dry seal" joint using the elastomeric seal provided with the shower stall.

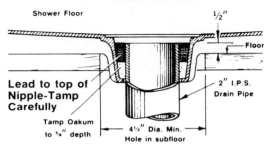

OAKUM AND LEAD

Shower Floor

½″

Floor

**Lead to top of
Nipple-Tamp
Carefully**

Tamp Oakum
to ⅛″ depth

2″ I.P.S.
Drain Pipe

4½″ Dia. Min.
Hole in subfloor

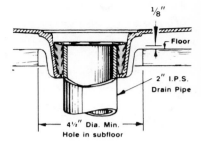

⅛″

Floor

2″ I.P.S.
Drain Pipe

4½″ Dia. Min.
Hole in subfloor

SHOWER DRAIN SEAL DS-2

example 3–16. Shower-stall drain installation.
Where required, use oakum
and lead; otherwise use the
dry-joint elastomeric seal
packed with each Fiberglas
prefab unit.

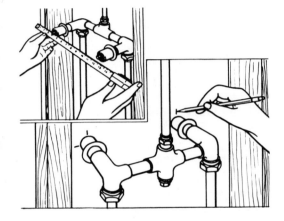

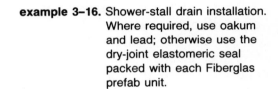

example 3–15. Marking Fiberglas panels for
plumbing holes. Careful
measurement is an absolute
necessity; either measure the
faucet centers from a common
reference (inset) or put the
shower panel in place and
mark centers on the reverse
side as shown.

Bathroom Accessories

Grab bars, soap dishes, towel racks, tissue dispensers, and the like come under the general heading of bathroom accessories. When they're improperly installed, they will offer a safety hazard that won't be worth the time you have saved in a quick-and-dirty installation.

In the HUD design guide for home safety, prepared by an engineering firm in Huntsville, Alabama, certain recommendations are made concerning installation of bathroom accessories. Example 3–17 is a listing of the most important of these with drawings that offer suggested dimensions for placement of grab

example 3–17. Government-recommended bathroom accessory placement for optimum safety of occupants.

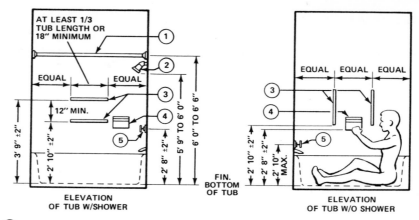

ELEVATION
OF TUB W/SHOWER

FIN.
BOTTOM
OF TUB

ELEVATION
OF TUB W/O SHOWER

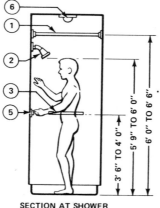

SECTION AT SHOWER

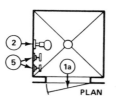

PLAN

1. SHOWER CURTAIN ROD: KEEP WITHIN INSIDE OF TUB OR SHOWER.

1a. ENCLOSURE DOORS: IF SWINGING DOORS ARE USED, PLACE HINGES ON THE SIDE OPPOSITE CONTROL VALVES.

2. SHOWER HEAD: SEE ELEVATION OF TUB AND SHOWER STALL FOR RECOMMENDED HEIGHTS.

3. GRAB BARS SHALL BE MANUFACTURED OF SHATTER-RESISTANT MATERIAL, FREE FROM BURRS, SHARP EDGES AND PINCH POINTS. KNURLING OR SLIP-RESISTANT SURFACE IS DESIRABLE.

4. RECESSED SOAP DISH SHALL BE FREE FROM BURRS AND SHARP EDGES. WHERE GRAB BAR IS AN INTEGRAL PART OF THE SOAP DISH, IT MAY HAVE A MINIMUM LENGTH OF 6 INCHES.

5. FAUCET SHALL BE MANUFACTURED OF SHATTER-RESISTANT MATERIAL, FREE FROM BURRS AND SHARP EDGES. ALL FAUCET SETS IN SHOWERS, TUBS AND LAVATORIES SHALL BE EQUIPPED WITH A WATER-MIXING VALVE DELIVERING A MAXIMUM WATER TEMPERATURE OF $110^{\circ} \pm 5^{\circ}$F.

6. SHOWER STALL LIGHT: SHALL BE OF A VAPOR-PROOF FIXTURE WITH THE ELECTRICAL LIGHT SWITCH A MINIMUM OF 72 INCHES AWAY FROM SHOWER STALL.

bars, faucet handles, shower head, and so on. In the example that follows (3–18), a variety of grab-bar installation schemes is pictured; use any of these, but be absolutely certain your installation can support the weight of a human body without pulling loose from the wall.

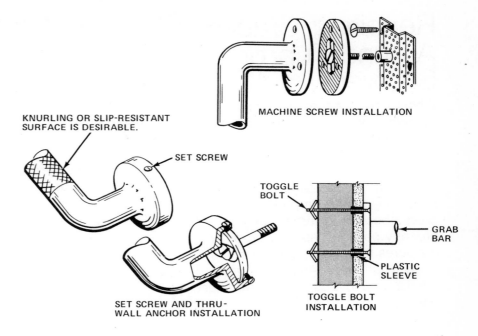

MACHINE SCREW INSTALLATION

KNURLING OR SLIP-RESISTANT SURFACE IS DESIRABLE.

SET SCREW

TOGGLE BOLT

GRAB BAR

PLASTIC SLEEVE

SET SCREW AND THRU-WALL ANCHOR INSTALLATION

TOGGLE BOLT INSTALLATION

WOOD FRAMING

PLASTER/WALLBOARD OR TILE

WOOD SCREW

GRAB BAR

EXPANSION SHIELD

CONCRETE OR SOLID MASONRY

WOOD SCREW OR EXPANSION SHIELD INSTALLATION

PLASTER OR WALLBOARD

CHROME PLATED BRASS PLATE

GRAB BAR

BLOCK CONSTR. (VARIES)

PERFORATED ANCHOR TEMP.

2 HALF NUTS

THRU BOLT

THRU-BOLT INSTALLATION

example 3–18. Typical grab-bar variations and their installation methods.

Repairing Cracks Around Tub or Shower

When a crack develops between the bathtub or shower and the wall, it should be filled as soon as possible to prevent water seepage. Water can damage the walls and house frame, and the crack serves as a dirt catcher, presenting sanitation problems. When cracks become enlarged, they provide an entry point for insects and other creatures that have made their home under your home. And, of course, cracks around tubs and showers just plain *look* bad. They should be sealed.

There are several types of waterproof crack filler, but the most common are waterproof grout and plastic sealer (such as General Electric's RTV compound). Grout comes in powder form and must be mixed with water to make a paste. This material is fairly inexpensive and can be mixed in small amounts as required for each job. Plastic sealer comes in a tube and resembles toothpaste. It's easier to use than grout but costs more. Both types are effective, but the tube sealant probably will offer the greatest longevity.

To prepare the surface, remove as much of the old crack filler as possible from the existing crack, then wash the surface thoroughly to remove soap, grease, and dirt buildup. Finally, dry the surface well and reinspect after it's dry to be sure all surface grime has been removed (see example 3–19).

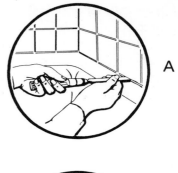

A

B

C

example 3–19. Filling cracks around tubs and showers.

Grouting

If you're using grout for the repair, put a small amount of it in a bowl, then slowly add water (example 3–20A), mixing until you have a thick paste. Put this mixture in the crack with a putty knife or similar spatula (as in B), then press it into the crack (as shown in C). Smooth the surface (as in D), then wipe the excess grout from the wall before it has the chance to dry and harden. It is a wise idea to keep your family from using the tub or shower until after the grout dries thoroughly. (Dispose of leftover grout in a rubbish container—not down the drain—and then wash the bowl and knife before the grout dries on them.)

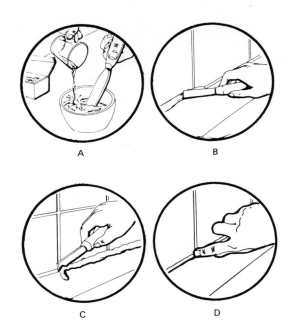

A   B

C   D

example 3–20. Applying grout to positioned tiles.

Plastic Sealer

If you're careful, you can apply plastic sealer directly from the tube as shown in example 3–21. Squeeze the sealer from the tube in a uniform ribbon along the crack. Use a putty knife or spatula to press it down and into the crack. Smooth the surface as you go, and work as fast as you can because the stuff dries rapidly. Be sure to put the cap back on the tube when you're finished with the job; otherwise the material will harden in the tube and you'll have to discard the portion you haven't used.

example 3–21. To seal tubs and shower corners with plastic compound, just squeeze the tube as you draw it along the crack to be sealed, trying to maintain a uniform bead size as you go.

Replacing Bathroom Tile

Often when you replace a fixture or repair a leaky pipe in the bathroom, you'll find it necessary to remove or break tiles in order to gain access to an area behind a wall. And sometimes you may have to replace tiles that have loosened as a result of excessive moisture seeping through cracks in the tile mortar.

There are two basic variations of tile: flexible and ceramic. Each has its own repair requirements. Generally, however, to make repairs in tile surfaces you'll need a bowl to mix grout (and a mixing stick), a tile adhesive that is applicable to the type of tile used in your bathroom, a putty knife or paintbrush, a knife or saw (such as the coping saw shown in example 3–22), as much new tile as you'll require for the repair, and grout.

Installing Flexible Tile

The first order of business is to remove all loose and damaged tiles. Application of a warm iron to the tiles will serve to soften the adhesive sufficiently to allow you to pull the tile

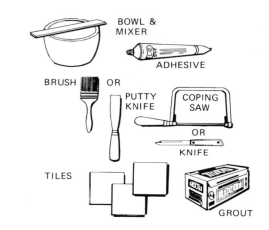

example 3–22. Equipment required to set new flexible-plastic or ceramic tile.

from the wall (place a damp cloth or paper towel under the iron's surface to avoid discoloration or marring of the tile, as shown in example 3–23A). When the tiles have been removed from the floor or wall, try to scrape away all the old adhesive (B) so that the tile will seat properly when it's reapplied. If you plan to reuse the original tile, you'll have to scrape it too.

Make sure the replacement tiles or tile portions fit exactly. Some tiles can be cut with a knife (C) or shears, but some of the early types require cutting with a saw. Flexible tile is less apt to break or chip if you keep it warm while you're working with it.

Now spread the adhesive evenly on the floor or wall with a paintbrush or putty knife (depending on the requirements of the tile you're using) as pictured in D. Once the adhesive begins to set and has a tacky feel to it, place the tile in position and press it down firmly. Use a rolling pin (E) to make sure all tile surfaces are even and smooth in relation to each other.

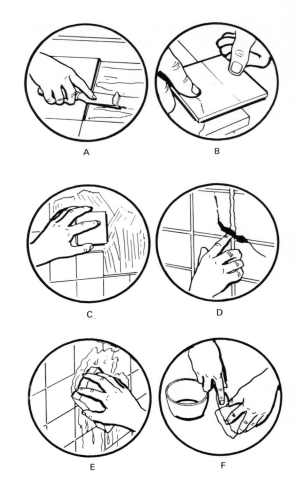

Installing Ceramic and Stiff Plastic Tile

Remove all damaged and loosened tiles, then scrape all old adhesive from the walls (and tiles that are to be reused). As you read the instructions here, check the alphabetized illustrations in example 3–24).

If you are using new tile and need to fit it to some size other than the as-purchased dimensions, mark it carefully to size and cut it with a saw. On heavy ceramic tile, you can make straight cuts by scoring the mark first. It will then snap along the scored line if you press it on the edge of a hard surface (example 3–24B).

Spread adhesive on the floor or wall where the tiles are to be placed as well as on the back of the tile itself. Let the adhesive momentarily set to provide a tacky surface, then place the tile firmly into position (C).

Tap each piece of new tile into place with a wood piece (hammer handle or similar tool), then roll across the newly placed portion with a rolling pin.

Joints on ceramic tile should be filled with grout after the tile has firmly set; mix grout (in powder form) with

example 3–23. Steps for positioning and affixing flexible bathroom tile.

example 3–24. Surface preparation and ceramic-tile cutting, placing, and affixing.

water to form a fairly stiff paste, then press the mixture into the joints with your fingers (as shown in D). Smooth the surface as best as you can, then remove the excess grout from the tile surface before it dries (E).

Empty the excess grout into a rubbish container (not down the drain, unless you want more plumbing problems than you bargained for) and clean all surfaces and tools. Try to avoid letting the grout get wet for at least 10 hours, even longer in humid environments.

Water-Closet Installation

It is not possible to include procedures for installation of every type of water closet available since so many varieties abound. The general procedure given here, however, will serve as a universal guide that will allow you to adapt accordingly.

In example 3–25, a sketch shows one of the older two-piece water-closet assemblies; the inset shows the modern equivalent. The older type is pictured in greater detail because it has a connecting water-closet elbow that is integral with the newer units. The only significant difference between the two types is the placement of the bowl horn; the older type mounts onto a floor flange and the pictured modern version mounts on a wall. Rough-in measurements of the traditional water closet are shown in example 3–26; in the drawing, the term "C/L" stands for *centerline*.

To install a new toilet (as opposed to replacing one that has been removed), make sure the floor flange fits the bowl. Slip the water-closet floor flange (example 3–27) over the in-floor

bend, then slide it down until the flange is level with the floor. (The floor must be finished for this operation, because the flange must seat flush with the surface of the finished floor.) Prepare the flange-to-bend joint for lead calking, then pour and calk the water-closet flange to the pipe bend. If any portion of the pipe protrudes above the surface of the water-closet flange, break it off using a cold chisel and a hammer; but be very careful to avoid damaging the closet bend below the flange. If the floor hole permits, secure the flange to the floor with four screws. But don't place the screws in the bolt slots (see top of example 3–27).

When the flange is secured to the floor, position the brass (usually) holddown bolts into the proper flange slots; these are to be placed so the threaded portion juts upward, as shown in the sketch. Slip a preformed sealing ring (puttylike compound available at building-supply and hardware stores) over the toilet bowl horn; this will form a watertight gasket for the water closet against the face of the flange. Don't use ordinary putty for this—it dries eventually and cracks. (Remember that sewage line leaks, as

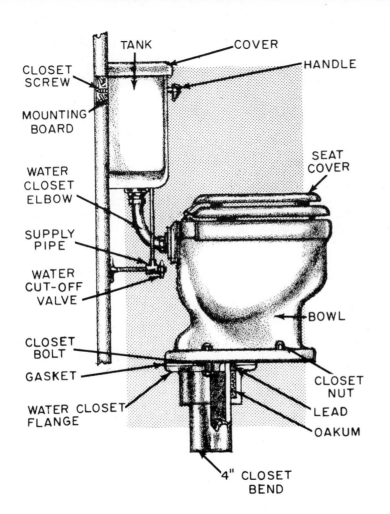

TANK

COVER

HANDLE

CLOSET
SCREW

MOUNTING
BOARD

WATER
CLOSET
ELBOW

SUPPLY
PIPE

WATER
CUT-OFF
VALVE

CLOSET
BOLT

GASKET

WATER CLOSET
FLANGE

SEAT
COVER

BOWL

CLOSET
NUT

LEAD

OAKUM

4" CLOSET
BEND

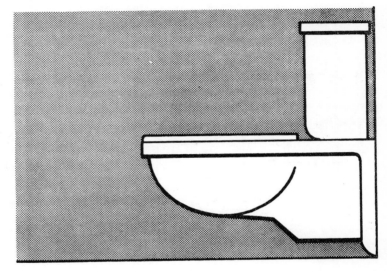

example 3–25. Water-closet mounting details. Inset shows modern wall-hung version.

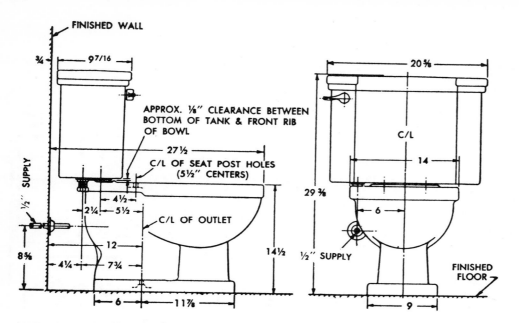

example 3–26. Typical water-closet rough-in measurements.

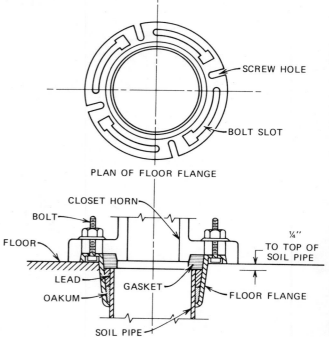

example 3–27. Connection details, water closet to floor flange and soil pipe.

from cracked bowl seals, allow the escape of gases that are toxic as well as unsanitary.)

Now set the water-closet bowl onto the flange with the horn of the bowl projecting down into the flange, guiding the holddown bolts up through the bolt holes on the base of the water closet as shown in example 3–27 (bottom). All you have to do now is secure the bowl with the brass nuts. (Some units are supplied with china nuts; others have plastic or enamel covers that must be cemented over the brass nuts. In any case, it is important that the nuts not be cinched down too tightly; otherwise, you can overtighten one of the nuts and crack the water closet.)

Water-Closet Removal and Replacement

When the water closet is obstructed beyond possible clearance with conventional in-place techniques, or when there is definite leakage around the bottom of the bowl, it will be necessary to remove the toilet to unclog the trap or replace the seal. Repeated gathering of moisture into puddles at the base of the bowl is a clear indication that the seal or gasket between the bowl and its outlet has ruptured. Don't rule out ordinary water condensation, however, or leakage from the tank—two common problems.

When you've determined that puddling is definitely attributable to a faulty seal, you must remove the bowl and install a new seal to prevent damage to your floor and the possible entry of sewer gases into the room. Ruptured gaskets are almost a certainty where (1) the original seal was formed with some compound other than water-closet sealing material, or (2) the closet's holddown bolts have remained loose long enough to allow the water closet to become wobbly.

Removal Procedure

There are several methods of flushing, and each represents a different water-closet configuration. For the discussion here, we'll assume that flushing is by means of a flush tank mounted on the wall behind the toilet. Newer units are one-piece assemblies (inset, example 3–25), and the same general instructions apply. Our discussion centers around older units because they are the most likely to be replaced or require repairs, and they present the most difficult repair challenge.

Before disconnecting any of the water lines serving the closet tank, make sure the water supply valve is shut off. If your unit is the type that has a plated elbow connecting the tank to the bowl, you have an advantage because the flush tank may be left in place while you remove the bowl. You *do* have to remove the elbow, however. Connection details are shown in example 3–28. Just

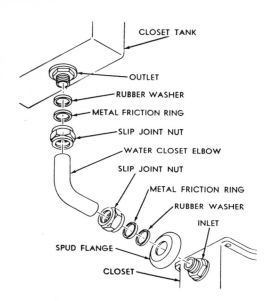

CLOSET TANK

OUTLET

RUBBER WASHER

METAL FRICTION RING

SLIP JOINT NUT

WATER CLOSET ELBOW

SLIP JOINT NUT

METAL FRICTION RING

RUBBER WASHER

INLET

SPUD FLANGE

CLOSET

example 3–28. Elbow connecting wall-mounting tank of older toilet to water closet.

unscrew the two slip-joint nuts on the elbow and slide them toward each other; the joints themselves are friction-fit and should break easily.

If the water closet you're removing is of comparatively recent vintage, it may be necessary for you to remove the tank and bowl as an assembly. In that case, first disconnect the water supply from the tank by unscrewing the union nut located immediately beneath the tank, and then flush the tank to remove all the water. If the water closet is secured to the wall as an integral assembly, the wall-holding screws have to be removed. If there's water or waste in the bowl, remove it by bailing or pumping.

Unscrew the nuts that fasten the bowl to the floor. Examine these first. If round or large crowned protrusions, they are simply decorative caps to conceal the brass nuts beneath. Don't use a wrench on these porcelain or plastic caps; they must be pried free with a broadblade screwdriver or similar tool. If they don't come free with a slight prying pressure, tap them lightly all around with a soft-faced mallet. Once these caps are removed, the brass nuts will loosen with ease.

Examine the water closet to be certain there are no more holding fasteners, and then ease the fixture from the floor. In many cases the fixture will itself be frozen into position and will require gentle knocks. Don't pull the water closet from its "mooring" by grabbing the tank if the tank and bowl are a one-piece assembly, especially if the connection between the tank and bowl is a relatively small area—a weak spot. Use the palms of both hands to jar the bowl from every angle; when the seal is broken at the floor flange, the bowl will be free.

In summary, here's the procedure for bowl removal:

1 Shut off the water.
2 Empty the bowl and tank of water by sponging, bailing, siphoning, or some other means.
3 Disconnect all water lines to the tank.
4 Disconnect the tank from the bowl if the assembly is a two-piece unit; if you have to remove the tanks (e.g., in the event the tank is not secured to the wall), be sure to place it where it won't be damaged.
5 Remove the seat and cover; if you plan to replace the unit with a new water closet, don't bother removing the old seat. But if you need to free the bowl of a stubborn obstruction, the seat must come off.
6 Pry loose the bolt covers and remove the bolts holding the bowl to the floor flange.
7 Jar the bowl enough to break the seal at the bottom, then remove the bowl gently.
8 Set the bowl upside down on something that has a surface that will not chip or break it.
9 Clear away the obstruction at the horn. If no obstruction is evident, and you're sure the pipe bend and flange are all right, the most likely blockage point in the water closet is the baffle or baffle trap (example 3–29), which occasionally manages to ensnare a child's toy or some other bulky object. With the closet separated from its mooring, you can reach every square inch of the toilet's innards through either the top or the horn.

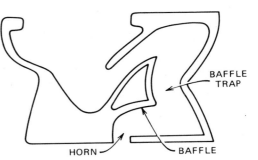

example 3–29. Cross section of water-closet bowl showing trap sections.

BAFFLE TRAP

HORN

BAFFLE

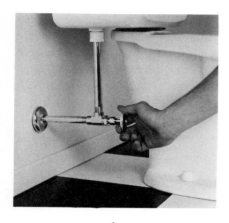

1

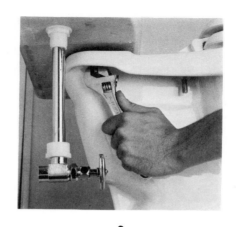

3

5

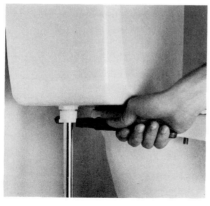

2

4

6

7

1. **Shut toilet water supply off completely, then drain all water from tank and bowl— by flushing, sponging, or bailing**

2. **Disconnect water line and drain all remaining water from tank**

3. **Disconnect two-piece tank from bowl**

4. **Remove seat and cover**

5. **Pry up porcelain or plastic bolt covers and remove holddown nuts beneath them**

6. **Jar bowl sharply using palm of hand to unseat water closet from floor flange**

7. **Remove bowl and place upside-down on carpeted surface; remove debris from horn with fingers. (If obstruction is inside bowl, use stiff wire as grappling hook.)**

Replacement Procedure

Never reuse gaskets or packing materials when replacing water closets. Use preformed sealing rings or specially designed wax seals. If the holddown bolts or nuts are badly worn or corroded, they must be replaced. Also, be sure that the floor area where the bowl is to be placed is clean and dry. Clean the bowl horn and dry it thoroughly, then place the new seal material around the horn and press it into position. (Suitable seals or gaskets are available from hardware and plumbing stores.)

Set the bowl in place and press it down firmly. Install the bolts that hold it to the floor flange. Tighten the nuts snugly, but be careful not to overtighten—porcelain bowls break when excessive pressure is applied in one spot.

Use a carpenter's level (Appendix) to seat the bowl; with the seat still on the bowl, place the level across the bowl opening and tighten the holddown bolts alternately. If the repair is being made in an older house, the floor itself may have sagged as a result of house settling; in this case, it may be necessary to shim up one side of the bowl to compensate for the slope of the floor.

Once the bowl is secured, connect the tank to the bowl and replace the water pipes. Replace all gaskets and make sure every fitting and pipe is immaculate. Test the installation for leaks by flushing a few times; when you're satisfied that the joints are all secure, reinstall the seat and cover.

Water-Closet Repairs

The mechanical portions of most water closets are contained within the tank assembly; to accurately assess a problem with a water closet, it is important to understand how the system functions. Unless you are a plumber already, the nomenclature used in plumbing will be unfamiliar to you; as you read the *principles of operation* presented here, refer to the tank illustration of example 3–30.

When you trip the lever outside the flush tank, the rubber or soft plastic tank ball (it may be a soft disk on some models) is raised from the flush valve seat to release water into the closet bowl. As this emptying process gets under way, the float ball moves downward as the water level recedes. Movement of the float arm causes a plunger in the intake valve assembly to open and admit a new supply of water through the supply pipe.

The tank ball (shown) or flat stopper disk seats as the water flows from the tank. Water coming into the tank creates sufficient pressure to hold the stopper disk or tank ball in position, thereby sealing the flush valve and preventing further flow. The new supply of water, then, is to be used for the next flushing operation.

As the water level rises during the refilling portion of the cycle, the float ball rises with it. The buoyancy of this float is considerable—enough to provide the force required to operate the intake valve. In the intake valve closeup shown in example 3–31, the rod lever lowers the plunger and causes the water to shut off when the tank fills to the proper level (in the pictured assembly, a thumbscrew allows adjustment of the shutoff point).

Notice the refill tube shown in example 3–30; this feeds water into the overflow pipe. During the refill operation, water flows through the refill tube into the overflow and down into the water-closet trap. If you've ever tried to make an adjustment in a malfunctioning valve assembly, you have doubtless observed that the toilet will operate whether the refill tube spouts water into the overflow or not.

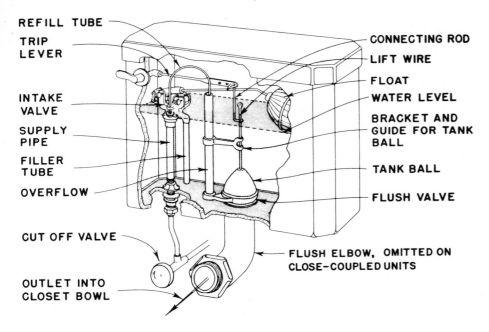

REFILL TUBE

TRIP LEVER

INTAKE VALVE

SUPPLY PIPE

FILLER TUBE

OVERFLOW

CUT OFF VALVE

OUTLET INTO CLOSET BOWL

CONNECTING ROD

LIFT WIRE

FLOAT

WATER LEVEL

BRACKET AND GUIDE FOR TANK BALL

TANK BALL

FLUSH VALVE

FLUSH ELBOW, OMITTED ON CLOSE-COUPLED UNITS

example 3–30. Flush tank cutaway with identification of components.

example 3–31. (A) Closeup of intake valve assembly; (B) plunger, washer, and cap in plan view.

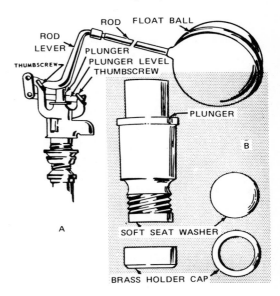

ROD   FLOAT BALL

ROD LEVER

PLUNGER

THUMBSCREW

PLUNGER LEVEL THUMBSCREW

PLUNGER

B

A

SOFT SEAT WASHER

BRASS HOLDER CAP

The reason for the refill tube is prevention of gas backflow—it is insurance that water in the trap will be replenished. Make sure the refill tube stays in place.

Some typical toilet maladies and their cures are described in the following text. Parts that most commonly require repair are the flush valve, the intake valve, and the float ball.

Tank Sweating

When cold water enters a water-closet tank, it may chill the tank enough to cause atmospheric moisture to condense on the sides of the tank. This sweating can be prevented by insulating the tank to keep the temperature of its outer surface above the condensation temperature (dew point) of the surrounding air. Insulating jackets or liners that fit inside water closet tanks are available from plumbing supply dealers. Another cure is to wrap a pipeline heating cord around the fresh water supply line that serves the water closet. Unless the problem is particularly severe, and water drips continuously from the tank to the floor, it's just as well to leave the tank alone; the trouble will disappear by itself when the humidity in the bathroom drops.

Water Running Continuously

You can be fairly certain that there is a problem with the intake-valve assembly when water continues to run into the closet bowl after the tank is full. If water runs into the bowl without the tank refilling, the problem probably lies in the flush-valve assembly.

When the water continues to run when the tank is full, the plunger has most likely failed to close the intake valve as it should. The excess water continues to flow and is discharged through the overflow pipe and into the toilet bowl. Common trouble sources for this problem are leaky or waterlogged float bowl, bent float-bowl arm, worn washer on the bottom of the plunger (see example 3–31B), a worn valve seat, or an improperly adjusted thumbscrew (or similar maladjustment).

Probably the best starting point is at the float bowl itself. With the tank full and the fresh water turned on so you can hear and see the results of your efforts to shut off the water flow, pull upward on the float ball lever as close to the ball as possible. If this causes the water flow to stop, and you haven't had to apply a great deal of pressure, you can be fairly certain the intake valve is all right.

Is the water level even with the mouth of the overflow tube? If so, the float-ball arm is bent out of alignment. You can fix this easily by holding the arm securely with the left hand at a point near the intake valve and then bending down the arm *slightly* at a point near the float bowl. Example 3–30 shows the proper bend position of the float-ball arm.

Intake-Valve Repair Procedure. Intake valves leak for a variety of reasons, some of which have been discussed previously. If a leak exists that can't be corrected with simple adjustments of the float ball or arm, the most likely culprit is a worn plunger washer, one version of which is pictured in example 3–31B. To replace this washer, perform these steps:

1 Shut off the fresh-water supply and drain the tank.
2 Remove the two thumbscrews that hold the rod lever (example 3–31), and push out the levers.
3 Lift out the plunger.
4 Unscrew the cup on the bottom of the plunger and insert a new washer (it's a soft material such as rubber or leather).
5 Examine the seat for nicks, scratches, grit, and particulate matter from the water. (If the surface is not smooth or if it's scratched, the washer seat will have to be refaced.)

Flush-Valve Repair Procedure. If you're experiencing trouble that can't be fixed with the routine examination and simple corrective procedure already discussed, you'll have to start at the top and eliminate possibilities one by one.

There are two principal difficulties associated with flush valves: (1) continuous operation and (2) failure to deliver the desired amount of water. Both problems can result in a large amount of water waste over extended periods. Once you understand the principle behind the flush valve's operation, you won't have too much trouble tracking down troubles with this cantankerous device.

Is the water level well below the mouth of the overflow tube? If continuous flow exists when the tank is full and the water level is at the proper point or thereabouts, there is a problem in the flush-valve assembly. The tank ball may not be making good contact with the flush valve; this can be caused by a variety of things: misaligned bracket and guide for tank ball, bent or corroded lift wire or connecting rod, worn out (excessively soft) tank ball, or damaged flush-valve receiver (the portion of the valve that seats the tank ball).

When the water continues to run when the tank is not full, it's highly probable that water is leaking out of the tank faster than the intake valve can fill it. Stop the intake flow by holding up the float ball or by supporting it in its raised position with a stick or wedge (but be careful to avoid damage to the intake-valve assembly). Drain the tank by raising the rubber or soft plastic stopper ball (example 3–32). Look at the stopper ball carefully to see if it is out of shape or cracked. Squeeze it to see if it has lost its elasticity.

Make sure the lift wire slides through the guide on the connecting arm (upper lift wire in example 3–32) without encountering excessive friction. If all seems satisfactory, remove the stopper ball by unscrewing it from the lift wire, and then examine the wire in relation to the valve seat; the lift wire should fall about in the center of the valve seat. If it does not, the lift-wire guide is likely out of alignment. If there is a thumbscrew or other setscrew adjustment point on the bracket, try loosening it and repositioning the

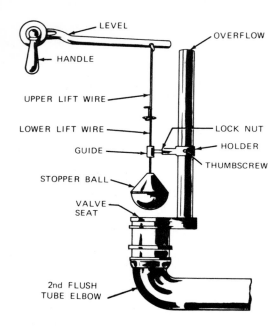

LEVEL

HANDLE

OVERFLOW

UPPER LIFT WIRE

LOWER LIFT WIRE

GUIDE

STOPPER BALL

VALVE SEAT

2nd FLUSH TUBE ELBOW

LOCK NUT

HOLDER

THUMBSCREW

example 3–32. Flush valve for low tank.

bracket and guide for the tank ball (*holder* in example 3–32). If the bracket and guide are welded or soldered in place to the overflow tube (as in example 3–30), you may have to bend the bracket slightly. Be extremely careful here, however; the slightest misalignment will create sufficient drag to keep the valve from closing properly.

Feel both lift wires to see if they are smooth; in areas where the water tends to be corrosive, copper flushing mechanisms deteriorate rapidly. If this is the case, you'd be well advised to replace the metal parts with plastic.

Sluggish or Ineffective Flushing

A tank should empty completely within 10 seconds. Because of insufficient rising from the ball's seat or elongation of the ball, the time may be longer than 10 seconds, which would make the flushing action correspondingly weak or sluggish. This problem may be corrected by shortening the loop in the upper lift wire. If this changes the resting position of the flush handle appreciably, bend the lever onto which

the upper lift wire connects—carefully! It's also a good idea to apply a few drops of lubricating oil to the lift wires and the lever (at both ends). Oiling the lever inside the tank where it goes through to mate with the handle will make the handle itself operate more freely; oiling the end where the upper lift wire connects minimizes the chance of binding.

Another cause of sluggish flushing is a small blockage at some point in the soil pipe or in the baffle of the bowl. If a sanitary napkin or other small object gets jammed across one of the baffle trap surfaces, it will restrict the flow to such an extent that emptying will be unreliable. Waste material won't be able to get by until it becomes liquefied. When this condition persists and cannot be corrected with the procedures used to unclog the bowl, removal is the only remedy.

Frequent Clogging of Bowl

When a bowl clogs often, it's usually accompanied by general sluggish operation. The cause is usually some foreign object in the toilet baffle or sewage line. Obstructions other than roots growing through one of the soil lines can ordinarily be cleared by chemicals or by manually operated devices designed for this service. The method of clearing an obstruction depends on the severity of the problem and the nature of the stoppage. It is extremely important to remove the obstruction completely rather than free it only to the extent where it can find another lodging place further downstream.

Everyone is familiar with the force cup, or "plunger," or *plumber's friend*, commonly used to clear stoppages in sinks and tubs (example 3–33). There are several versions of this, but the operational concept is the same: the cup is compressed over the drain area, then pulled up sharply to create a vacuum; the resulting displacement in the drain often frees the obstruction. Force cups aren't usually as effective at clearing blocked toilets as they are at unstopping sink drains, because they don't form an airtight seal in a toilet bowl.

To use the force cup properly, partly fill the water closet with fresh water. Position the force cup as squarely as possible over the throat of the bowl at its smallest diameter, then work the handle up and down to

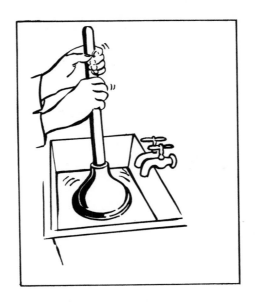

example 3–33. The force cup, universally known as the *plumber's friend*.

provide alternate compression and suction in the drain. If the downward thrust doesn't clear the stoppage by forcing air behind it, the pulling action sometimes will dislodge the obstruction enough for the water to flush it away. The force cup that has a bell-like extension (example 3–34) generally works more efficiently in a toilet bowl because it allows a better seal—the bell portion goes further into the bowl throat than the simple force cup shown in example 3–33.

When plunging action doesn't do the job, a *closet* or *trap auger* is required. The trap auger shown in example 3–35 is popularly known as a *plumber's snake;* it resembles the brake cable of a bicycle in that it consists of a tube with an inner flexible rod. A twisting force applied to one end is transferred to the other end, which is inserted to the obstruction point. Perhaps the most effective auger device is the *closet auger,* shown in example 3–36. This is a cane-shaped tube with a coiled-spring "snake" inside; it's equipped with a handle for rotating a coiled hook at the opposite end.

To insert the closet auger into the trap of a water closet, retract the

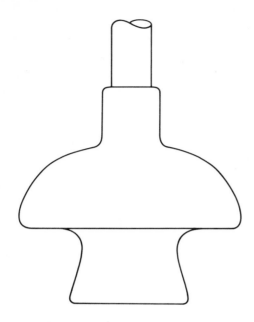

example 3–34. Water-closet force cup with a flanged bell, which provides a tighter seal than the conventional plumber's friend.

example 3–35. The trap auger, known generally as a *plumber's snake*.

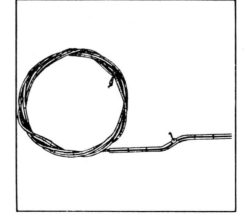

coiled spring up as far as possible into the sheath; then hook the "cane" portion around the protruding bowl section as shown in example 3–36. Hold the sheath or cane with one hand and crank the snake with the other; as you turn the crank the snake pays out, clearing its way as it goes. Turn the handle slowly until you reach the obstruction, which should snag on the coiled hook of the auger.

Once you reach the major obstruction, start pulling back on the auger while continuing to rotate the handle; this procedure serves to bring the snagged object up into the closet bowl, where it can be removed by hand. And *don't* assume the bowl is cleared after removing an obstruction; as often as not there will be more debris further downstream. Repeat the clearing operation until the snake passes without undue resistance all the way down into the horn of the bowl. Then flush a couple of wads of tissue just to be sure.

Modernizing Old Tanks

Several companies manufacture devices you can use to update worn-out valves in toilet tanks. When these devices are installed in an old tank, they make the entire flushing arrangement seem a lot simpler than it should be. It can make you wonder why the water-closet manufacturers haven't been doing the same kind of design improvements and simplification over the years. Let's look at a couple of these modernizing approaches.

Flush-Valve Replacement. The Fluidmaster Company markets a deceptively simple looking flush-valve replacement kit called the *Flusher Fixer.* As example 3–37 shows, there isn't much to it. And when you start ripping out the old tank parts the kit is designed to replace, you'll very likely do a little worrying, because you'll remove more parts than you replace. The parts to be removed can be seen at a glance in example 3–30: the connecting rod, both upper and lower lift wires, the bracket and guide for the tank ball, and the tank ball itself.

Installation requires no tools and takes no more than a few minutes. But there is a waiting period of two hours for the sealant to set thoroughly. The sealant is an interesting variation of the messy two-tube epoxy sets we

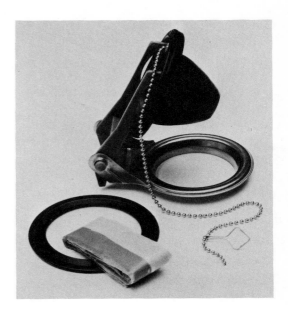

example 3–37. The complete Flusher Fixer kit. The striped ribbon is kneadable two-part epoxy, which is used to seal the flush valve in place.

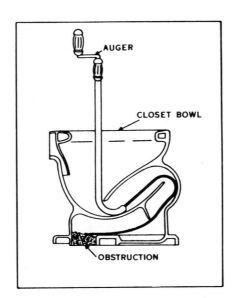

AUGER

CLOSET BOWL

OBSTRUCTION

example 3–36. The closet auger, identifiable by its cane-shaped sheath, is the best all-around device for clearing stubborn obstructions in toilets.

have seen in the past, but this one isn't messy at all. See the "candy ribbon" in example 3–37? That's a pliable two-part epoxy tape; all you have to do to get the epoxy mixed is to knead the tape. The epoxy tape comes with two colored stripes, one blue and one yellow. When you knead the ribbon, it turns green; and when the entire ribbon is green, you know the epoxy has been mixed properly. Your hands never get goopy.

To install the Flusher Fixer, just follow a few simple steps:

1 Prepare the tank by turning off the water and flushing the toilet. Then remove the old tank ball or flapper, the lift wires, and guide bracket. Don't remove the trip lever or handle, though; you'll need these to control the new assembly you're installing.
2 Determine the best position for the new flush valve by setting the Flusher Fixer in place on the old valve seat without sealant. Rotate the new flush valve a bit one way and the other to make sure you have good clearance. The new flapper and hinge frame cannot be allowed to come into contact with the overflow pipe or the side of the tank. The position of the valve is not critical, so long as no part of the unit touches anything during the flushing operation. Once you have the position established, remember the orientation.

3 Now set the new flush-valve assembly aside and clean off the old flush-valve seat with a scouring pad—and steel wool, if you have it. Then rinse it well. Knead the epoxy ribbon until it turns green, then apply it to the underside of the stainless-steel ring of the new flush valve. Don't let any of the adhesive come into contact with the flapper or hinge—they must be free completely.
4 Carefully center the ring onto the old flush-valve seat, placing it in the position established in step 2. Press it down firmly to distribute the epoxy sealant equally all around. Place a weight on the assembly—the maker suggests a 9-ounce can of vegetables—and add just enough water to cover the top rim of the Flusher Fixer. Let this arrangement sit undisturbed for two hours.

5 At the end of the two-hour period, remove the weight and connect the ball-chain clip to the hole in the trip lever. Adjust the chain for a bit of slack when the lever is down. Then cut off the excess chain with scissors or wire cutters. Turn on the water and allow the tank to fill normally.

Inlet-Valve Replacement. In the original inlet-valve assembly of your toilet tank (see example 3–31), a long rod with a float ball attached to it provides the leverage required to shut off the valve and stop the incoming water flow. But as we have seen, as the old assembly wears, it requires increasing leverage to keep things working. One effective solution to this problem is to replace the whole assembly with a new one of radically different design.

Fluidmaster manufactures a unique inlet-valve assembly that doesn't even use a float ball. The entire assembly is shown in example 3–38. The "rifling" on the shaft of the replacement unit allows the valve portion (at the top) to be positioned as may be required in individual tanks. This concept has allowed Fluidmaster to market one device that can be fitted to any toilet tank of any style or configuration. The ballcock is adjustable simply by sliding a clip on the assembly—this allows effective and complete control over the amount of water entering the tank.

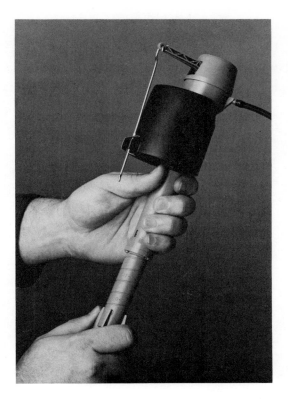

example 3–38. Fluidmaster's radical but simple replacement inlet valve does away with your old float-ball and rod assembly. A simple adjustment allows the unit to be tailored to any tank.

This unit is no more difficult to install than the flush-valve assembly. What you have to do is shut off the water at the cutoff valve (example 3–30), then remove the entire supply pipe with its intake-valve assembly, including the float ball and rod. Once the old assembly has been removed entirely, you'll be left with a hole at the bottom of the tank where the old supply pipe was mounted. Just mount

the new plastic flush valve in its place (example 3–39). Adjust the water level ring for the correct height, then tighten the assembly in place. Connect the supply piping to the plastic replacement on the underside of the tank, then clip the new refill tube to the existing overflow. Example 3–40 shows the complete installation in a tank that also contains the Fluidmaster Flusher Fixer kit.

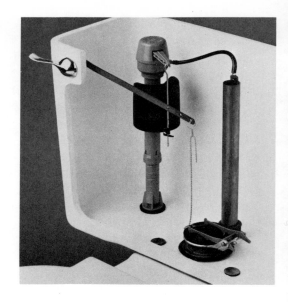

example 3–39. The plastic replacement assembly fills the void left by the removed inlet valve. The water supply connects to the threads after the unit is tightened into position.

example 3–40. This photo shows how the new refill tube is clipped to the overflow pipe. In this installation both the inlet and flush valves of an old tank have been replaced with improved versions.

Water Heaters

There aren't many homes these days without a hot-water heater, a "fixture" second in importance only to the water closet. Early water heaters required a great deal of attention; they had to be lighted every time hot water was required and turned off manually when the water reached the proper temperature. Today's heaters are virtually maintenance-free; they come on automatically when the water temperature drops below a specified point and shut off when the temperature rises to its specified upper limit. They have built-in safety devices to keep the unit from overheating or generating steam.

The heating elements in water heaters may be electric, gas, or oil, each with its own peculiarities and unique operational characteristics. There is little resemblance between the systems, even though they all perform the same basic function: keeping the temperature of your hot water supply within a certain range constantly.

Hot-Water Service

Instructions for connecting water heaters to plumbing systems come with the units; the tanks have the necessary internal piping already installed, so that all you have to do is connect the hot- and cold-water lines and the fuel supply. Unless you have the equipment to check temperatures and water pressure, coupled with knowledge of the fuel system being employed, you must have your unit installed by a qualified professional.

Every modern heater has certain features in common—pressure relief valves and thermostats. Pressure and temperature relief valves vent the pressure in the heater storage tank and associated pipes if other control equipment in the system fails and the water temperature climbs to the point where dangerous pressures could exist. As water heats, it expands, and the expansion will become sufficient to rupture the tank or pipes if the water can't be forced back into the cold water line or discharged through a relief vent (valve).

The size of the hot-water storage tank, and the capacity of the associated burner, depends on the size of the family being served in the dwelling—the prime determinant as to the volume of hot water required during periods of peak use—and the recovery rate of the heater. The recovery rate is proportional to the size of the heating element in *British thermal units* (Btu) and is measured in gallons per hour; the higher the Btu rating, the better (more gallons per hour) the recovery rate.

The recovery rate of water heaters varies also according to the type of fuel required. In conventional units intended for home installations, oil and gas heaters tend to have a higher recovery rate than electric heaters of similar size; however, some manufacturers of electric water heaters do have "quick recovery" units that compare favorably with oil and gas types. These involve the use of several high-wattage heating elements. Electric water heating does have advantages over gas and oil: the latter types require special flues to vent the products of combustion.

According to information acquired by the U.S. Government and released by the Department of Agriculture, a family of four or five requires a water-tank size of about 35 gallons for oil or gas heaters, not less than 40 gallons for quick-recovery electric heaters, and about 50 gallons for standard electric water heaters. For larger families, or where unusually heavy use will be made of hot water, correspondingly larger capacity heaters should be installed. It's probably a good idea to consider the government figures as generally conservative.

In comparing water heaters that require different fuels for their operation, it may be convenient to convert kilowatts to British thermal units and vice versa. One British thermal unit represents the amount of heat required to raise the temperature of one pound of water one Fahrenheit degree when the ambient temperature is 39.2° F. But the Btu figure doesn't mean much unless it's linked to a time designator; otherwise there would be no way of knowing how long heat would have to be applied to the pound of water before its temperature increases one degree. Gas and oil heaters are rated generally at Btu *per hour*. To get a fair idea of the Btu rating of an electric heater, multiply the heater's total input wattage by 3.41.

Electric water heaters are rated in kilowatts, or thousands of watts (10 kilowatts is the same as 10,000 watts and 27.4 kilowatts is the same as 27,400 watts).

If you want to replace an oil or gas heater with an electric type, you can approximate the power consumption of the replacement unit (assuming it's to be the same capacity as the old one) by dividing the old heater's Btu/hr rating by 3.41. The result will be watts, which you can convert to kilowatts by moving the decimal point three positions to the left. To approximate the power output of a 96,000 Btu gas water heater, for example, you'll need an electric water heater with more than 28 kilowatts of burner capacity.

Swimming-Pool Heaters

The swimming-pool heater is really no different from the home water heater, but the selection criteria are unique and important. The size of heater you need for your pool depends on several factors: pool size, the climate of the area you live in, and the water temperature you prefer. And, of course, an indoor pool requires a heater of less capacity than an outdoor pool.

A pool heater that is too small, even though it operates flawlessly, won't be able to keep your pool at the temperature you select. It will not be able to overcome the considerable cooling effects of evaporation on dry days, for example, nor will it heat at a rate that equals the natural cooldown on days that aren't sunny and warm.

A pool heater that is "too large" means unused capacity (and concomitant waste), which might cost you more in upkeep dollars than you are getting in swimming fun. The Ruud Manufacturing Company, which manufactures home and pool water heaters, uses a sizing guide to assist customers in selection of the proper heater for various pool types and sizes. The method involves four fairly uncomplicated steps, which are reproduced here with Ruud's kind permission:

1 Calculate the area of your pool by multiplying its width by its length. If your pool is other than rectangular, refer to the sizing guides for round and oval pools (examples 3–41A and B).

2 The area figure obtained in step 1 must be adjusted according to the surface heat loss in your area. If your pool area receives typical daily breezes of between 3.5 and 5 mph, increase the area figure by 25%; if the pool area sees frequent winds of 10 mph or so, increase the calculated area by 50%. Write the total so that you can refer to it when you use the heater selection guide of example 3–41C.

SWIMMING POOL SIZING GUIDE

TYPE POOL	DIA.	AREA SQ. FT.
ROUND POOL	8'	50
	10'	78
	12'	113
	15'	177
	18'	254
	21'	346
	22'	380
	24'	452
	27'	572
	28'	615

A

example 3–41. Pool heater selection guide. Use chart A to calculate dimensions of circular pools, B for oval pools. Chart C (see text) gives you recommended heater sizes for your pool after considering wind loss for your area. (Courtesy Ruud Manufacturing Company.)

SWIMMING POOL SIZING GUIDE

TYPE POOL	W	L	AREA SQ. FT.
OVAL POOL	8' x 12'		82
	10' x 15'		128
	12' x 18'		185
	12' x 24'		257
	15' x 25'		327
	15' x 26'		342
	15' x 27'		357
	15' x 30'		4092
	15' x 32'		432
	16' x 24'		329
	16' x 32'		457
	16' x 40'		585
	18' x 34'		542

FOR RECTANGULAR POOL MULTIPLY LENGTH BY WIDTH.

B

3 Look at the pool-heater selection guide (C) and determine the column that most closely describes the number of degrees higher than ambient (39.2° F) you'd like the water to be on cool swimming days. The first column, for example, is applicable if you would like the water temperature to be about 10° higher on the average than the cloudy-day air temperature around the pool. Naturally, this is an extremely rough estimate. The Ruud people have a procedure you may adopt to arrive at a more precise temperature differential figure: First, determine the pool temperature you'd like. Then determine the average daily temperature during the coolest month of your planned swimming season. When you have both these figures, just subtract the second from the first to obtain the temperature differential.

4 Read down the column and spot the square containing the figure nearest to or greater than the area figure you jotted down in step 2. (If your home is served by liquid petroleum gas, use the parenthesized numbers; if you are served with natural gas, use the other.) The Btu requirement can be read in the two left-hand columns.

TEMPERATURE DIFFERENTIAL BETWEEN AIR and POOL WATER		10°F	15°F	20°F	25°F	30°F
INPUT BTU/HR NATURAL	(L.P.)	POOL SURFACE AREA IN SQUARE FEET**				
80,000	(69,000)	533 (460)	355 (306)	266(230)	213(184)	178(153)
160,000	(160,000)	1067(1067)	711 (711)	533(533)	426(426)	355(355)
240,000	(225,000)*	1600(1500)	1067(1000)	800(750)	640(600)	533(500)
300,000	(240,000)*	2000(1600)	1333(1067)	1000(800)	800(640)	667(533)

* PROPANE ONLY ** SURFACE AREA IN PARENTHESIS ( ) IS FOR L.P. GAS

C

Section 4

Drains and Vents

Section 4

The drainage system includes all piping that carries sewage or other liquid waste to the house sewer line, which in turn carries it to the disposal facility. Since the escape of sewage or sewer gases can be a serious health hazard, it is extremely important to make installations and pipe repairs properly. The information in this section should be all you need to remove and replace pipe sections above ground, add drains and disposal line branches, and vent waste lines according to accepted plumbing practice.

Bear in mind always that polluted water or sewage may carry disease such as typhoid and dysentery. When you do your own plumbing work, be especially careful to avoid leaks, cross connections, and back siphonage. Leaks in drain systems are an invitation to sewer gases, which are not just a health hazard but at times explosive and toxic.

As long as cross connections are avoided, you can be reasonably certain that all water sources that are safe remain so; and when there is trouble in one line, the absence of any cross connection will allow quick isolation so that repairs can be initiated.

Once a pipe has become polluted, it is usually difficult to clean; for this reason building codes do not normally permit the use of secondhand pipe in newly installed plumbing systems or partial systems. Use new material throughout.

In this section, we attempt to steer clear of needless specifications and tedious charts and tables. You don't *have* to know how fast wastes travel down pipes of certain materials, dimensions, and slopes if you follow some of the basic rules of good plumbing practice. A master plumber *does* have to know these things before he can qualify for his license. This book, however, is written not for the master plumber or the apprentice seeking master status, but for you and others like you who want to be capable of making additions, modifications, and repairs to existing plumbing systems.

Vent and Drain Principles

If you have ever tried to pour a gallon of gasoline from an unvented can into a stalled car, you'll have little trouble understanding the principle behind venting. And your understanding will be even more complete if you have ever found it necessary to siphon gasoline from a tank to a can. Let's take the example of an unvented gas can first.

When the can of gasoline is right-side up, there is an air space at the top that is of the same air pressure as the atmosphere around the can—about 14.7 pounds per square inch at sea level. When you invert the can so that the spout, the only opening into the can in our example, protrudes downward as in example 4–1, the liquid, being heavier than air, goes to the bottom of the can and the air becomes trapped. Immediately, if the spout is open, gasoline will begin to pour from the can, which creates a larger air space inside the can as some of the liquid falls from the spout. But this causes the air pressure inside the can to drop substantially.

If there is an air pressure of 14.7 pounds per square inch with an air-space volume of 100 cubic inches, it stands to reason that the air pressure per square inch would be less if the air-space volume were to increase suddenly. And drop it does. But the instant the air pressure in the can drops below the air pressure around the can (atmospheric air pressure), we say that a *vacuum* is formed. At first the sides of the can will begin to be pushed in around the air space; but the can is metal, so the sides will only permit a certain amount of flexing in this way. In order for the flow of liquid to continue from the can to the gas tank, some air must be introduced into the air space in the can.

At this point, the only entryway for air is through the spout itself; when the pressure differential is great enough, outside air pressure coming up through the spout pushes the liquid out of the way (momentarily defeating the flow of gravity and stopping liquid flow from the can) and rushes into the air space, equalizing the pressure in the can once more. When the pressure in the can is *equalized,* or made the same as the pressure on the

example 4–1. This unvented gasoline can can't discharge liquid effectively. For every cubic inch of gasoline removed from the can, one cubic inch of air must be introduced to displace it. With no vent, the air has to stop the gasoline flow momentarily while air goes up the spout; the result is a cyclic gurgling and uneven discharge of gasoline.

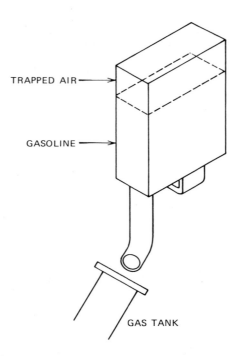

TRAPPED AIR

GASOLINE

GAS TANK

outside of the can, the process repeats itself. This occurs again and again as the can is emptied: the sides of the can push in as liquid comes out the spout, then air goes up the spout as gas flow is stopped. Air entering the can forces its sides back to the original configuration, and flow is permitted again. This happens quickly in repetitive cycles so that what you hear as you pour are gurgles and glugs, and what you feel is the can distorting and regaining its shape. Inside the can, the liquid burbles undecidedly as air pressure shifts back and forth between atmospheric and below-atmospheric. All that is needed to smooth the flow and make it continuous is a small air vent above the level of liquid when the can is in its pouring position. This would keep the air pressure inside the can always the same as the air pressure on the outside of the can, and the liquid would come from the can in one continuous stream because of gravity.

In a plumbing system, the drain and soil pipes are the gas can and spout. And drain pipes must be vented to allow the smooth, fast flow of waste. But the vent in a drain system has yet another purpose, as we can

see from the siphon example. With a siphon hose you can make liquids flow uphill. Examine the classic gasoline siphoning arrangement shown in example 4–2. For liquid to flow from the automobile tank to the gas can, the fuel must first "defy gravity" by flowing up the hose to the gas-tank opening. If you were to place all the proper components in their respective "correct" positions, you could sit and watch till the end of eternity and no fluid would flow from the tank to the can. Before the flow can begin, the air pressure at the lowest point of the hose must be substantially less than the pressure above the fluid line in the tank (which, of course, is roughly 14.7 pounds per square inch). It's not very pleasant to taste, but you can start the flow in example 4–2 by sucking on the end of the hose as if it were a soda straw while the other hose end remains immersed in the tank. *After all air has been removed from the hose,* air pressure on the fluid level in the tank will continue to force fluid through the tube, but it isn't really defying gravity at all. As a matter of fact, the flow continues *because* of gravity. It works this way:
When the tube is completely full

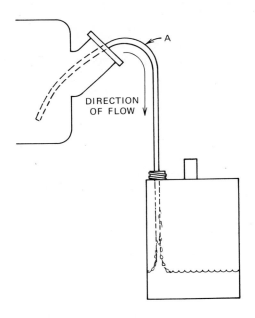

DIRECTION OF FLOW

A

example 4–2. When air pressure at the lowest hose mouth is less than that pushing on the gasoline in the tank, the fluid will continue to flow once all air has been removed from the hose. Note that flow occurs despite the fact that point A is at a greater elevation than the inlet and outlet. A vent at point A would make all liquid in the tube fall according to the law of gravity.

of gasoline, gravity causes all the liquid at the highest point (point A in the sketch) to fall. If there were a hole in the hose at or near A, the liquid in the vertical portion of the tube would fall into the gas can and the liquid on the tank side of the loop would rush back into the tank. The hole would be the vent needed in the first example. With no hole in the hose, though, and the flow actually started by reducing the air pressure at the lowest point, it must continue as long as the outlet end remains below the fuel level of the tank. Gravity makes the fuel drop from A to the gas can; but something must displace the liquid at A as it falls. The only thing that *can* displace it is the liquid at the upper mouth of the tube; and it will continue to do so as long as the fuel level in the tank is subjected to atmospheric air pressure. Seal off the tank, and flow will cease. Introduce air to the hose and flow will cease.

Remember these two examples and your plumbing troubleshooting tasks will be simplified greatly when problems develop. And it will help you appreciate the importance of a lot of piping that might otherwise seem useless to you.

One of the most pesky sources of trouble in a household drainage system is a clogged main-stack vent, not because birds tend to build nests in rooftop vent openings (even though it can and does happen), but because people who install TV antennas tend to anchor masts inside roof vent openings. When you plug a vent or obstruct it, you're asking for sluggish drainage, waste siphonage from fixture traps, and recurring drain headaches that can't be cured with an Anacin tablet.

Drainage Piping

Over the years, plumbing hasn't changed much. Like the automobile windshield wiper that's been modified only slightly since its inception in the early 1900s, plumbing has remained a successfully developed art that relies on but a few scientific principles for its operation. One area of plumbing that has seen some innovative changes, however, is materials. And these changes are so recent that many local codes still don't recognize their existence. Because of the slow responsiveness of local governments to technological advances, you may find it mandatory to melt lead and use it with oakum to seal certain pipes in your house drainage system while other localities permit use of plastic pipe and chemical adhesive sealants.

Drainage piping may be made of plastic, cast iron, galvanized wrought iron or steel, copper, lead, or brass.

Cast iron is commonly used for building drains that are buried under concrete floors or underground.

Plastic Piping

Plastic pipe and tubing, when carefully selected and properly installed, offer several advantages over conventional piping materials such as galvanized steel and copper conduit. But bear in mind that there are no "perfect" plumbing materials, and all must be installed with knowledge of their physical properties and limitations.

Even when plastic piping isn't the most economical type you can use, it is light and easy to install, which gives it a leg up on some of the other perhaps less costly competitors.

Different types of plastic pipe require different substances for sealing. If you want to use plastic for your add-on or replacement drainage piping, determine first if your locality permits it, and then use the sealing agent recommended by the manufacturer of the pipe you select. Your plumbing-supply dealer will prove extremely helpful here.

Here are the elements to remember when shopping for plastic pipe to be used in drain systems: make sure the material is marked with the manufacturer's name or trademark, pipe size, the plastic material type or class code, pressure rating, the standard to which the pipe is manufactured (ASTM), and the seal of approval of an accredited testing laboratory such as the National Sanitation Foundation. Plastic pipe is easier to cut, carry, fit, seal, and work with all around; if you *can* use it, *do* use it.

This book is not intended as a "plumber's handbook," and it does not cover rough-in plumbing as such. There may be occasions, however, when you want to install a laundry facility in a basement or add a bathroom where no rough-in plumbing is provided. Consequently, there are some basics of plastic piping that will be of benefit for you to know. What size of drain pipe do you use for waste line? Does plastic pipe have a tolerance for chemicals and solvents? How difficult is it to incorporate plastic fittings into an existing plumbing system?

The fact is, there's a plastic pipe or fitting for just about every soil-pipe function you can imagine. There are no universal guidelines that all localities and codes agree on, unfortunately. If your plumbing installation requires inspection, you should check beforehand as to whether or not plastic pipe is acceptable in your area. If plastic *is* acceptable, you should choose your materials according to the requirements of your household. Drain piping, for example, doesn't have to handle liquids under pressure, so you needn't be concerned about the specifications relating to pressure.

Sizes and Grades

Polyvinyl-chloride (plastic) pipe to be used in drain lines is called DWV, for *drain-waste-vent*. You can buy DWV piping in various thicknesses, but the grading system for those thicknesses isn't readily identifiable to nonplumbers according to inches or millimeters; rather, there are specifications or *schedules* that identify the thicknesses to the initiated. Schedule 120 pipe, for example, is very thick. Schedule 40 pipe is relatively thin-walled. Thicknesses greater than schedule 40 are pressure-rated and are used for water service; schedule 40 and below are typically good for DWV lines. But if you need to thread your pipe for fittings (some codes require it), you should opt for schedule 80 or thicker

because manufacturers of plastic pipe don't recommend threading schedule 40; threading would reduce the thickness of the wall in the threaded area excessively. When you have to thread plastic pipe, you should follow the guidelines in the section on water service (Section Two), which covers the subject extensively.

PVC (polyvinyl chloride) pipe is "faster" than comparable metal conduit. As a consequence, when you make an installation you can use one size smaller than the recommended metal pipe—assuming that local codes permit. Local plumbing codes should be checked for the sizes of drain pipe required, of course, but when no codes are applicable, the U.S. Department of Agriculture recommends pipe diameters of:

- $1\frac{1}{4}$ inches for lavatories
- $1\frac{1}{2}$ inches for bath, dishwasher, sinks and laundry trays, and fixture branches
- 2 inches for floor drains and shower stalls
- 3 inches for soil stacks, water closets, and beyond-soil-stack connections

But these recommendations are for rigid metal pipe; when permissible, you can use $2\frac{1}{2}$ inch diameters where 3-inch diameters are called for, $1\frac{1}{2}$ inches where 2 inches are called for, and so on, when you use plastic pipe.

Cutting, Fitting, and Joining

The installation instructions provided here, provided by R & G Sloane, a company engaged in the manufacture and sale of plastic DWV pipe, are applicable to nonthreaded joints; threading and joining details are covered in Section One (piping materials). Sloane representatives state that 95% of all plumbing failures (leaks, premature disconnects, inability to seal, and other problems) are caused by poor cementing practice.

The first step is to make sure the fittings and pipe are the *same temperature;* it's the only way to assure a good fit because plastic pipe has an unavoidable tendency to expand with higher temperature. If the fitting and the pipe are at two separate temperatures, the one with the highest temperature will be stressed when it

CUTTING

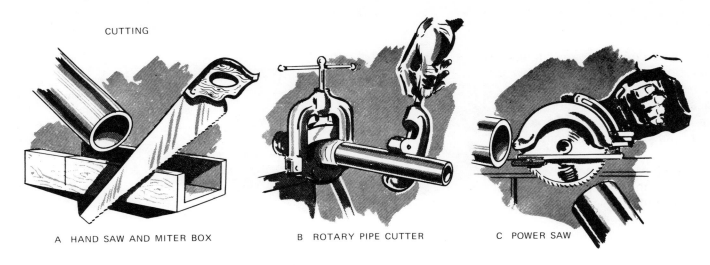

A HAND SAW AND MITER BOX   B ROTARY PIPE CUTTER   C POWER SAW

example 4–3. The important thing to remember when cutting plastic pipe is to make the cut as square as possible; you can accomplish this with a hand saw and miter box (A), a rotary pipe and tube cutter (B), or a power saw (C). (Courtesy Harvel Plastics Inc.)

begins to regain its original configuration.

Cutting isn't critical by any means, but it must be done properly. To be sure you're cutting the pipe square, use a miter box or tubing cutter (example 4–3). You can use a power saw so long as you can maintain a square cut.

The next step, beveling, is where many go wrong; they simply tend to forget to do it. But beveling is extremely important because it provides a large contact area for the cement used to make the weld. As shown in example 4–4, the bevel should measure not more than $\frac{1}{8}$ inch nor less than $\frac{1}{16}$ inch; you should try for a 10- to 15-degree bevel at the cut.

Before the fitting and pipe are mated, be absolutely sure both are clean and dry. Use a soft, clean cloth to wipe both the end of the pipe and the inside of the fitting to be mated. Then check the fit (without solvent). The pipe should enter the fitting socket from half- to full-socket depth.

Once you're satisfied that the fit is right, scrub in an even coat of primer on the pipe end and fitting socket with brush or applicator until the surface starts to soften or dissolve. (The right

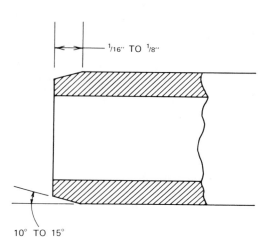

$\frac{1}{16}$" TO $\frac{1}{8}$"

10° TO 15°

example 4–4. Sealing a joint in a plastic pipe isn't difficult if the operation is done properly; the main point to remember is to allow as much contact surface as possible for the binding cement. Ideally, the bevel will be 10 to 15 degrees from the axis of the pipe, and the width of the bevel will be between $\frac{1}{16}$ and $\frac{1}{8}$ inch, depending on wall thickness and the diameter of the pipe.

primer for your pipe is available from the pipe distributor.) Then apply a full but even coat of cement to the pipe end and the fitting socket, all the way back to and including the pipe stop in the fitting. Do it again. Don't be sparing in the use of the cement; as the Sloane people point out, the cement you use here is probably the least expensive item in your inventory.

Then, while the surfaces are soft and wet with solvent, force the pipe into the fitting socket while giving the pipe a quarter turn. The pipe must be bottomed in the socket and held in place for a minute or so. It's very important that you hold the connection well during this brief period since the pipe will have a definite tendency to push out from the fitting.

If you've done a good job, your properly made joint will show a bead of cement around the perimeter of the joint. If there's a gap (no cement showing) it may indicate a poor joint due to insufficient cement or the use of a cement that is too light-bodied. If you're satisfied with your connection, wipe away the excess cement; but don't disturb the joint until it has had the chance to set. At normal room temperatures, the recommended

setting time is 30 minutes; at the cold end of the spectrum (0 to 20° F) you have to wait 4 hours before you can use the system.

General Requirements

Even though PVC and other plastic pipes are tough, resilient, and durable, they are softer than metal and are susceptible to damage by abrasion and gouging. For this reason, plastic pipes and fittings should be stored separately from metal pipes and fittings.

When you use plastic pipe, be sure to support long runs with hangers and brackets at every 6-foot interval. Avoid installation of PVC pipe near hot objects such as steam lines or radiators, hot-water-service piping, and so on. Be absolutely sure that all hangers and brackets are mounted straight so they won't cut into the pipe they are intended to support.

The expansion coefficient of PVC pipe is something like ten times that of metal, but this won't prove troublesome for home plumbing installations. On very long runs of the kind you might find in an industrial situation, however, some joints must

be made with a special expansion fitting, which permits expansion and contraction of pipes without compromising any of the seals at the fittings.

Cast-Iron Piping

Cast-iron soil pipe and fittings are normally used in and under buildings, protruding some 3 to 5 feet from the building. Here the cast-iron line connects to the house sewer line (Section Six). Cast iron is used both above and below the ground. If the soil is unstable, it will be to your advantage to use cast-iron pipe, but you should avoid cast-iron pipe in soil containing cinders or ashes because this is an indication of sulfuric acid content, which encourages corrosion and general deterioration of iron.

Cast-iron drainage pipe sections and fittings are usually of the hub-and-spigot type and are joined by packing with hemp, wool, or oakum and sealing with lead. These sections come in 5- and 10-foot lengths with inside diameters of 2, 3, 4 inches, and larger. They are available as double- or single-hub lengths, as shown in example 4–5. The hubs are enlarged sleevelike fittings cast as part of the pipe itself.

When joined properly with oakum and lead (Appendix), the joint must be fitted and packed so that the sections are concentric, leaving no obstructions

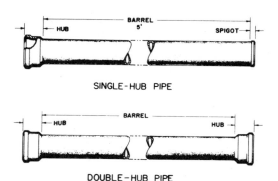

SINGLE-HUB PIPE

DOUBLE-HUB PIPE

example 4–5. A single-hub pipe section has a bell at one end; a double-hub section has one on each end.

example 4–6. Hub-and-spigot joint in cast-iron soil pipe; note that the direction of flow is always from hub to spigot. Some local plumbing codes still require the use of oakum and lead; others recognize the advantages of speed-seal connections and alternative pipe materials. But when you have to repair or replace a section in an existing line, you don't have a great deal of choice in the matter.

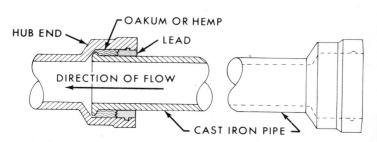

to the flow of liquid or projections against which solids can lodge. The direction of waste flow must always be as shown in example 4–6.

There is another system for connecting cast-iron pipe sections, but not all localities permit its use. Briefly, this comparatively recent technique involves the use of no-hub pipe (pipe that has spigot openings on both ends). Pipe sections are joined by a Neoprene sleeve gasket held in place by a wraparound stainless-steel shield fastened by steel bands with worm-drive clamps (example 4–7). The absence of hubs allows you to install 2- and 3-inch pipe sections within the confines of a standard wall framed with two-by-fours.

Measurement

Cast-iron soil pipe sections are generally referred to as being 5 or 10 feet in length, but this is not precise. Reference to a 5-foot length applies to the laying length rather than the overall dimensions. The pipes fit one into the other, and when they are all fitted, the total length of each section is 5 feet, or 10 feet, depending on the pipe size

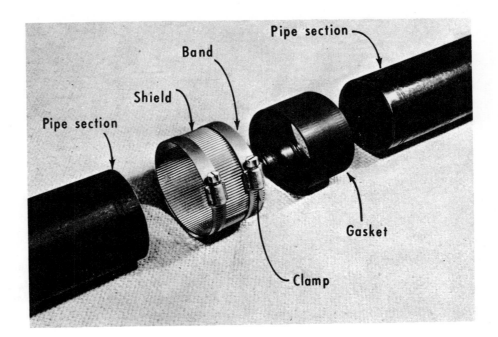

example 4–7. Hubless pipe can be assembled by butting the spigots together and sheathing the butted joint with a Neoprene gasket and a stainless-steel shield, which is held in place with a pair of hose-clamp bands.

used. The length of the hub for a 3-inch diameter pipe is 2¾ inches.

The most common measurement of cast-iron soil pipe for a shorter length than 5 feet is the overall measurement. When making this measurement for, say, 4-inch pipe, take the desired length of pipe for the installation and add 3 inches to it for the bell.

Cutting

Before joining cast-iron soil pipe, you will often have to cut the pipe to provide the length desired for the job at hand. Cast-iron soil pipe can be cut with an abrasive cutter, a band saw, a special soil-pipe cutter, or a hammer and a cold chisel. Since the hammer and the cold chisel are the most common, our cutting instructions will involve these only.

To cut cast-iron soil pipe using a hammer and cold chisel, you will need the following equipment:

- 6-foot folding rule
- Piece of soapstone or crayon
- 18-inch wraparound
- 2 two-by-fours (approximately 3 feet long)

- Hammer
- Cold chisel
- Clear goggles

1 Place the cast-iron soil pipe on the wood blocks, one under each end of the pipe on the floor or ground, so it can be steadied with your knee.
2 Measure the length of the piece to be cut and mark this length on the pipe with an arrowhead (see example 4–8A).
3 Make a line around the pipe using a wraparound and a piece of soapstone or crayon. This will aid you in cutting the pipe squarely.
4 Place one two-by-four under the pipe where the cut is to be made.
5 Hold the cutting edge of the chisel firmly and squarely against the pipe at the mark and strike with the hammer, scoring the cast-iron soil pipe lightly (see example 4–8B).
6 Rotate the pipe slightly and place the cold chisel against the mark and strike it again with the hammer.

7 Repeat steps 5 and 6 until the pipe has been evenly scored on its whole circumference. Give very light blows with the hammer until the cast iron has been scored all the way around, to prevent the pipe from breaking jaggedly, rendering it valueless.
8 Gradually increase the force of the blows with the hammer while rotating the pipe, deepening the score until the pipe breaks evenly around the scored mark.

If you have to cut a short piece of cast-iron soil pipe, cut it from a double-hub pipe, because the remaining pipe still has a hub good for future use. However, when you cut the hub off a single-hub pipe, the remaining portion of the pipe is usually wasted.

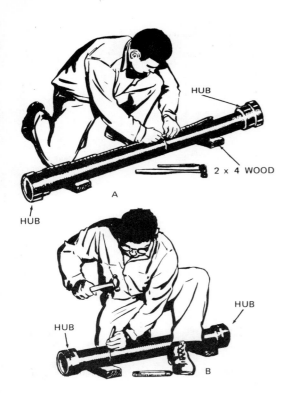

HUB

2 x 4 WOOD

A

HUB

HUB

HUB

HUB

B

example 4–8. Cutting and measuring cast-iron soil pipe. First, place the pipe section across a couple of two-by-four blocks to protect the hub and keep the pipe horizontal; then mark the entire circumference of the pipe where the cut is to be made (A). Finally, score the line with a hammer and cold chisel evenly all the way around the pipe, starting with light blows and gradually increasing the force of blows until the section separates. Always wear goggles when performing this operation!

Fittings

Cast-iron soil pipe fittings are used for making branch connections or changes in the direction of a line. Both pipe and fittings are brittle, so exercise care to avoid dropping them on a hard surface. Some cast-iron fittings are described below.

Bends. A number of different types of bends are generally used on jobs involving cast-iron soil pipe. Among common types are the $\frac{1}{16}$, $\frac{1}{8}$, short-sweep $\frac{1}{4}$, long-sweep $\frac{1}{4}$, and reducing $\frac{1}{4}$ bend. Example 4–9 shows the shape and appearance of each of these types of bends.

The $\frac{1}{16}$ bend is used to change the direction of a cast-iron soil pipeline $22\frac{1}{2}$ degrees. A $\frac{1}{8}$ bend is used to change the direction of a line 45 degrees.

The *short-sweep* $\frac{1}{4}$ bend is a fitting used to change the direction of a cast-iron soil pipeline 90 degrees in a close space. The *long-sweep* $\frac{1}{4}$ bend is used to change the direction of a pipeline 90 degrees more gradually than a quarter bend.

The increasing $\frac{1}{4}$ bend gradually changes the direction of the pipe 90

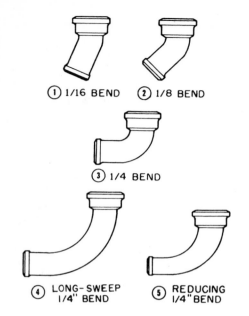

example 4-9. Some cast-iron soil-pipe bends.

① 1/16 BEND ② 1/8 BEND

③ 1/4 BEND

④ LONG-SWEEP 1/4" BEND ⑤ REDUCING 1/4" BEND

degrees; and in the sweep portion it reduces one size. A 3-by-4 increasing long-sweep $\frac{1}{4}$ bend, for instance, has a 3-inch spigot on one end, increasing in 90 degrees to a 4-inch hub on the other end. Note that for all fittings the spigot end always is listed first.

Tees. Tees are used to connect branches to continuous lines. Learn to recognize the four designs of tees shown in example 4-10.

For connecting lines of different sizes, *reducing* tees often are suitable.

The *test* tee is used in stack and waste installations where the vertical stack joins the horizontal sanitary sewer. It is installed at this point to allow the plumber to insert a test plug and fill the system with water while testing for leakage. (The test tee also is used in multistory construction.)

The *tapped* tee is frequently used in the venting system, where it is called the main-vent tee. The sanitary tee is commonly used in a main stack to allow the takeoff of a cast-iron soil-pipe branch.

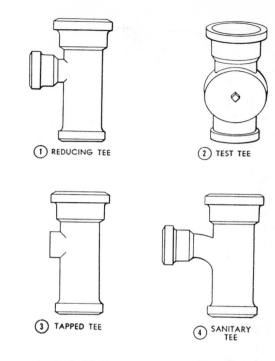

① REDUCING TEE ② TEST TEE

③ TAPPED TEE ④ SANITARY TEE

example 4-10. Some cast-iron soil-pipe tees.

90-degree Wye Branches. Four types of 90-degree wye branches generally used are illustrated in example 4–11. These are normally referred to as *combination wye* and $\frac{1}{8}$ *bends.*

The *straight* type of 90-degree wye branch has one straight-through section and a takeoff on one side. The side takeoff starts out as a 45-degree takeoff and bends into a 90-degree takeoff. This type is used in sanitary sewer systems where a branch feeds into a main; it is desirable to have the branch feeding into the main as nearly as possible parallel to the main flow.

The *reducing* 90-degree wye branch is similar to the straight type. But as indicated in example 4–11, the takeoff of the 90-degree branch is of a smaller size than the main straight-through portion. Its general use is the same as the straight type except that the branch coming into the main is a smaller size pipe than the main.

The *double* 90-degree wye branch is easy to recognize because there is a 45-degree takeoff bending into a 90-degree takeoff on both sides of the fitting. It is especially useful as an individual vent.

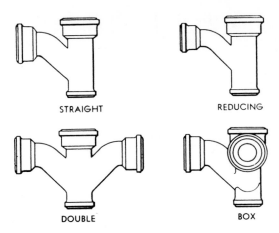

STRAIGHT   REDUCING

DOUBLE   BOX

example 4–11. Some cast-iron soil-pipe 90° wye branches.

The *box* type of 90-degree wye branch has two takeoffs. It is designed so that each takeoff forms a 90-degree angle with the main pipe. The two takeoffs are spaced 90 degrees from each other.

45-degree Wye Branches. Two types of 45-degree branches are the reducing and the straight types, both shown in example 4–12.

The *reducing* type is a straight section of pipe with a 45-degree takeoff of smaller size branching off one side. You will use different sizes of this fitting. As an example, a 4-by-4-by-3 reducing 45-degree branch would have a 4-inch straight portion with a 3-inch 45-degree takeoff on one side.

The *straight* type of 45-degree wye branch is the same as the reducing type except that both bells are the same size. It is used to join two sanitary sewer branches at a 45-degree angle.

Cleanouts. Cleanout plugs are installed to allow the removal of stoppages from waste lines. Example 4–13A shows one type of cleanout plug. It consists of an iron ferrule that is calked into the hub of a pipe or fitting. Its top opening is tapped and threaded to accommodate a pipe plug.

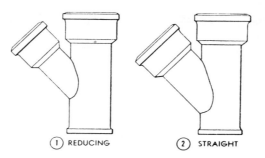

example 4–12. Two types of cast-iron soil-pipe 45° wye branches.

① REDUCING ② STRAIGHT

example 4–13. Cast-iron soil-pipe adapter fittings.

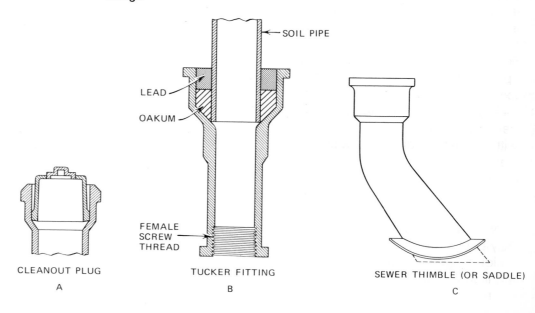

SOIL PIPE

LEAD

OAKUM

FEMALE SCREW THREAD

CLEANOUT PLUG
A

TUCKER FITTING
B

SEWER THIMBLE (OR SADDLE)
C

Adapters. A *Tucker* cast-iron soil-pipe drainage adapter (example 4–13B) is a specialized fitting used to connect a bell-and-spigot pipe section to a threaded pipe section. It is recommended especially for pipe joints where unions are not desirable. The adapter has a bell at the top and a female screw thread on the bottom.

To insert the adapter in a line, slide the bell up on the spigot of the bell-and-spigot pipe until it clears the male thread on the threaded pipe. The adapter should be screwed onto the threaded pipe with a wrench; the bell is then calked with oakum and lead.

Another type of adapter is a sewer thimble or *saddle*. This is a specialized fitting used to tie into an existing tile sewer line (example 4–13C). It has a hub on one end, bending around to almost 45 degrees, with a flange near the opposite end. You may also cut one hub off of a double-hub pipe and use that when no sewer thimble is available.

To install, cut a hole halfway between the top and centerline in the clay tile about the same size as the outlet portion of the thimble beyond the flange. Slip the thimble into the opening until the flange seats on the sewer pipe. Using oakum and concrete, grout around the thimble to make a watertight joint.

Joining Cast-Iron Soil Pipe

Various methods are used in joining pipe. You should know the procedure to follow in making various types of joints required for the kind of pipe to be joined. Joints in cast-iron hub-and-spigot pipe have traditionally been made by calking with oakum and lead. The Appendix includes complete instructions for doing this.

Construction Fundamentals

The drainage system in every house that contains conventional plumbing should include these basic parts: fixture drains, fixture branches, soil stack, and building drain. The *fixture drain* is the piping through which a fixture drains; from Section Three you'll recall that each fixture must be properly trapped and vented. A *fixture branch* is a drain pipe that connects several fixture drains. The *soil stack,* which is discussed in greater detail later in this section, is the vertical soil pipe into which the water closet and other fixtures drain. It connects to the building drain and is vented up through the roof to the outside air. The vent portion is referred to as the *vent stack.* The *building* drain is the main horizontal (or at least *apparently* horizontal) drain that receives the discharge from soil, waste, or other drainage pipes inside the building and carries it outside the building to the building sewer, which carries it to the disposal facility.

The wastes from all fixtures flow through the drainage system by gravity to the sewer line, which is the lowest physical portion of the household plumbing system. When the sewer line is above the level of the lowest drain line, a covered sump is used as a waste gathering place, and a sewage pump or ejector empties the sump as necessary by forcing the waste to flow against gravity to the house sewer line. The drainage piping must be of the proper size and slope to insure good flow.

The standard slope of a horizontal drain-pipe run using pipe that is 3 inches or less in diameter is $\frac{1}{4}$ inch or more per foot of pipe length. (Larger pipes needn't have quite that much slope, but you're not apt to find drain lines with inside diameters greater than 3 inches.)

Vents

Every house drainage system must include at least one main vent through which gases can escape to the outside air. As noted in the discussion on *drain and vent principles,* a well-vented drainage line prevents siphonage of traps and it promotes the smooth flow of wastes through the line.

Example 4–14 shows the drainage line for a typical bathroom and the manner in which the vent is implemented. The soil stack must always be vented by means of the stack vent, which extends up through the roof of the house with the same size pipe as the soil stack itself. Plumbing codes in different localities vary, so it would be wise to check for any special requirements before adding a new fixture branch to a finished house; it may be that you'll have to install an additional vent rather than a vent line leading to an existing stack vent. (If no local codes apply, of course, you may use the guidelines in this section to determine the vent requirements applicable to your specific situation.)

Vent Types

There are a variety of vent types used in the venting of fixture and floor drains. The selection of a particular type depends largely on the manner in which the plumbing fixtures are to be located or grouped. Some of the more common vent types are back vents, unit vents, and circuit vents.

The *back vent* (also called *individual vent*) connects the main vent with the individual trap under or behind a fixture. This method of venting is illustrated in example 4–15. When installing two or more fixtures on an individual vent basis, it is important that the leg (illustrated) connecting the vent to the main vent be of sufficient size to carry the combined load.

When you want to vent two traps to a single vent pipe, you should use a *unit vent* (example 4–16). This vent should be used when a pair of lavatories are hung back to back on both sides of a partition or wall. The waste from both fixtures discharges into a *double sanitary tee.*

The *circuit vent* (also called *loop vent*) serves a group of fixtures. As indicated in example 4–17, a circuit

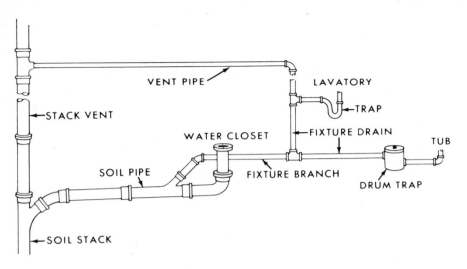

VENT PIPE

STACK VENT

LAVATORY

TRAP

FIXTURE DRAIN

WATER CLOSET

TUB

SOIL PIPE

FIXTURE BRANCH

DRUM TRAP

SOIL STACK

example 4–14. Method of venting a group of bathroom fixtures. The vertical fixture drain for the lavatory is also called a *wet vent* because it serves the dual purpose of draining and venting.

LEG

IND. VENT

WALL

IND. VENT

CLOSET

MAIN VENT

FLOOR

STACK

example 4–15. The individual or back vent connects fixtures to a main-vent leg separately.

example 4–16. A unit vent is ideal for venting back-to-back fixtures when two bathrooms are served by a single waste line.

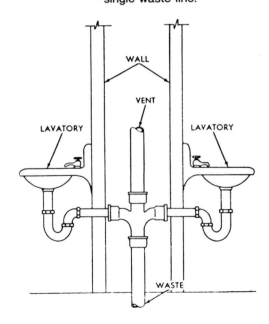

WALL

VENT

LAVATORY

LAVATORY

WASTE

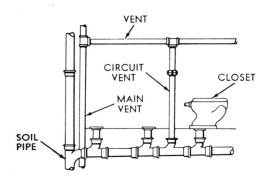

example 4-17. A circuit vent is like a back vent except that it serves a group of fixtures on a branch rather than a single fixture. It links the main vent with the group and is connected to the branch at a point between the final two fixtures on the branch.

vent extends from the main vent to a position on the horizontal branch between the final two fixture connections. In this type of vent, water and waste discharged by the last fixture will tend to scour and flush the vents of the other fixtures on the same outlet line.

When liquid flows through a portion of a vent, that portion is referred to as a *wet vent*. This type may be used on a small group of bathroom fixtures such as a lavatory, water closet, and shower. Its inside diameter should never be less than 2 inches.

As indicated in examples 4-14 and 4-18, a lavatory should be individually vented. This is necessary to prevent loss of the trap seal through indirect siphonage. The relatively clean waste water discharged by a lavatory tends to scour wet vents, thus preventing buildup of waste materials in the vent line.

Installation

A venting system must be properly installed if it is to serve its intended purpose and require a minimum amount of repair and upkeep.

Referring again to example 4-18, notice that a main-vent tee is used to form a junction between the vent and the main soil-and-waste line. This is a tapped tee, having an outlet on the side of the stack. It should be installed by calking in the vertical stack at least 6 inches above the overflow level of the highest fixture connected. After this has been done, the vertical stack should be extended, full size or larger, through the roof to form the vent terminal.

The vertical portion of the soil stack above the drainage line should extend at least 6 inches above the roof, and even more than 6 inches if there is any chance that snow could accumulate and block the vent. But you can't just let a pipe stick up through the roof of your house; otherwise, you'd have problems of rain water seeping down the sides of the pipe and into the innards of your home. What you have to do is *flash* the protruding portion of the soil vent to make it fully watertight.

The manner in which flashing is accomplished for both flat-roof and pitched-roof houses is shown in example 4-19. Flashing is constructed usually of galvanized iron, steel, or

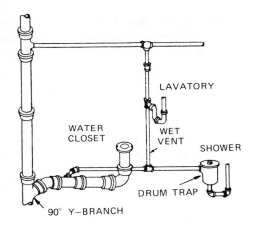

example 4-18. A lavatory drain also used as a vent for a companion fixture such as a shower or tub is called a wet vent.

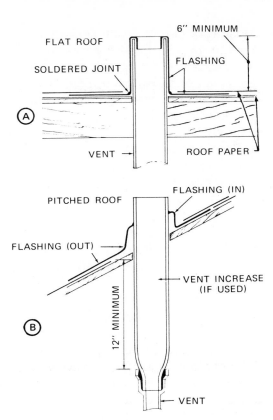

example 4-19. Flashing seals protruding vents against water entry around vent stacks. A flat roof is flashed with roofing material overleaving the metal (A); a vent in a pitched roof requires a different flashing, with the upper portion under the roofing material (B) and the lower portion over it.

copper; various types are available at hardware stores, building-supply dealers, and even electrical distributors (because electrical conduits also require flashing when wire entry is through the roof).

When flashing is installed on a house that has a slanted or pitched roof, the flashing must itself be considered the same as a roofing shingle. In example 4–19B, note that that portion of the flashing that is on the lower side of the vent is placed above the roof paper or shingle, and that portion on the higher side of the vent is placed under the shingle or paper. As you can see, this allows water to drain off the house efficiently without trapping it between flashing and roofing material.

When installing the roof flashing on a shingled roof, extend it under two courses of shingles above the pipe. On a flat roof, place it between layers of the roofing material and have the finishing layer over the top of the flashing. To complete the installation on either type roof, always apply a coat of roofing cement as added protection against leakage.

In freezing climates, suitable provision must be made to prevent closure of the main soil-and-waste vent at the roof outlet by freezing. The air discharged by this vent is humid, causing condensation in cold climates. Unless proper precautions are taken, this condensation will freeze if exposed to extremely low temperatures.

One method by which you can prevent this kind of freezing is to increase the pipe to a size or two larger than the vertical vent passing through the roof. As the rising air condenses and freezes on the surface of the conduit, the enlarged pipe diameter allows the ice to build up without choking off the air.

Another method involves the installation of high-lead flashing, which will provide an insulating pocket of air between the flashing and the end of the main soil-and-waste vent above the roof. Being open to the heat of the building, the air pocket makes possible an intermediate warming area for gases leaving the main soil-and-waste vent.

As noted, each fixture in your house has to be vented. The only exception is a drain located right at the building drain. Vent piping for each fixture should be installed between the trap for that fixture and the sewer line, and it should be the same size as the drain pipeline. If the vent piping is connected to the soil stack, it must be connected above the highest fixture drain. The only option is to vent each fixture directly to the outside (above the roof) with a separate line.

The distance (trap arm) from the trap of each fixture to the vent is governed by the size of the fixture drain. The following list relates the inside diameter of a drain in inches to the maximum permissible distance from the trap to the vent in feet:

$1\frac{1}{4}$ inches—$2\frac{1}{2}$ feet

$1\frac{1}{2}$ inches—$3\frac{1}{2}$ feet

2 inches—5 feet

3 inches—6 feet

4 inches—10 feet

Materials used in vent piping ordinarily include galvanized pipe, cast-iron soil pipe, brass, copper, and plastic piping. Asbestos-cement pipe can also be used for venting soil-and-waste pipe. A single length of this pipe is often sufficient for venting a stack. For such an installation, pipe is available with a machined end, which is placed in the hub of the soil or waste pipe, the connection being made by proper calking.

The use of proper piping is important. A good point to remember is that the diameter of the vent stack or main vent must be not less than 3 inches. The actual diameter depends on the developed length of the vent stack and on the number of fixture units installed on the soil or waste stack. The diameter of a stack vent should be at least as large as that of the soil or waste stack.

Branches

Soil-and-waste pipe branches are horizontal branch takeoffs used to connect various fixtures and the vertical stack. One method of installing a branch takeoff from the vertical stack is by using a wye branch with a $\frac{1}{8}$ bend calked into it. Another method is by using a sanitary tee. Of these two methods, preference usually is given to the sanitary tee, which eliminates one fitting and an extra calked joint.

Under some local codes, you may be permitted to connect more fixture units to a given size stack when a combination wye and $\frac{1}{8}$ bend are used.

In such a case, the combination wye and $\frac{1}{8}$ bend may be more desirable than the sanitary tee. Once either fitting is calked into place, however, the horizontal branch can be extended as necessary with lengths of soil pipe. They, too, are joined by calking.

Waste pipes, as noted earlier, should be graded downward to insure complete draining. Horizontal vents should be pitched slightly to facilitate drainage of condensate. Normally this pitch is $\frac{1}{4}$ inch per foot.

Traps and Cleanouts

Traps are covered extensively in Section Three, so the information included here should serve as no more than a capsule summary, which may help you relate all portions of a drain system to one another. Traps—one for each fixture drain—prevent gases from backing up through open drains or overflows and escaping into the house. Each trap should be the same size as the drain pipe with which it is used, and it should be positioned as close as possible to the fixture outlet. Remember that a water closet has its own trap built in, so no other is required for this fixture. *Never double-trap a fixture.*

Even floor drains have to be trapped (floor drains are used in showers, laundry rooms, basements, and utility rooms). If the building drain is laid under a floor as in example 4–20, it must be far enough below the floor to accommodate a trap for a floor drain. You shouldn't have to vent this drain, though, because it will probably be very close to the building drain, which is adequately vented already.

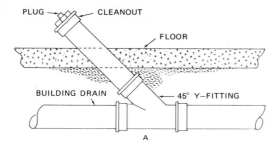

PLUG — CLEANOUT
FLOOR
BUILDING DRAIN
45° Y-FITTING
A

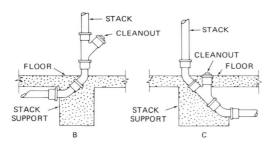

STACK
CLEANOUT
FLOOR
STACK SUPPORT
B

STACK
CLEANOUT
FLOOR
STACK SUPPORT
C

example 4–20. Soil-and-waste pipe cleanouts and support. Though access is good, the arrangement in A is not as satisfactory as those in B and C because it presents a hazard to occupants: it's easy to trip over a cleanout like this in the dark.

P-traps are typically used for lavatories and sinks. S-traps are commonly used for laundry drains. Drum traps (see example 4–14) are used commonly for tub drains; drum traps are 3 or 4 inches in diameter and have a removable top or bottom to allow cleaning of both the trap and the line in which it is used.

Cleanouts most often consist of 45-degree wye fittings with removable plugs. Cleanouts are used, as the name implies, to permit you to remove obstructions and debris from drainage pipes. Cleanouts, like traps, should be the same diameter as the pipes with which they are associated.

As shown in example 4–20A, cleanouts should be installed where they are readily accessible and where cleanout tools can be easily inserted in the drain pipe. Place one cleanout at or near the floor of the soil stack (example 4–20B and C) and install others at intervals of not more than 50 feet along horizontal drain lines less than 4 inches diameter.

Drain-Line Problems and Cures

By far the greatest majority of disposal-line problems can be traced to a blockage in a trap, drain, soil line, or vent. But occasionally other problems develop: a pipe joint will loosen, a pipe section will corrode and leak, or perhaps you'll break a fitting during an attempt to remove a plug from a cleanout. This subsection should give you all the information you need to solve these and other drain line problems by yourself.

Learning the Layout

Before you can apply a logical troubleshooting approach to plumbing problems that develop in an existing installation, you have to know where the drain lines are and how to get at them. Without this information, you become a traveler in a strange land without a roadmap. In the well planned house, you won't find long circuitous paths between fixtures; rather, you'll see a built-in economy of pipe routing that may not be obvious immediately because of the walls, partitions, and possibly stories separating the rooms containing fixtures.

If your house has a basement, determining its plumbing layout is somewhat simplified. Just walk down the stairs and spend a few minutes examining the exposed overhead pipes. Pay particular attention to those pipes immediately beneath bathrooms and the sink area of the kitchen. Probably the first thing you'll notice is the larger-diameter piping for the main soil-pipe run with branches that you should be able to identify easily, such as water-closet drain, tub-and-shower drain, sink drain, and so forth.

Now inspect the drain line for cleanouts—plugs at the ends of pipes or in wye branches. These are your drain access points. When a drain clog occurs that you can't remedy at the fixture drain itself, you'll have to remove the appropriate cleanout plug and clear the obstruction from that point.

If your house has no basement, you'll still be able to "psyche out" the system. One very effective technique that is enlightening as well as practical is to map out your own household plumbing system. This isn't as tough as it might sound. All you'll need is a carpenter's collapsible tape measure and piece of quadrille (graph) paper about the size of a standard sheet of typewriter paper. (Quadrille paper has light-blue or green vertical and horizontal lines evenly spaced at $\frac{1}{8}$- or $\frac{1}{4}$-inch intervals over its entire surface, so that the sheet contains equal-size squares over its complete face.)

If you let each square on the paper represent one square foot, you can quite easily make a very accurate "map" of your house. Simply measure the distance from one corner of the house (outside) to the next, and transfer that measurement to the paper with a line; continue doing this until you have a plan view of your home's shape. From this point, it's easy to draw in the partitions separating the rooms. And when you've done that, all you have to do is sketch in the plumbing fixtures using the symbols shown in the final portions of the Appendix.

It might not have been obvious to you at first, but when you have mapped out the fixtures in your house, you'll likely see the rationale used in the original layout. If your house has several bathrooms, they'll most likely be back to back so that one drain line in the separating wall can serve both facilities. Even if there is only one bathroom, you are not likely to find it in a corner diagonally opposite the kitchen. Instead, you'll see a logical grouping so that one short pipe run can serve the entire household. Example 4–21 illustrates a floor plan of a typical one-bedroom home; in this plan, notice that the plumbing fixtures are strung along one straight line.

Once you have drawn a simple plan of your house and sketched in the plumbing fixtures as in example 4–21, you'll be able to do most of your drain-line troubleshooting right

from your easy chair. A clogged drain? Determine if there's more than one drain involved. If not, the trouble lies in or near the single drain. If more than one drain is involved, how does the second clogged drain relate to the first? Chances are it's on the same branch. Are all the drains clogged? Then look for trouble in the main soil-and-waste pipe downstream from the lowest fixture. As you can see, it's relatively easy to troubleshoot problems if you have a good idea of the routing of your drain lines.

Drain-Line Troubleshooting

You can get a fair idea of the pipe routing in your bathroom by studying the phantom view depicted in example 4–22, which plainly shows the soil stack and vent, the vent pipe, and all traps, drains, and fittings. As you troubleshoot problems in your own bathroom, study this sketch until you have a clear image of your own plumbing scheme. The kitchen plumbing arrangement will be similar.

Clogged Drains

Water closets are the most commonly clogged plumbing fixtures because they traditionally receive the most solids. Water closets are covered extensively in Section Three, so refer to that section if you have a blockage problem that can be isolated in this fixture.

Obstructions in Traps. As mentioned earlier, every drain in your house is *trapped*. Most often the P-trap is used, though you'll occasionally find S-traps, drum traps, and even pipe fittings arranged in such a way as *to form a*

example 4–21. Typical plumbing layout for a small house.

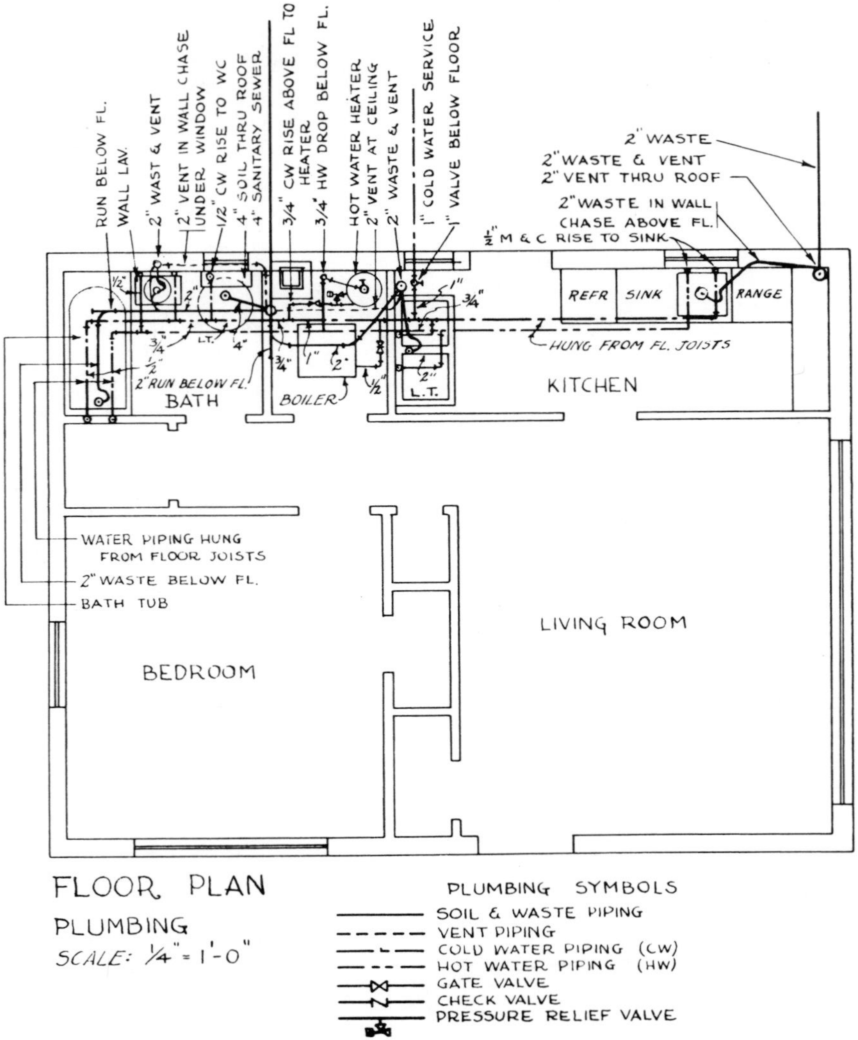

RUN BELOW FL.
WALL LAV.
2" WAST & VENT
2" VENT IN WALL CHASE UNDER WINDOW
1/2" CW RISE TO WC
4" SOIL THRU ROOF
4" SANITARY SEWER
3/4" CW RISE ABOVE FL TO HEATER
3/4" HW DROP BELOW FL.
HOT WATER HEATER
2" VENT AT CEILING
2" WASTE & VENT
1" COLD WATER SERVICE
1" VALVE BELOW FLOOR
2" WASTE
2" WASTE & VENT
2" VENT THRU ROOF
2" WASTE IN WALL CHASE ABOVE FL.
½" M & C RISE TO SINK
REFR
SINK
RANGE
HUNG FROM FL. JOISTS
KITCHEN
BATH
BOILER
L.T.
2" RUN BELOW FL.
WATER PIPING HUNG FROM FLOOR JOISTS
2" WASTE BELOW FL.
BATH TUB
LIVING ROOM
BEDROOM
FLOOR PLAN
PLUMBING
SCALE: ¼" = 1'-0"
PLUMBING SYMBOLS
SOIL & WASTE PIPING
VENT PIPING
COLD WATER PIPING (CW)
HOT WATER PIPING (HW)
GATE VALVE
CHECK VALVE
PRESSURE RELIEF VALVE

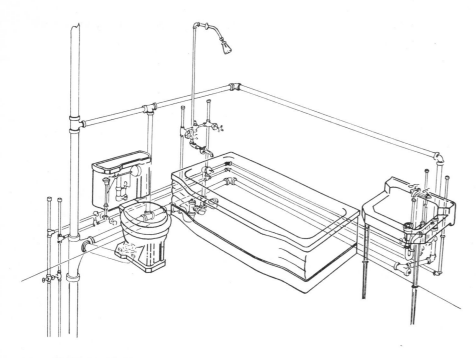

example 4-22. If your bathroom were constructed of transparent fixtures and walls, it would look something like this one, which shows all the piping, traps, cleanouts, and valves and the manner in which they relate to the fixtures they serve.

trap. When you can't free a blockage in a fixture drain with a force cup (see discussion under "Water Closets" in Section Three) or chemicals, the logical next step, so long as no more than a single drain is involved, is to disconnect the trap or remove it altogether. Traps that have built-in cleanouts, such as that pictured in example 4-23, probably won't have to be removed or disconnected at all. All you have to do is loosen the cleanout cap with a suitable wrench, then pull the obstruction out through the cleanout port. Usually, though, you won't find a cleanout on a trap.

To avoid loosening other pressure connections in the drain line, you must be careful not to upset the concentricity of the existing pipework. The correct procedure for disconnecting a P-trap is to loosen both the slip joint and the crown packing nuts (example 4-23), then slide the slip-joint nut out of the way of the connection and swivel the trap so that the connection is severed but the axis of the inlet relative to the outlet is maintained. When you do this, protect the finish of the packing nut with friction tape or wrap a cloth around the jaws of the wrench.

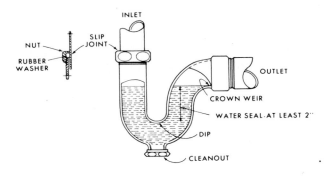

INLET

SLIP JOINT

NUT

RUBBER WASHER

OUTLET

CROWN WEIR

WATER SEAL-AT LEAST 2"

DIP

CLEANOUT

example 4-23. When a P-trap is equipped with a cleanout, the unclogging operation involves nothing more complicated than removing the cap and draining the trap. But without the plug, you'll have to loosen the packing nuts at the slip joint and crown. The slip joint is a pressure fitting typically; a gasket compresses when the slip-joint packing nut (shown in the inset sketch) is tightened accurately.

Remember when you remove a P-trap that it contains water—it's required to maintain a seal against gases in the sewer line. So use a container to catch the overflow when you drop the trap. Nine times out of ten, your problem will be solved by removing and replacing the trap associated with a drain. But once in a while, you may have to use a probe. Don't use a heavy steel-spring coil snake to clear traps under lavatories, sinks, or tubs. Rather, use a flexible wire or spring snake that will follow the bends in the drain line.

When you've cleared a drain that's been clogged, it's a good idea to scour it. This doesn't mean that you have to swab down the inside of a pipe with a rag or auger. You can scour a drain by pouring something abrasive in it. Contrary to popular belief, coffee grounds are excellent for this. As a matter of fact, you can keep drains in pretty good working order all the time (except for obstructions, of course) by pouring coffee grounds down the drain every day. Just remember to flush with water for 15 or 20 seconds after doing so, because this prevents the grounds from becoming lodged in solids that have been accumulating along the inside surface of the drain pipes.

Obstructions beyond Traps. If the drain-clogging problem is recurring, it's very likely that there is a buildup of semisolid material that can't be flushed away with a single scouring with coffee grounds. But grease and soap clinging to a pipe can sometimes be removed by flushing with cold water and a chemical intended for this purpose. Lye or lye mixed with a small amount of aluminum shavings may also be used. When cold water is added to the mixture, the gas-forming reaction and simultaneous production of heat loosens the grease and soap so that they can be flushed away. Use *cold water only* unless you're using a chemical that specifically requires hot water.

Chemical cleaners should never be used in pipes that are completely blocked because chemicals require a flushing action at the blockage point to be effective. And once you have cleared a sluggish drain pipe with a chemical, let water flush the drain for several minutes.

Waste-Line Obstructions

Sometimes an obstruction will be particularly stubborn; you can't clear a drain pipe regardless of how much you try. When this happens, use the logical troubleshooting approach: determine whether the problem is isolated in a single branch or located in the main soil-disposal line. (If the main line is at fault, you'll have trouble with all drains rather than just those on one branch.)

The first step—after checking the appropriate trap—is to locate the cleanouts in the problem line. In example 4–22, for example, you'll see a cleanout plug in the drain line immediately below the lavatory. An auger inserted at this point would allow access to the drain all the way to the tub drain.

If you have mapped your plumbing fixtures using the quadrille-paper method described earlier, you should have a fair idea where the cleanouts are located throughout your disposal line. If the cleanout locations aren't obvious at first, try looking in inappropriate places. Kitchen cleanouts, for example, are often outside the house—in the vicinity of the sink, but protruding from the house just above ground level.

Blocked Pipes. Even when the auger fails to do the job, it *will* let you know where the obstruction is. And when you know this, you can at least remove that section of the drain line causing the trouble and replace it. In galvanized pipe installations where the fittings on both sides of the problem section are not immediately accessible, the bad section can be cut out. After cutting, the procedure requires two people, however—one person to hold the remaining drain pipe with a wrench and the other to cut new threads on it while it's in place. The cut-out section should be replaced with a coupling, a pipe section of the proper length, and a union.

Roots in Lines. One of the most common problems with clogged lines underground is root encroachment. With trees growing close to sewer lines, you can expect root trouble eventually. Roots can cause all manners of disposal line problems—breaks in the pipe as well as clogging. Such trees as elms, poplar, and willows are among the worst offenders; when these trees are growing within 100 feet of a buried line, you can look for problems sooner or later. Example 4–24 shows how roots actually penetrate a line.

The problem of root growth is compounded when the pipe is cracked or has defective joints. You can clear the stoppage temporarily by using a root-cutting tool (available at equipment-rental stores) or by application of blue vitriol. It's generally advisable to try the vitriol first, since it's the easiest and most economical approach. But use extreme caution when working with this chemical because it is poisonous.

Blue vitriol is copper sulfate. It comes in small crystals. The procedure for root killing is to throw the copper sulfate crystals into the line at the nearest upstream access point *while the sewer line is flowing*. Since it is necessary to kill only the small root that is entering the line, and not the whole tree, the important consideration

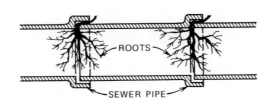

example 4–24. Roots exhibit inexorable persistence and they're in no hurry. If your pipe sections have a weak spot or if they have deteriorated to any extent, roots find an entry; and once they gain a foothold, they are going to cause you ceaseless trouble because they keep on growing.

is reaching the root with the chemical; it isn't essential to use a large quantity.

There is no established rule as to the exact quantity of copper sulfate crystals required for effective results, but experience indicates that with a flow of 5 gallons per minute through a large sewer pipe, a handful of crystals will stop the obstruction. If the flow is greater, larger quantities and repeated application might be necessary.

You must also remember that chemicals can't work in a line that is completely blocked; there has to be at least some movement of waste through the line to carry the chemical to the root.

If copper sulfate crystals won't cure the problem, you'll have to decide whether to replace the pipe section or use a power-driven scraping-and-cutting tool.

Clogged Vents

There was a time when a clogged vent was a real rarity—not so any more. As pointed out earlier, an open vent on the roof seems too much of a temptation for do-it-yourself antenna installers. If you've been plagued with sluggish drains and can't find the obstruction, start looking up instead of down. Never ever allow a roof vent to be used as an anchor point for a TV or radio mast.

Normally a clogged vent isn't particularly difficult to clear. You can use the same technique as employed in clearing a drain with an auger. When the lodged object is too long to remove after being forced down to the cleanout at the bottom of the soil stack, the problem isn't quite so simple.

One family had drain problems for a long time, with frequent toilet overflows and sewage backup into fixtures. Several plumbers made concerted efforts to find the trouble, not suspecting that an obstructed vent was the culprit. Very likely the problem could have been cured the first time around had the family given the plumber the complete history of the house's drains. But each call was on a case basis: a line was clogged, and when the plumber managed to get waste flow, he presented his bill and left. Family members didn't offer information as to repeated sewer backup, which itself is indicative of siphonage attributable to a blocked

vent. It wasn't until one of the young men in the family became a plumber's apprentice that the recurring problem was finally identified—and solved.

What had happened was that someone had, a long time previously, attempted to make a temporary antenna installation for a new TV set. He inserted a mast into the vent opening and wedged it into place with a broom handle and some other long sticks. Eventually, whether from the wind or human carelessness, those wedges dropped into the vent. They were finally removed with a long pole that had been fitted with a protruding 6-penny nail at the end (example 4–25). The nail was used to spear the objects, which were then withdrawn.

Most clogged vents can be cleared with a good deal less difficulty; the house in question had no cleanout that would allow access to the main soil-and-waste vent.

Leaky Lines

Occasionally a pipe may start leaking when materials needed to repair it permanently are not readily available. In such cases it might be necessary to make a temporary or emergency

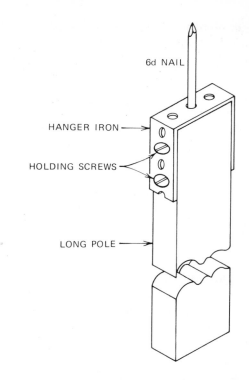

example 4–25. You can make a rudimentary spear like this to retrieve wood objects from roof vents when you can't reach them from a cleanout. To make the spear, insert a large-head 6-penny nail up through an 8-inch length of $\frac{1}{2}$-inch plumber's tape, then bend the tape down over the end of a long stick and secure it with a couple of wood screws on each side.

repair. But bear in mind that a *permanent repair* should always be made as soon as the tools or materials become available.

One method of repairing a leaky pipe involves the use of sheet rubber and sheet-metal clamps (example 4–26). Place the two sheet-metal clamps over the sheet rubber as shown, then secure the clamps with nuts and bolts.

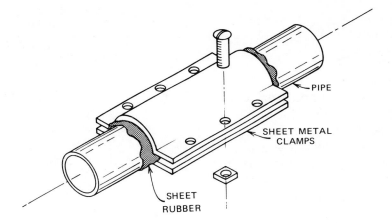

PIPE

SHEET METAL CLAMPS

SHEET RUBBER

example 4–26. If you have a couple of sheet-metal clamps like this, with a radius of curvature that approximately matches that of the pipe, you can effect a temporary repair by wrapping sheet rubber around the break and tightening the clamps as shown.

Waste-Line Add-Ons

When you have to add a branch to an existing disposal line and there's no readily available tie-in point, you'll have to install a fitting. There are several ways to do this. If you want to install a fitting in a short space or if the existing pipework can't be moved easily, you can cut off the spigot end of a pipe section whose inside diameter is slightly larger than the outside diameter of the existing drain line. Example 4–27 shows the procedure for hub-and-spigot pipe.

First, cut a section of the existing pipe completely out. The cut-out portion must be of such length that the cut end of one piece of the existing pipe can slip into the hub of the add-on fitting while the fitting butts against the end of the other pipe section. Before installing the fitting, slide the coupling (actually the spigot end discussed above) over the piece of the existing pipe that will butt against the spigot end of the fitting; then put the fitting in place and slide the coupling up over the butt joint. Calk the sleeve at both ends and calk

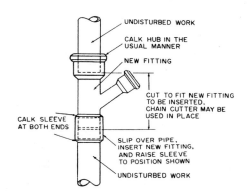

UNDISTURBED WORK

CALK HUB IN THE USUAL MANNER

NEW FITTING

CUT TO FIT NEW FITTING TO BE INSERTED. CHAIN CUTTER MAY BE USED IN PLACE

CALK SLEEVE AT BOTH ENDS

SLIP OVER PIPE, INSERT NEW FITTING, AND RAISE SLEEVE TO POSITION SHOWN

UNDISTURBED WORK

example 4–27. Installing a hub-and-spigot fitting in a restricted space.

the hub connection in the usual manner.

To replace a fitting or insert one in an existing line that can accommodate it, follow this procedure (refer to example 4–28):

1. Melt the lead from joints 1 and 2.
2. Cut pipe A to correct length.
3. Assemble pipe A and fitting B, and insert assembly into the line.
4. Straighten the line carefully, then calk joints 1, 2, and 3.

To add a connection to an outside vitrified-clay sewer line, follow this procedure (refer to example 4–29):

1. Remove a section of existing pipe that is long enough to accept a new wye fitting.
2. Break half of the hub rim of the new wye fitting (A).
3. Insert the spigot end of the wye fitting into the hub of the existing pipe. At the same time, place the remaining half of the hub end of the wye fitting over the cut end of the existing pipe—the wye branch pointing away from the new inlet (first position, B).

example 4–28. Method for insertion of new fitting into an existing line.

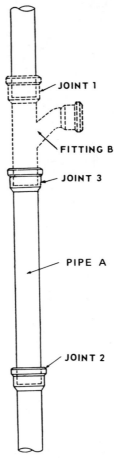

JOINT 1

FITTING B

JOINT 3

PIPE A

JOINT 2

STEP 1. MELT LEAD FROM JOINTS 1 AND 2.
STEP 2. CUT PIPE **A** TO CORRECT LENGTH.
STEP 3. ASSEMBLE PIPE **A** AND FITTING **B**
AND INSERT INTO LINE.
STEP 4. STRAIGHTEN LINE AND CALK JOINTS
1, 2, AND 3.

4. Rotate the wye fitting so that the broken half of the hub is up and the wye branch is in the correct position to receive the new inlet connection (final position, B).
5. Pour the joint carefully. Round over the broken hub half with concrete or bitumastic compound (C).

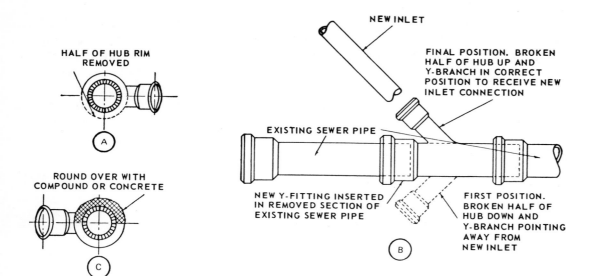

HALF OF HUB RIM REMOVED

(A)

ROUND OVER WITH COMPOUND OR CONCRETE

(C)

NEW INLET

FINAL POSITION. BROKEN HALF OF HUB UP AND Y-BRANCH IN CORRECT POSITION TO RECEIVE NEW INLET CONNECTION

EXISTING SEWER PIPE

NEW Y-FITTING INSERTED IN REMOVED SECTION OF EXISTING SEWER PIPE

FIRST POSITION. BROKEN HALF OF HUB DOWN AND Y-BRANCH POINTING AWAY FROM NEW INLET

(B)

example 4–29. Adding connections to vitrified-clay sewer line.

Sump Pumps

When the sewer-line outlet is higher in elevation than the basement floor level, a sump may be installed in the floor at the lowest point to accept liquid drain wastes. Sumps normally are not used for water closets because of the sanitation problems posed by their construction; rather, sumps are typically open or grated and are used only to accumulate water runoff during rains and the input from a floor drain. The information on pumps presented in the section dealing with fresh-water supply is equally applicable to sumps. In this subsection we concentrate on the system that controls the pump's operation.

Example 4–30 shows the classic mechanical sump pump arrangement, which consists of an electric motor on a shaft that terminates in a submersible impeller system. It works like this: water enters the sump area, either from foundation seepage or gravity drainage. As the water level in the sump rises, it lifts the float ball (see example). An adjustable stop on

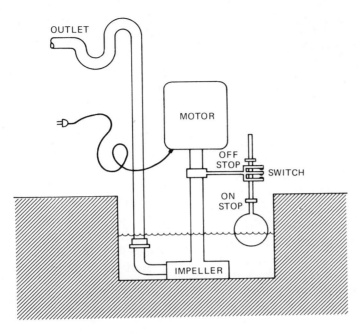

OUTLET

MOTOR

OFF
STOP

SWITCH

ON
STOP

IMPELLER

example 4–30. Standard sump-pump
arrangement.

the float-ball spindle makes contact with a microswitch when the water level rises to a predetermined level, causing the pump motor to turn on. When the water level drops to almost empty, a second stop shuts off the pump.

Sump pumps are always designed for automatic operation; if they are correctly installed and not abused, they'll require very little attention. Dirt, lint, and other waste particles should, of course, be kept out of the pit. The most common troubles with sump pumps are the following:

- Motor and shaft alignment become disturbed, causing outlet line to break or leak. When this happens, the pump will operate continuously as the liquid pours from the drain line back into the sump.
- The float ball rubs against side of the pit, preventing the pump from turning on.
- Spindle shaft of float ball gets bent, preventing normal on/off operation of pump.
- Stops on spindle shaft become loose, causing the pump to start too late or stay on too long.

As you can see, most of these difficulties can be overcome with a good secure installation.

If you're building a new sump for a pump, be sure to line the pit with a length of large drain tile or with concrete and metal to prevent cave-ins. Inlets or holes should be provided in the lining material to admit ground water.

Gutters and Downspouts

Houses should have gutters and downspouts to take care of roof water from rain and snow. These should be considered very much the same as other pipelines in the drainage system, and kept free of debris and toys. And, of course, a rain gutter should never be used as a grapple bar to gain access to the roof.

Where leaves and twigs from nearby trees tend to collect in a gutter, install a basket-shaped wire strainer over the downspout outlet. And make it a point to repair gutters and downspouts as soon as they appear to require it. It's also a good idea to keep them painted, since water tends to corrode bare metals of all types.

Downspouts usually have an elbow or shoe on the lower end to discharge the water slightly above the ground and away from the wall. If your house has a basement, this elbow is absolutely essential to prevent water buildup inside. To prevent water buildup at the discharge point, use a concrete splash block or trough to carry the water away. This trough should slope away from the house at the rate of about an inch per foot of distance, and its edges should be flush with the grade (see example 4–31).

Disposal of roof water as shown in example 4–31 makes it easy to clear clogged downspouts. Roof water can also be piped underground to a storm drain, dry well, or surface outlet that is 15 feet or more from the house; the arrangement for accomplishing this is shown in example 4–32. Of course, the bottom of a dry well should be lower than the basement floor and terminated in earth or rock that drains rapidly.

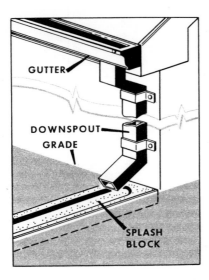

GUTTER

DOWNSPOUT

GRADE

SPLASH BLOCK

example 4–31. Properly installed gutters and downspouts prevent roof water from forming in puddles around the house and serve to keep water from entering a basement through wall seepage.

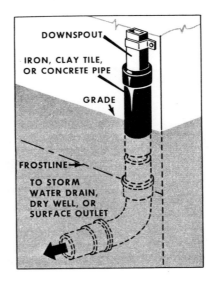

DOWNSPOUT

IRON, CLAY TILE, OR CONCRETE PIPE

GRADE

FROSTLINE→

TO STORM WATER DRAIN, DRY WELL, OR SURFACE OUTLET

example 4–32. Roof water can be routed to a storm drain or other outlet by terminating a downspout as a conventional pipeline. The pipe should lie below the frostline for a given area.

Section 5

Private Water Systems

Section 5

The water for your household may come from any one of a variety of sources: a community reservoir, a local well, an underground or overground water storage tank, a stream, or perhaps even an artificial pond. In most metropolitan areas of the United States, Americans don't need to concern themselves with water sources; water is delivered under pressure to their homes, where it passes through a metering device that indicates the quantity of water used and the rate of consumption. If your house is served by such a system, you can skip this section; but if your house is fed by an individual water supply, or if you're contemplating construction of a private water-supply system, this section should give you the kind of information you need to plan the system, put it into operation, and keep it working in good order, delivering fresh water in a volume that is right for your needs.

The water sources in example 5-1 all receive their supply from rain or snow. As water falls, it may collect to form lakes, rivers, or ponds (known as surface waters). The water that seeps below the surface of the earth may be absorbed by the plant roots or

172

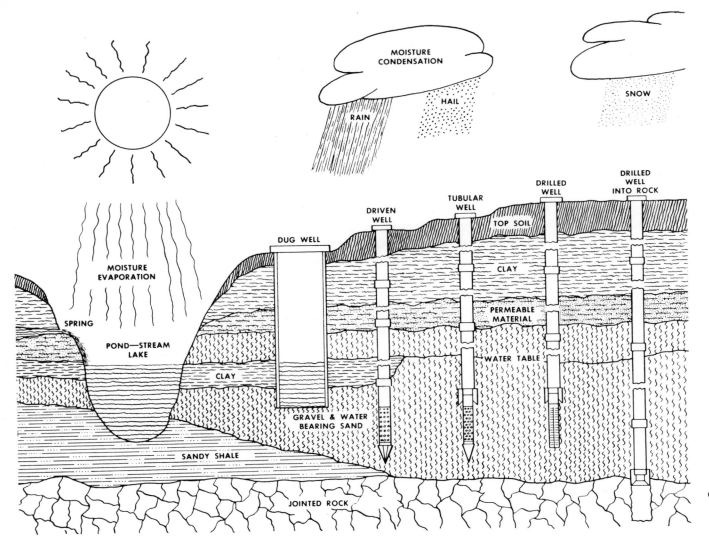

example 5-1. Water sources and various well types.

Labels within the figure:

MOISTURE CONDENSATION

HAIL

SNOW

RAIN

DRILLED WELL INTO ROCK

DRILLED WELL

TUBULAR WELL

TOP SOIL

DRIVEN WELL

CLAY

DUG WELL

MOISTURE EVAPORATION

PERMEABLE MATERIAL

SPRING

POND—STREAM LAKE

WATER TABLE

CLAY

GRAVEL & WATER BEARING SAND

SANDY SHALE

JOINTED ROCK

continue until it meets solid strata and accumulates, where it is available as natural spring or well water. Each well shown in example 5–1 represents a different method of obtaining water from underground sources.

Cisterns and reservoirs used for storage of rain water offer a limited source of soft water, which should not be used for human or animal consumption unless it is thoroughly boiled or properly treated with chlorine or iodine.

Water from springs and pools for domestic use must be carefully checked for purity periodically because the chances of contamination are very great and the purity of such sources is not dependable during periods of drought. Springs not located sufficiently near the homesite or the point of use require extensive piping systems and construction of a basin to collect the water against future need. The basin should be large enough to impound most of the flow and should have slots or openings to admit the water flowing from the spring.

The *dug well* permits the flow of water into the excavation, furnishing a relatively large quantity of water from a shallow source. The walls of a dug well should be curved or lined with rocks, bricks, wood, or concrete to prevent the entry of surface water and caving. Because of their necessary location in low points of the terrain, close to ground level, dug wells are generally subject to contamination from surface seepage and subsurface drainage.

A high percentage of dug wells are polluted. Every possible precaution should be taken by those using this type of well to guard against pollution. Driven or drilled wells are preferable because they give greater assurance of safety against pollution.

Drilled wells are installed when greater volume and depth are needed. In the construction of a tubular well, the drilling operation stops at the water-bearing and sand stratum; in the drilled well the casing or pipe is driven down as the well is drilled and extends either to water-bearing sand or on to rock (example 5–1).

There are other types of driven wells, such as the *open-end* variety, which are less easily clogged with fine particles of sand in areas adjacent to large lakes, but such wells require special apparatus and equipment for proper development and should be installed by experienced well drillers competent to undertake such work and have the knowledge of the ground strata.

When the well goes down to rock, the casting or pipe is driven until rock is reached; then drilling continues into the rock to the water supply. This type of well often does not require a screen. As a guide to drilled well installations, study the typical tubular and drilled wells shown in example 5–1.

Planning the System

Before a water system can be planned successfully, there must be an adequate supply of safe water. Consult local health authorities or your county agent for help on such problems as where to drill new wells, how to protect wells from surface drainage, and how to determine water purity.

If you live in an area where it is difficult to obtain a good supply of well water, you may have to resort to using a farm pond. Pond water should be filtered and purified with chlorine or iodine before it is used for household purposes.

Since a complete water system may cost more than some families wish to spend at one time, it may be advisable to install only a basic system first. Later, more piping and water-using appliances can be added.

The type and depth of the well determines the type of pump to use. If the water in the well is always less than 25 feet below the pump, including drawdown during pumping, plan to use a *shallow-well jet* or *piston pump.* If this level is more than 25 feet but less than 100 feet, use a *deep-well jet* or *piston pump.* If the water level is more than 100 feet deep, use a *deep-well piston pump* or a *submersible pump.* Do *not* use jet or submersible pumps if the water contains heavy sand deposits.

A pressure tank should be used with the pump. Such a tank allows you to draw small quantities of water without causing the pump to start. It is also helpful in evening out the flow of water. The most common size is 42 gallons, which permits you to use about 8 gallons of water between the stopping and starting of the pump. A pressure switch attached to the tank provides automatic control of the motor and pump.

The capacity of the pump should not be greater than the capacity of the well to supply water. If the water flow is strong enough, choose a pump that will deliver at least 5 gallons of water per minute. This capacity enables you to use a $\frac{3}{4}$-inch hose for fighting fires and for watering gardens. A 10-gallon-per-minute pump with special hose and nozzle is recommended by some fire-protection agencies for effective fighting of small fires or for protecting adjacent buildings during a larger fire.

Aside from fire-protection needs, the amount of water you require should determine pump capacity. It has been estimated that each person in a household uses about 100 gallons of water per day. Milk cows need about one third this amount, and other animals need about a tenth.

Generally, you will want your pump to deliver most of this water in a few short periods during the day. Therefore, select a pump size that will furnish the entire daily water needs in 2 hours or less of actual pumping time. If the well capacity does not meet the daily requirements in 2 hours of pumping, use a reservoir or a large pressure tank. The capacity of the reservoir or tank plus the amount of water delivered by 2 hours of pumping should equal at least the daily water requirements.

Pump-Selection Criteria

Unfortunately, selecting the right pump for a given application is more complicated than the foregoing information might suggest. A long length of pipe reduces water flow considerably; and the fittings used in our pipelines, the age of the pipes, and the number of appliances we want to use all contribute toward muddying any simplistic formula we might otherwise use to estimate the proper pump size. There's no reliable way to specify a pump for a given application without learning some of the principles of flow and drawdown.

The function of most pumps is *to create a vacuum.* When the pump is completely installed, the suction line that connects the well to the pump is filled with air at atmospheric pressure (14.7 pounds per square inch). But when the pump is started, it sucks air out of the line; if the well is vented so that outside air is allowed to push it at the continuous 14.7 psi rate, water will begin to flow into the vacuum line and on to the pump. However, it is very important to realize that there is a practical limit to the number of feet we can raise a column of water using this principle.

Suction Lift

How high can water be raised or lifted by suction? The answer is a height equivalent to the pressure differential established between the 14.7 pounds of atmospheric pressure (on the surface of the water to be pumped) and the suction chamber of the pump.

This raises the problem of converting pounds of pressure to feet. It has been proved that 1 pound of pressure from any source will raise a column of water 2.3 feet. If we can remove *all* the pressure of atmosphere from the pump suction chamber, the pressure difference becomes 14.7 pounds; therefore, a perfect vacuum would cause water to rise 14.7 times 2.3, or 33.8 feet. However, due to hydraulic and mechanical losses in a vacuum pump, 25 feet is considered the maximum practical total suction lift at sea level. At altitudes higher than sea level, total suction lift will be reduced 1 foot for each 1000 feet of elevation. (As a practical matter, it is advisable to deduct 1.2 feet from the pump manufacturer's specifications for each 1000 feet above sea level.)

Head Loss in Pipes

The movement of water through a pipe develops friction similar to that of a brake. The amount of this friction depends on the diameter of the pipe, its length, and the rate of flow of water in gallons per minute. (The *age* of steel pipe also has a bearing on the amount of friction loss, because the inside diameter becomes smaller with rust and scale.) The degree to which the flow of the water is retarded by the pipe is known as *head loss,* explained in Section One.

The use of a pipe that is too small will result in an unsatisfactory flow of water, in pressure loss, and in higher pumping costs. Table 5–1 shows the head loss in feet due to friction in hundred-foot lengths of 17-year-old steel pipe. It is very easy to determine the friction loss through any of the pipe sizes shown for any flow of water. The figures on the left show gallons per minute and the figures in the table show the friction loss for each 100 feet of pipe of the various sizes.

For example, 5 gpm through 100 feet of 1-inch pipe results in a friction loss of 3.3 feet. The same 5 gpm through 100 feet of $\frac{3}{4}$-inch pipe will result in 10.5 feet of friction loss, and 41.0 feet of friction loss through 100 feet of $\frac{1}{2}$-inch pipe. If the pipe lengths were doubled, the losses would also be doubled.

Head loss *must* be taken into consideration when pipe is selected for a water system.

Conversion Factors

In the selection of a water system, we must convert pounds of pressure to feet, and vice versa. To do this, we use *conversion factors.* They are as follows:

- To reduce pounds of pressure to feet of head, multiply by 2.3.
- To reduce head in feet to pounds of pressure, multiply by 0.434.

To see how pounds of pressure relate to feet of head, consider a weightless container one cubic foot in volume and filled with water to the brim. (See sketch, example 5–2.) A cubic foot of water weighs 62.5 pounds, which works out to a weight of 0.434 pound per square inch. If the container were twice as large on all sides there would be eight times as much water (2 ft × 2 ft × 2 ft = 8 cu ft); but only twice the water pressure at the bottom; this is because *height* rather than *volume* determines the pressure per square inch. Since we know that a column of water which is one foot high exerts a pressure of 0.434 pound per square inch regardless of the diameter or width and breadth of the column, we can calculate the amount of pressure we have to apply to raise water in a column. If we were to apply a pressure of 0.434 pound per square inch, the foot-high column of water would be weightless or neutral; increase the pressure and the water column will begin to rise. For every 0.434 psi we apply, the column will be elevated 1 foot. If we apply 1 pound of pressure per square inch, we can raise the column 2.3 ft. And this is where the conversion factors (2.3 and 0.434) come from. Use of these factors is necessary to determine the pneumatic tank pressures required to overcome pipe-friction loss or elevation. When the necessary pressure is known, the water system can be selected.

The following is a typical example of the application of conversion factors in selecting a water system.

TABLE 5–1. HEAD LOSS PER 100 FT IN 17-YEAR-OLD STEEL PIPE
(For new pipe, multiply readings by 0.6; for older pipe, by 1.2.)

U.S. G.P.M.	$\frac{1}{2}$	$\frac{3}{4}$	1	$1\frac{1}{4}$	$1\frac{1}{2}$	2	$2\frac{1}{2}$	3	4	5	6	8	10
1	2.1												
2	7.4	1.9											
3	15.8	4.1	1.3										
4	27.0	7.0	2.1										
5	41.0	10.5	3.3										
6	57.0	14.7	4.6	1.2									
7	76.0	19.5	6.0	1.6									
8	98.0	25.0	7.8	2.0									
9		31.2	9.6	2.5	1.2								
10		38.0	11.7	3.1	1.4								
11		45.0	13.3	3.5	1.7								
12		53.0	16.4	4.3	2.0								
13		62.0	18.7	4.9	2.3								
14		71.0	22.0	5.7	2.7								
15		80.0	24.2	6.4	3.0	1.1							
16		91.0	28.0	7.3	3.4	1.2							
17			30.5	8.0	3.8	1.3							
18			35.0	9.1	4.2	1.5							
19			38.2	10.0	4.6	1.7							
20			42.0	11.1	5.2	1.8							
21			45.5	11.9	5.5	2.0							
22			50.0	12.9	6.2	2.1							
23			54.0	14.0	6.6	2.3							
24			59.0	15.2	7.3	2.5							

TABLE 5–1. HEAD LOSS PER 100 FT IN 17-YEAR-OLD STEEL PIPE
(For new pipe, multiply readings by 0.6; for older pipe, by 1.2.)
(continued)

U.S. G.P.M.	$\frac{1}{2}$	$\frac{3}{4}$	1	$1\frac{1}{4}$	$1\frac{1}{2}$	2	$2\frac{1}{2}$	3	4	5	6	8	10
25			64.0	16.6	7.8	2.7							
26			68.0	17.8	8.4	2.9							
27			73.0	19.0	9.0	3.1							
28			78.0	20.2	9.7	3.3	1.1						
29			83.0	21.7	10.0	3.5	1.2						
30			89.0	23.5	11.0	3.8	1.3						
35				31.2	14.7	5.1	1.7						
40				40.0	18.8	6.6	2.2						
50				60.0	28.4	9.9	3.3	1.4					
60				85.0	39.6	13.9	4.7	1.9					
70					53.0	18.4	6.2	2.6					
80					68.0	23.7	7.9	3.3					
90					84.0	29.4	9.8	4.1	1.0				
100						35.8	12.0	5.0	1.2				
120						50.0	16.8	7.0	1.7				
140						67.0	22.3	9.2	2.3				
160						86.0	29.0	11.8	2.9				
180							35.7	14.8	3.6	1.2			
200							43.1	17.8	4.4	1.5			
220							52.0	21.3	5.2	1.8			
240							61.0	25.1	6.2	2.1			

TABLE 5–1. HEAD LOSS PER 100 FT IN 17-YEAR-OLD STEEL PIPE
(For new pipe, multiply readings by 0.6; for older pipe, by 1.2.)
(continued)

U.S. G.P.M.	$\frac{1}{2}$	$\frac{3}{4}$	1	$1\frac{1}{4}$	$1\frac{1}{2}$	2	$2\frac{1}{2}$	3	4	5	6	8	10
260							70.0	29.1	7.2	2.4			
280							81.0	33.4	8.2	2.8	1.1		
300							92.0	38.0	9.3	3.1	1.3		
325								43.5	10.7	3.6	1.5		
350								50.0	12.2	4.2	1.7		
375								56.0	14.8	4.6	1.9		
400								65.0	16.0	5.4	2.1		
425								72.0	17.2	5.8	2.4		
450								79.0	19.8	6.7	2.6		
475								87.0	21.6	7.3	2.9		
500								98.0	24.0	8.1	3.2		
550									28.7	9.6	3.8		
600									33.7	11.3	4.5	1.2	
650									39.0	13.2	5.3	1.3	
700									44.9	15.1	6.1	1.5	
750									51.0	17.2	6.8	1.7	
800									57.0	19.4	7.7	2.0	
850									64.0	21.7	8.7	2.2	
900									71.0	24.0	9.8	2.4	
1000									88.0	29.2	11.9	3.0	1.0
1100										33.5	13.7	3.6	1.2
1200										39.3	16.1	4.2	1.4
1300										45.6	18.6	4.9	1.6
1400										52.3	21.4	5.6	1.9
1500										59.4	24.3	6.4	2.1

Based on Williams & Hazen Formula with Constant C—100

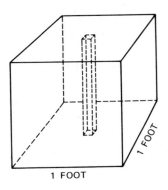

1 FOOT

1 FOOT

example 5–2. A weightless container measuring 12 X 12 X 12 inches would hold 1 cubic foot of water, which would weigh 62.5 pounds. Each square-inch column of water in the container would weigh 0.434 pound, since there are 144 of these square-inch columns in the container (62.5 pounds divided by 144 equals 0.434 pound per column of 1 square inch and 1 foot high).

A water system, including the tank, is installed at the well in a well-ventilated frost-proof pit. The pump has a capacity of 9 gpm. The dairy barn is located 250 feet from the well, with an elevation of 25 feet above the water system location. After referring to Table 5–1, the 1¼-inch pipe has been selected, having a friction loss of 2.5 feet per 100 ft (250 feet times 2.5 equals 6.25 feet of friction loss); 6.25 feet of friction loss plus 25 feet of elevation equals 31.25 feet of friction loss and elevation. It now becomes necessary to find the required pressure to overcome 31.25 feet of extra head. We do this by multiplying 31.25 feet by 0.434, the factor used in converting feet to pounds of pressure per square inch. This gives us 13.56 pounds pressure (per square inch), which will be required to move the water to the dairy barn and nothing more. When it arrives there, it will have no further pressure.

In most cases when water leaves the pressure tank, a minimum pressure of 20 pounds and a maximum pressure of 40 pounds is desired. In our example, therefore, we would have to add 13.56 pounds pressure to the 20 and also the 40 pounds to determine the corrected setting of the pressure switch. (The pressure switch starts the pump motor when the pressure in the tank has declined to the minimum and stops the pump motor when the pressure has reached the maximum.) Therefore, our pressure switch setting for the example would be, to use round figures, 34 pounds minimum and 54 pounds maximum.

Now let's assume that in our example, instead of using 1¼-inch pipe, we used 1-inch pipe. Referring to our friction table, we note that 1-inch pipe at 9 gpm would create 9.6 feet friction per 100 feet of pipe. Multiplying this by 2.5 gives the friction loss in 250 feet of 1-inch pipe as 24 feet.

With ¾-inch pipe, the friction loss would increase to 78 feet, more than three times the elevation!

Obviously the proper size for the above installation is 1¼ inches. This example shows very clearly that when there is excessive pipe-friction loss (as would be the case with ¾-inch pipe), the pumping cost would be much more. Always use a pipe size large enough to keep friction losses to a minimum.

Total Head

The *total head* is the load on the pump designated in feet. When reciprocating or plunger pumps are to be used, it is necessary to know the total head in order to determine the size of electric motor to place on the pump. The loading of motors on submersible and ejector pumps is discussed later.

The total head is the sum of the vertical lift from the water to the pump, plus friction loss, plus the distance the water is to be elevated after leaving the pump, plus the friction loss in the discharge lines (pipes leading from the pump to the storage tank or point of discharge), plus the desired pressure at the point of discharge.

Here again the pipe-friction loss must be taken into consideration. The more friction loss in both suction and discharge lines, the greater the total head of our pump load.

Horsepower

Reciprocating or plunger pumps are available in different horsepower ratings. The more horsepower, the more work the pump will do in terms of feet of head and gallons per minute. So, in selecting our water system, we must determine the horsepower needed to operate the pump.

Here is a simple rule for arriving at required horsepower: pump capacity in gallons per minute, multiplied by the total head in feet, then divided by 2000 equals horsepower. This rule applies to shallow-well reciprocating pumps.

Shallow and Deep Wells

In general, a well of 25 feet or less in depth is known as a *shallow well;* one over 25 feet is called a *deep well.*

Actually, the distance down to the pumping level in the well dictates whether a shallow or a deep-well pump must be installed. If a well were, say, 200 feet deep and the water, in sufficient quantity, were available at only 20 feet below ground level, a shallow-well pump could be used.

Remember the limitations of shallow-well pumps. Lifts in excess of 25 feet will require deep-well equipment. There are no general rules for the best type of pump for a given installation; each type of pump has its own advantages and limitations. As new pumps are developed and old ones improved, the ranges of best usefulness change.

Tank-Selection Criteria

Selection of a pressure tank for a water system will involve a tradeoff: the larger the tank, the less often the pump will start and stop; the smaller the tank, the lower the cost and space requirements. However, the operating cost tends to drop when a pump cycles less frequently but for longer periods, so it's generally better to avoid being too conservative in estimating the size requirement for a particular installation.

Water is an incompressible liquid; air, of course, is compressible. Because of the incompressibility of water it is necessary to maintain a supply of air in the tank at all times; this is done with an *air-volume* control. Water entering the tank compresses the air and keeps the water under a relatively constant pressure; expansion of the compressed air drives the water through piping to the points of use.

The amount of air in the tank is important since it determines the amount of water the tank can release before the pump restarts. When a tank contains $\frac{1}{3}$ air and $\frac{2}{3}$ water at 40 pounds pressure per square inch, the tank will release approximately 20% of its total capacity before the pressure drops to 20 psi and the pump restarts (illustrated in example 5–3). On the basis of 200 gallons of water used daily, the pump would start 25 times a day if a 42-gallon tank is employed. This would not be excessive cycling of the system; the pump should operate satisfactorily for many years if the fresh-water demand remains relatively static.

The drawbacks of a pressure tank of too small a size are undesirable from several standpoints, one of which is the too-rapid cycling of the electric motor, resulting in inordinately high electric bills as well as excessive motor wear and tear. Another tangible disadvantage is the extreme wear on the delicate rocker arms and diaphragm in the pressure switch, which would almost certainly cause premature failure. And of course there's the practical consideration of reserve capacity. A tank sized too small would not be able to provide sufficient water for peak-load periods, when several faucets are being used at one time. Unless the water system is planned to be used only for limited

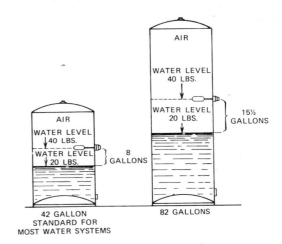

example 5–3. Two common pressure tanks and the capacity in gallons between minimum and maximum pressure.

service, as in a summer cottage or mountain cabin, the pressure tank should have a capacity of at least 42 gallons.

Associated Controls and Switches

The controls associated with a pump and tank depend on the service—whether a shallow-well or a deep-well installation—and the type of pump employed.

Air-Volume Control

An automatic air-volume control maintains the correct volume of air in a pressure tank and is essential to successful water-system operation. Unless the correct amount of air is maintained in the tank, it will become waterlogged, which works a hardship on the motor, the pump, and the associated controls in the same way an undersized pressure tank would. Leakage of water at a faucet will then cause the system to start and stop too frequently.

The amount of air in a pressure tank gradually diminishes because of the gradual absorption of the air by the water in the tank. If the air is not occasionally replenished, the tank will actually fill with water, and whatever

little air under pressure remains will be insufficient to force the water through the piping. That is what is known as *waterlogging.*

Shallow-Well Float Controls. The float-type air volume control (example 5–4) automatically maintains the proper air content by operating in accordance with the amount of water in the tank. The tank is initially set so that maximum air pressure exists when the float arm just touches the water surface. If the water level rises beyond that point, the air pressure is falling below maximum; when this happens, the float begins to rise with the water surface, opening a valve connected to the opposite end of the float arm. This valve, on the outside of the tank, allows the pump to secure air from the atmosphere. The air travels downward through a tube, enters the pump body, is mixed with water in the pump, and then forced into the pressure tank.

While the air content in the tank is sufficient, the float never rises, so no new air is introduced. This characteristic, the introduction of air only when it is needed in the pressure tank, is applicable to shallow wells.

example 5–4. Float air-volume control for shallow-well pump system. (Courtesy F. E. Myers Company.)

Deep-Well Float Controls. Example 5–5 shows an air-volume control typically used on tanks with submersible and deep-well reciprocating pumps, or with any pump where a venting control is needed.

In operation, the proper amount of air under pressure is maintained in the tank by means of a float connected to a shutoff valve (the opening and closing of which is controlled by the position of the float in the tank) and a release air valve (the opening and closing of which is controlled by the air pressure in the tank).

When water in the tank raises the float to its maximum height, the shutoff valve closes. As the water level lowers, allowing the float to drop, the shutoff valve opens. Air in the tank then passes through the release valve and is vented to the atmosphere.

The pressure in the tank at which the release valve opens is established by adjusting the tension on the spring that holds the release valve to the valve seat. This is typically preset to prevent the air-venting operation at tank pressures below 30 psi. An adjustment screw permits resetting so that air cannot escape below a

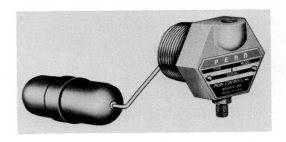

example 5–5. Float air-volume control for deep-well pump system. (Courtesy F. E. Myers Company.)

pressure sufficient to force water through the service piping.

Diaphragm Control. An air-volume control commonly used for both shallow- and deep-well installations where ejector pumps are employed is the diaphragm type of floatless construction. This control cannot be used with reciprocating pumps.

In service, a small charge of air is put into the tank each time the pump stops, but only if air is needed. When the tank contains sufficient air, air entering the diaphragm control (example 5–6) simply works back and forth in the control body and is not permitted to be passed into the tank.

Pressure Switch

As the name implies, the *pressure* switch opens and closes an electrical circuit in response to changes to the pressure applied to the switch. This switch controls the operation of the pump. By starting the pump motor when the pressure reaches a predetermined low point and stopping it after the pressure in the system is built up again, a constant supply of water is kept in the tank at suitable

example 5–6. Diaphragm-type air charging control, typically used on shallow- and deep-well jet pumps. (Courtesy F. E. Myers Company.)

pressures. Usually this is 20 pounds (psi) cut-in and 40 psi cut-out or 30 pounds cut-in and 50 psi cut-out. The 20-pound differential between cut-in and cut-out settings is normal.

Example 5–7 shows a two-pole pressure switch. Mechanically it consists of a pipe flange, a flexible diaphragm, range spring, differential spring, linkage, and electrical contacts and connections. A small movement of the rubber diaphragm caused by the pressure against it is multiplied sufficiently by the mechanical linkage to open and close the electrical contacts.

Pressure Gage

The pressure gage (example 5–8) is an inexpensive device that indicates water pressure in a complete water system. With the ejector system the gage is properly mounted on the pump case or the pump side of the pressure regulator valve. This aids in correctly setting the minimum operating pressure to make the deep-well jet pump work most efficiently. On most all other types of systems, the pressure gage is mounted on the pressure tank.

example 5–7. Pressure switch. (Courtesy F. E. Myers Company.)

Pressure-Relief Valve

A relief valve is a spring-loaded safety valve with adjustable spring tension (example 5–9). It is installed on the *discharge* or pressure side of a pump. The spring tension is adjusted usually to about 20 pounds higher than the cut-out setting of the pressure switch. If the pressure switch fails to stop the motor at the proper pressure, the relief valve will open to prevent dangerous pressure in the system.

The relief valve should be manually opened at least once yearly for inspection and for assurance of free operation.

example 5–8. Typical pressure gage. (Courtesy F. E. Myers Company.)

example 5–9. Pressure relief valve. (Courtesy F. E. Myers Company.)

Foot Valve

A foot valve is a combination check valve and strainer used on the submerged end of a suction pipe (example 5–10). The check valve holds the suction line priming when the pump is not running. The strainer prevents large particles and other foreign matter from entering the suction pipe.

The strainer on the foot valve is not designed to keep sand from entering the suction pipe. A mesh screen small enough to exclude sand would plug up quickly and prevent water from entering.

In the case of a driven well, where the well diameter is small, the check valve (less strainer) is used to maintain priming.

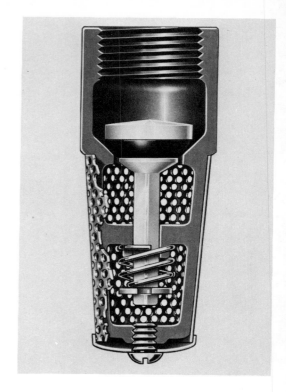

example 5–10. Cross section of conventional foot valve. (Courtesy F. E. Myers Company.)

Pumps

Simply defined, a pump is a device that uses an external source of mechanical power (prime mover) to move liquid from one point to another.

To accomplish this, the pump may employ a pushing, pulling, or throwing motion, or a combination of these effects. In this connection, pumps are often named or classified according to the type of movement that causes the pumping action: *diaphragm, rotary, centrifugal,* or *airlift,* for example.

Regardless of its design or classification, every pump has a power end and a liquid end. The *power end* is some form of prime mover, such as an electric motor or internal combustion engine. The basic purpose of the power end is to develop the mechanical motion or force required by the liquid end.

The liquid end is that part or portion of the pump where the mechanical motion or force developed by the prime mover is actually exerted on the liquid. This part of the pump must have provisions for suction (where the liquid enters) and for discharge (where the liquid leaves the pump). The liquid end is most often referred to as the *pump end.*

Finally, every pump must be equipped with devices for controlling the direction of flow, the volume of flow, and the operating pressure of the pump. A device that performs one or more of these control functions is called a *valve.* A valve that permits liquid flow in one direction only is classified as a *check valve.* In most cases, check valves open and close automatically; that is, they are kept closed or seated by spring tension or the force of gravity until the liquid pressure above or below the valve overcomes the spring or gravity resistance and causes the valve to open. Check valves of this type are used in *centrifugal* pumps to automatically control the suction and discharge of liquid in the pump end at the proper time. Example 5–11 shows a *vertical check valve.* In this case, the valve is kept seated by its own weight or the force

of gravity (but it could also be kept closed by a spring).

Another type of valve found in very large pump systems is the *stop valve.* Stop valves are usually opened or closed manually by means of a handwheel. They are used primarily to start or stop the flow of liquid through the pump during certain phases of operation. Thus, stop valves are often placed on suction and discharge lines so that the pump may be isolated or sealed off from the rest of the water system. Example 5–12 illustrates the operation of a *gate valve,* which is a type of stop valve. A gate or wedge is raised or lowered by turning the handwheel. Some types of stop valves are used for throttling purposes; that is, to regulate the flow of liquid. Gate valves, however, are never recommended for throttling service since the flow of liquid past the partially opened gate can rapidly corrode the gate face. Instead, the gate can be replaced with a *tapered needle valve* (another type of stop valve), which gradually opens or closes through the valve seat.

A third type of valve generally found on water pumps is the *relief valve.* As you can see from example 5–13, most relief valves are quite

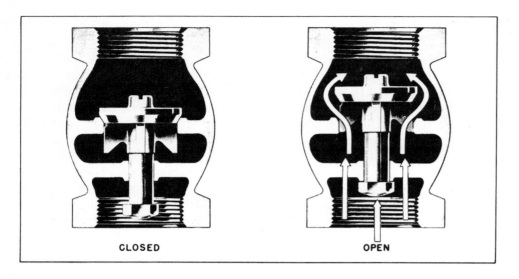

CLOSED   OPEN

example 5–11. Cutaway view of typical vertical check valve.

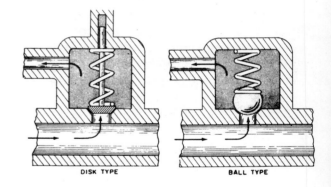

DISK TYPE   BALL TYPE

example 5–13. Disk-and-ball relief valves.

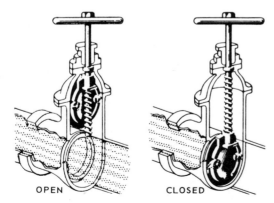

OPEN   CLOSED

example 5–12. Gate valve operation.

similar in their design to check valves. These valves are designed to open when the water pressure in the pump becomes dangerously high. In most cases, the outlet of the relief valve is connected to a recirculating line that passes the excess liquid back to the suction side of the pump. Almost all pressure relief valves will be fitted with an adjusting nut or screw that permits the spring tension to be regulated. In this way, the pressure at which the valve will open may be varied. (Valve repairs are covered in Section Two.)

Reciprocating Pumps

In any consideration of the several types of reciprocating pumps used for moving water, a division into two categories may be made. *Shallow-well pumps* are used to raise water (at sea level) from as much as 25 feet (total head, including well depth, friction losses, and so on.) *Deep-well pumps* are those used to lift water from pumping levels greater than 25 feet below the pump. As we might suspect, any pump capable of raising water from deep wells can also be used for shallow-well service.

Shallow-Well Reciprocator

The shallow-well pump is referred to as a *piston pump* because it relies on a moving piston for its operation. Some pumps of this class employ a piston that moves a flexible diaphragm rather than a movable diaphragm; these are called *diaphragm pumps,* but they are not as common for well service as they are for sewer pumping. Our discussion centers on the type of reciprocator shown in example 5–14. In the rendering,

portions of the pump have been cut away to simplify the explanation and to make the pump's principle of operation clearer.

When the plunger moves to the right, the plunger plates draw a vacuum in the left side of the cylinder body. Atmospheric pressure on the surface of the water in the well drives water into the cylinder through a check valve. When the plunger moves to the left, the water in the cylinder is forced up through the valve on the left as fresh water enters the cylinder on the right. As the plunger returns to the rightmost position the water trapped in that portion of the cylinder body is forced up through the associated valve. The pump has a *double-cycle action* as it discharges water at each forward and backward stroke of the plunger. The inlet and outlet connections for this pump are on the portion of the pump head that is cut away.

example 5–14. Popular shallow-well reciprocating pump, the design of which has remained stable for many years. The water end of this assembly has been cut away to show part detail. (Courtesy F. E. Myers Company.)

Deep-Well Reciprocator

The deep-well reciprocating pump (example 5–15) is often referred to as a *working head* because it acts merely as a power unit to operate the pumping barrel or cylinder in the well, force the water where desired, and supply air when needed. The most general usage is for securing water from lifts in excess of that obtainable with a shallow-well reciprocator.

Years ago, these pumps were manufactured in many stroke lengths and for loading to 15 or 20 horsepower. But the submersible pump has so many advantages that it has all but displaced heavy versions of this unit entirely. Some of the earlier deep-well reciprocators weighed as much as 2000 pounds. Today pumps of this type aren't being manufactured with a rating above 1 horsepower.

As with the shallow-well reciprocator, the rotating motion of the motor or prime mover is changed to reciprocating movement through the mechanical linkage of the pump. Because it is necessary to lower the pumping barrel down into the water, you also have to use a *drop pipe* and a *sucker rod* for attaching the barrel to

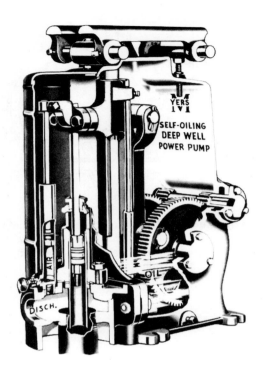

example 5–15. Reciprocating-pump "working head" for deep-well installations. (Courtesy F. E. Myers Company.)

the pump. For serviceability, the most desirable arrangement is the use of a wood sucker rod and a large-diameter drop pipe (see example 5–16).

Reciprocating-Pump Characteristics

Reciprocating pumps are usually employed for *low capacities* and *high-head* conditions, that is, where the total gallonage per minute is relatively low but the pressure requirement is high. Shallow-well models are available in capacities up to 65 gallons per minute and for pressures ranging up to 250 pounds per square inch. Deep-well models are intended for capacities up to 8 gallons per minute and pressure not exceeding 75 psi.

The general performance characteristics of reciprocating pumps are as follows:

1 Capacity is uniform and constant regardless of discharge pressure, suction lift, or lift from a deep well.

2 Horsepower varies directly with increase in discharge pressure.

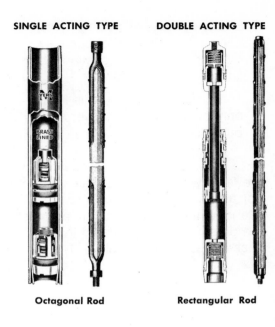

SINGLE ACTING TYPE   DOUBLE ACTING TYPE

Octagonal Rod   Rectangular Rod

example 5–16. At left, single-acting octagonal rod; at right, double-acting rectangular rod.

3 Efficiency remains more or less constant regardless of capacity, discharge pressure, suction lift, or lift from a deep well.

4 The discharge pressure is limited only by motor horsepower and pump construction.

These pumps cannot economically be fitted with one size of electric motor for all installations. From a cost standpoint alone, the total head must be determined as indicated in the loss-of-head chart (Table 5–1).

Rotary Pumps

All rotary pumps employ the principle of entrapment and displacement of fluid by *rotating* elements of various design. These rotating parts, which may be gear teeth, screws, lobes, or vanes, trap the fluid at the suction inlet and remove it to the discharge outlet. Instead of ''throwing'' the water as in a centrifugal pump, a rotary pump traps it, pushes it around inside a closed casing, and discharges it in a continuous flow. Since rotary pumps move liquid according to this method, they are often classified under the broad heading of *positive-displacement pumps.*

The simplest type of rotary pump is the *gear pump,* shown in example 5–17. This type of pump employs two spur gears that rotate in opposite directions and mesh together at the center of the pump. One of the gears is coupled to the prime mover (usually an electric motor) and is called the *driving gear.* The other gear, which receives its motion by meshing with the driving gear, is called the *driven gear.*

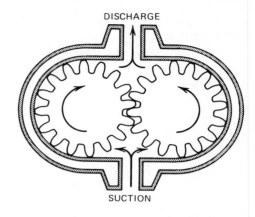

example 5–17. The gear-type or rotary pump is the ultimate in simplicity; it slings water by the meshing of gear teeth.

The actual movement of liquid is accomplished as the gear teeth rotate against the casing of the pump, thereby trapping the liquid and pushing it around to the discharge outlet. The meshing together of the two gears does not in itself move or pump liquid. Rather, the meshing of the gear teeth, in effect, forms a constant seal between the suction and discharge sides of the pump and thus prevents liquid from leaking back toward the suction inlet.

Very small clearances are permitted between the meshing gears, and between the gear teeth and pump casing in order to avoid unnecessary friction. This also allows the liquid being handled to act as a lubricant for the rotating parts. It should be evident, however, that if excessive clearances are allowed to develop between the gear teeth and casing, or between the gears where they mesh, the efficiency of the pump will be considerably reduced. For this reason, rotary pumps are rarely, if ever, used to handle corrosive or abrasive waters.

Most rotary pumps have stuffing boxes provided at the rotor shafts to prevent excessive leakage at the shaft joint. In addition, various types of bearings may be fitted at the ends of the rotor shaft to minimize friction.

There are many different types of rotary pumps, the classification generally being made according to the type of rotating element employed. Whatever form of rotating element is used, the basic principles of pump operation remain the same.

Generally, rotary pumps are considered to be *self-priming;* that is, the pump end need not be filled with liquid in order to initiate pumping action. Instead, the movement of the rotating elements will create a partial vacuum sufficient to lift or draw liquid into the pump and initiate the pumping process. Self-priming and good suction lift are actually characteristics of the whole class of positive displacement pumps.

Rotary pumps have the added advantage of being less expensive and considerably simpler in their construction. They can be used for pumping chemical feed in water purification systems and for priming larger pumps.

Centrifugal Pumps

When a body or a liquid is made to revolve or whirl around a point, a force is created that impels that body or fluid outward from the center of rotation. This phenomenon is called *centrifugal force,* and it is from this force that the *centrifugal pump* takes its name. An easy way to describe the operation of a centrifugal pump is to compare it with swinging a pail partially filled with water. The water stays in the bottom of the pail; centrifugal force keeps it there.

If a hole is punched through the bottom of the bucket, as in example 5–18, a stream of water would be released. The distance the stream would carry depends entirely on the speed at which the bucket is revolved. The faster the bucket is swung around in the circle, the greater the velocity at which the water would leave the hole in the pail. This is also true of the centrifugal pump.

However, instead of being swung in a pail, the water enters the *hub* or *eye* of the impeller and is thrown out at the *rim* of the impeller. This means

example 5–18. A swinging bucket with a hole in it explains how a pump can operate by centrifugal force.

that the diameter of the impeller and its revolutions per minute determine the velocity of the water as it leaves the rim of the impeller.

Basic Structure

The basic centrifugal pump has only this one moving part—the wheel or *impeller* that is connected to the drive shaft of a prime mover and which rotates within the pump casing. The design or form of the impeller may vary somewhat. However, whatever its form, the impeller is designed to impart a whirling or revolving motion to the liquid in the pump. When the impeller rotates at relatively high speeds, sufficient centrifugal force is developed to throw the liquid outward and away from the center of rotation. Thus, the liquid is sucked in at the center or *eye* of the impeller (center of rotation) and discharged at the outer rim.

By the time the liquid leaves the impeller, it has acquired considerable velocity. In this connection, there is a fundamental law of fluid physics that states, in part, that as the *velocity* of a fluid *increases,* the *pressure,* or pressure head, of that fluid *decreases.*

We can therefore characterize the liquid discharge from the impeller as having high velocity but low pressure.

Before the liquid can be discharged from the pump, however, some means must be found to *increase* the pressure. In other words, the primary concern in practically all pumping systems is that the discharge pressure be maintained so that liquid can be distributed effectively throughout the system. In the case of centrifugal pumps, what is needed is some device that will decrease the velocity of the impeller discharge and thereby increase the liquid pressure at the discharge outlet.

Example 5–19 shows a cross section of an impeller, hub in the center, and the several rotating vanes through which the water is discharged.

Example 5–20 again shows cross sections with the impellers cut in half. Both are the same diameter. The impeller shown in B will pump much more water because of a larger hub, or intake, and larger cross section for the water to pass through the impeller. However, both impellers are the same diameter, and when operated at the same speed will develop approximately the same pressure.

example 5–19. Cross section of impeller with arrow showing direction of rotation.

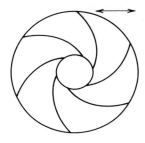

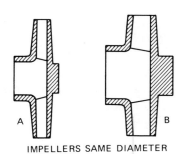

IMPELLERS SAME DIAMETER

example 5–20. Even though both impellers shown here have the same diameter, the one at the left is thinner so it will pump less water despite the fact that its pressure will be the same as the thicker one at the right.

example 5–21. A 4-inch-diameter impeller rotating at 3500 rpm will deliver the same pressure as an 8-inch impeller operating at half the speed (1750 rpm).

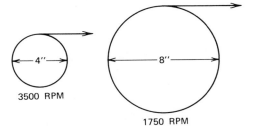

3500 RPM

1750 RPM

Example 5–21 shows a 4-inch impeller operated with a 3500 rpm motor and an 8-inch impeller operated by a 1750 rpm motor. Even though the 4-inch impeller is only half the diameter of the larger, it is operated at twice the rpm, causing both impellers to have the same rim speed, resulting in the same pressure.

Basic Types

Generally speaking, there are two basic types of centrifugal pumps. Example 5–22 shows the *volute* type, in which water leaves the impeller rim at high velocity into the volute case. The volute case increases in cross section until it has reached its maximum at the pump outlet. The purpose of this case construction is to slow up the high velocity and increase the low water pressure, converting to low velocity and a useful high water pressure.

The *turbine* or *diffuser pump* in example 5–23 changes the high velocity of the water as it leaves the rim of the impeller to low velocity and high pressure by discharging water into numerous vanes on the diffuser plate. These vanes have a narrow

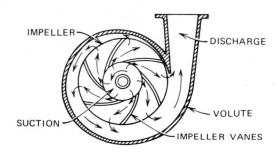

example 5–22. Volute centrifugal pump.

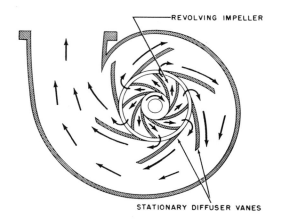

example 5–23. Diffuser-vane centrifugal pump.

cross section where the water enters from the impeller and is much wider at the rim of the diffuser plate. This causes the high-velocity water to be changed into useful pressure.

An analogy would be dropping a baseball from a height of 200 feet. The velocity of the ball would be $113\frac{1}{2}$ feet per second as it hits the ground. To throw the ball back to the height of 200 feet would require that the ball leave the thrower's hand at the same $113\frac{1}{2}$ feet per second. The force with which the ball is thrown determines the height or distance the ball will travel. The velocity of the water as it leaves the rim of the impeller determines the pressure the pump will develop.

Increasing Discharge Pressure

Because of the smooth, nonpulsating flow from a centrifugal pump, it is ideally suited for booster-pump service. An example of booster-pump service is shown in example 5–24. Because the first pump has a capacity of 16 gpm at 20 pounds pressure, water will enter the inlet to the second pump at 20 pounds pressure. The

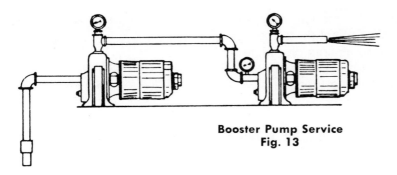

**Booster Pump Service
Fig. 13**

example 5-24. Impeller pumps in tandem allow additive pressure buildup.

second pump will use the incoming pressure and add it to its own pressure. The water then leaves the second pump at 40 pounds pressure. If a third pump were added, the final discharge pressure would be 60 pounds. There would be no change in the *amount* of water discharged, which would remain at 16 gallons per minute.

Another method of increasing the discharge pressure of centrifugal pumps is to provide additional impellers. Pumps having only one impeller are classified as *single-stage*. Pumps with two or more impellers are referred to as *multistage*. In multistage pumps, two or more impellers are placed on a common shaft (within the same pump housing), with the discharge of the first impeller being led into the suction of the next impeller, and so on. As the liquid passes from one stage to the next, additional pressure is imparted to it. In this fashion, the final discharge pressure of the pump can be increased considerably.

Maintenance and Repair

Generally, centrifugal pumps must operate at relatively high speeds in order to produce the necessary centrifugal force. Because of the high rotational speed of the impeller, its machine finish, or smoothness and balance, must be carefully preserved to minimize friction and avoid vibration.

Close clearances must be maintained between the impeller *hub* (the central portion of the impeller which fits around the shaft) and the pump casing in order to minimize liquid slippage between the discharge side of the pump casing and the suction side.

Because of the high rotational speed of the impeller and the necessarily close clearances, the running surfaces of both the impeller hub and the casing at that point are subject to relatively rapid wear. To avoid renewing an entire impeller and pump casing solely because of wear in this location, centrifugal pumps are fitted with replaceable metal *wearing rings.* One ring (impeller wearing ring) is fitted on the hub of the impeller and rotates with it; a matching ring (casing wearing ring) is attached to the casing and is therefore stationary. These metal wearing rings perform essentially the same function as *packing;* that is, they seal a moving joint and prevent excessive friction and wear. Like packing, the wearing rings are normally lubricated by the liquid handled in the pump. Also like packing, the rings must be replaced from time to time when they become worn in order to avoid excessive slippage or leakage.

Advantages and Disadvantages

The advantages of centrifugal pumps include simplicity, compactness, reduction of weight, and adaptability to high-speed prime movers. Turbine well pumps, to produce sufficient discharge pressure, are provided with multistage impeller arrangements referred to as *bowls.* To insure satisfactory suction, the impellers and bowls are set below the lowest drawdown or pumping level that the water in a well is expected to reach.

One of the disadvantages of centrifugal pumps is their relatively poor suction power. When the pump end is dry, the rotation of the impeller, even at high speeds, is simply not sufficient to lift liquid into the pump. The pump must, therefore, be *primed* before pumping can begin. For this reason, the suction lines and inlets of most centrifugal pumps are placed below the source level of the liquid pumped. The pump can then be primed by merely opening the suction stop valve and allowing the force of gravity to fill the pump with liquid. The static pressure of the liquid above the pump will, of course, also add to the suction pressure developed by the pump while it is in operation.

Another disadvantage of centrifugal pumps is that they are susceptible to the phenomenon known as *cavitation.* Cavitation, or boiling, will occur when the velocity of a liquid increases to the point where the consequent pressure drop reaches the *pressure of vaporization* of the liquid. When this happens, vapor pockets or bubbles form in the liquid and then later collapse when subjected to higher pressure at some other point in the flow.

The collapse of the vapor bubbles can take place with considerable force. This effect, coupled with the rather corrosive action of the vapor bubbles moving at high speed, can severely pit and corrode impeller surfaces and sometimes even the pump casing. In extreme instances, cavitation has been known to cause structural failure of the impeller blades. Whenever cavitation occurs, it is frequently signaled by a clearly audible noise and vibration (caused by the violent collapse of vapor bubbles in the pump).

There are several conditions that can cause cavitation, not the least of which is improper design of the pump or pumping system. For example, if the suction pressure is abnormally low (caused perhaps by high suction lift or friction losses in the suction piping), the subsequent pressure drop across the impellers may be sufficient to reach the pressure of vaporization. A remedy in this case might be to alter the pump design by installing larger piping to reduce friction loss, or by installing a *foot valve* to reduce suction lift.

Cavitation may also be caused by improper operation of the pump. For instance, cavitation can occur when sudden and large demands for liquid are made on the pump. As the liquid discharged from the pump is rapidly distributed and used downstream, a suction effect is created on the discharge side of the pump. One might think of this effect as a pulling action on the discharge side, which serves to increase the velocity of the liquid flowing through the pump. Thus, the pressure head on the discharge side, which serves to increase the velocity of the liquid flowing across the impellers, increases to the point where cavitation takes place. Perhaps the easiest way to avoid this condition is by regulating the liquid demand. If this is not possible, then the only other practical alternative is to increase the suction pressure by some means in order to maintain pressure in the pump under varying conditions.

Airlift
Pumps

The use of airlift pumps is confined entirely to well pumping. Unlike other pumps, the airlift pump needs no moving or rotating mechanism to produce liquid movement. Instead, the pump utilizes compressed air to move or lift the liquid.

The airlift pump operates on the principle that water mixed with air has less weight, or is more buoyant, than water that contains no air. When compressed air is introduced, a mixture of water and air is formed in one leg of the U-shaped pipe, as shown in example 5–25. The solid column of water in the other leg now has greater weight, or is exerting a greater static pressure, than the column containing air. Thus, the air-and-water column is forced upward to the point where it discharges over the top of the U-shaped pipe.

In actual practice, of course, wells are not dug in U-shape. Example 5–26 shows a central airlift pump. Compressed air is led down an air pipe to a nozzle or footpiece submerged well below the water level.

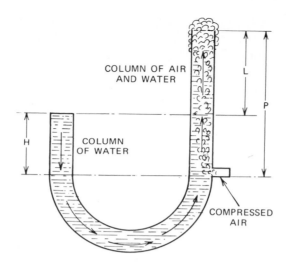

example 5–25. The principle of the airlift pump is explained in this drawing of a U-shaped column of water into which air is introduced. In practice, of course, the column isn't shaped like this, but the piping used works the same from the theoretical point of view.

example 5–26. Central airlift pump.

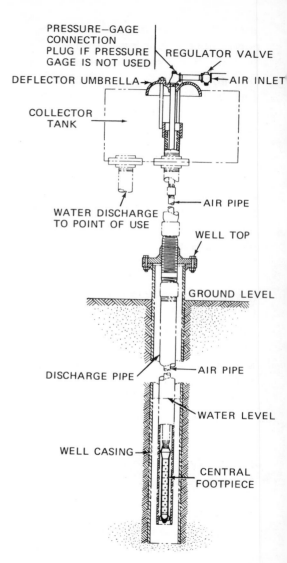

You will notice that the footpiece is suspended within a discharge pipe, which, in turn, is contained within the well casing. Note, also, that the discharge pipe is open at the bottom, directly beneath the footpiece. When compressed air is discharged through the footpiece, a column or mixture of air is formed above the footpiece in the discharge pipe. The solid column of water in the well casing, resting high above the footpiece and discharge pipe inlet, now has greater weight or static pressure. This forces the air-and-water mixture upward in the discharge pipe to the point where it is vented to the atmosphere through an open discharge outlet. In effect, the flow of water has a U-shape: down the well casing, around to the footpiece, and up the discharge pipe. The air-and-water discharge then strikes a separator or deflector that serves to relieve the water of air bubbles and entrained air vapor. Finally, the discharge settles in a collector tank.

The airlift pump is capable of delivering considerable quantities of water in the manner just described. The discharge pressure at which water is delivered, however, is relatively low. For this reason, airlift pumps cannot be used to discharge directly into a water-distribution system; that is, they do not develop sufficient pressure to distribute water horizontally above ground for any appreciable distance. Instead, the discharge is collected at the well for ground storage.

Air Pressure and Water Capacity

The capacity of the airlift pump will depend to a large extent on the submergence of the footpiece; that is, the greater the submergence of the footpiece below the water level in the discharge pipe, the greater the volume (column) of water the pump can deliver per unit of time. However, the deeper the footpiece is submerged, the greater the compressed air pressure must be to lift the column of water. In other words, a higher column of water (in the discharge pipe) above the footpiece will exert a greater pressure at the footpiece. The greater the static water pressure at the footpiece, the greater the air pressure must be in order to infuse air with the water properly.

It is also true that starting air pressure is always greater than working air pressure. When the pump is started, the *static* (at rest) *level* of water is drawn down somewhat to a *pumping* or *working* level. In effect, the column of water above the footpiece is decreased or lowered, and this in turn decreases the air pressure required to infuse the water with air. In wells where the drawdown is rather large, the pump is sometimes equipped with an auxiliary air compressor connected in series with the main compressor for starting purposes. Once the pump has been started, and the pumping level reached, the auxiliary compressor is no longer required.

Disadvantages and Advantages

The primary disadvantages of airlift pumps are low discharge pressure and the added depth required to obtain the proper submergence of the footpiece. Moreover, the entrained oxygen in airlifted water tends to make it more corrosive. In spite of these drawbacks, airlift pumps have several important advantages, not the least of which is their simplicity of construction and consequent lack of maintenance problems. They are particularly useful in emergency deep-well pumping, using a portable air compressor for power. They can be used to pump crooked wells and wells containing sand and other impurities. They can also pump hot-water wells with little difficulty.

Jet Pumps

Pumps that utilize the rapid flow of a fluid to entrain another fluid and thereby move it from one place to another are known as *jet pumps.* A jet pump contains no moving parts.

Jet pumps are generally considered in two classes: *ejectors,* which use a jet of steam to entrain air, water, or other fluid; and *eductors,* which use a flow of water to entrain and thereby pump water. The basic principles of operation of these two devices are identical.

Ejector Pump

The basic principle of operation of a simple jet pump of the true ejector type is illustrated in example 5–27. Steam under pressure enters chamber C through pipe A, which is fitted with a nozzle, B. As the steam flows through the nozzle, the velocity of the steam is increased. The fluid in the chamber at point F, in front of the nozzle, is driven out of the pump through the discharge line, E, by the force of the steam jet. The size of the discharge line increases gradually beyond the

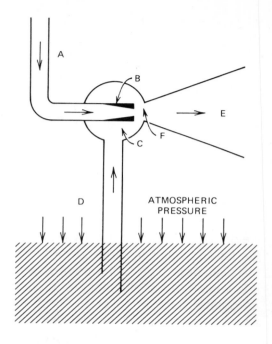

example 5–27. Principle of operation of an ejector-type jet pump. The text explains the operation of each alphabetically designated part.

chamber in order to decrease the velocity of the discharge and thereby transform some of the velocity head into pressure head. As the steam jet forces some of the fluid from the chamber into the discharge line, pressure in the chamber is lowered and the pressure on the surface of the supply fluid forces fluid up through the inlet, D, into the chamber and out through the discharge line. Thus the pumping action is established.

Single-Stage Centrifugal Eductor Pump

The centrifugal eductor pump combines two principles of pumping, that of the centrifugal pump and that of the eductor pump.

Shallow-well units are limited to a maximum pumping depth of approximately 25 feet, which is the practical depth limit for lifting water into an evacuated space by the use of atmospheric pressure. These shallow-well units have the *ejector* (it's actually an eductor, but pump makers refer to it as an ejector, so we'll use that nomenclature) placed inside or next to the casing of the centrifugal pump.

Deep-well units are used when the pumping level of the well is more than 25 feet below the pump location. In this system the ejector is immersed in the well water and connected to the pump with two pipes, the *pressure* pipe and the *delivery* pipe.

The pressure pipe carries pressure water from the discharge section of the centrifugal pump down to the nozzle in the ejector assembly. The delivery pipe is threaded to the top of the venturi assembly on the ejector and carries the recirculating water plus the well water up to the pump. This operation is smooth and continuous.

High velocity of the water has been mentioned in describing the operation of the centrifugal pump. Where there is high velocity there is low pressure. Example 5–28 shows a fire-hose nozzle. When this nozzle is in action and the high-velocity stream leaves the nozzle tip, most people think of high pressure. Actually, there is a low pressure and high velocity in the nozzle stream. If it were possible to have a pressure gage on the hose where the nozzle is attached and this gage registered 100 pounds when the water entered the tapered nozzle, the

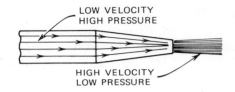

example 5–28. In a fire hose nozzle, the entering water is relatively low-velocity but under high pressure. The stream that leaves the nozzle is at high velocity but under low pressure.

velocity would increase and the pressure would dissipate. If it were possible to have a pressure gage at the nozzle tip, the pressure would be practically at zero.

Example 5–29 shows a venturi tube as used with Myers' *Ejecto* pumps. (Ejecto is that company's tradename for its line of centrifugal eductor pumps.) The water enters the nozzle, which is just below the throat of the venturi tube. This entering water is at high pressure, but it leaves the nozzle at high velocity and low pressure. Because of the high velocity of the water as it leaves the nozzle tip, there is a low pressure or partial vacuum created. Atmospheric pressure then drives water through the foot valve and into the venturi tube with the nozzle stream. Since the venturi tube now flares out, just the reverse of the fire nozzle, there is a change back to low velocity and useful pressure. This carries the recirculating water and the well water up to within suction distance of the centrifugal pump, when the pump is operating at its minimum operating pressure. Maximum pump flow is now possible.

A distinctive feature of the centrifugal eductor pump is that the

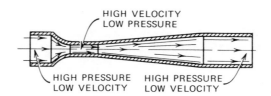

example 5–29. The venturi principle is illustrated by this eductor tube.

horsepower requirement remains practically constant regardless of discharge pressure or well lift. A given size of pump and motor will furnish generous amounts of water from well lifts of from 30 to 180 feet. This is accomplished by using different ejector assemblies only. Each ejector assembly is designed for a certain maximum lift.

Example 5–30 shows an eductor or ejector assembly and throat-to-nozzle area ratios. The relatively high ratio of 2.75 (case A) is for a maximum setting of 60 feet; at full operating pressure it will deliver a gallon of water for every 1.3 gallons circulated. Case B, in which the throat-to-nozzle area ratio is just slightly greater than 2 to 1, represents a system intended for applications where the setting will be between 60 and 90 feet; for every gallon of water delivered, 1.7 gallons are circulated. The third case, for lifts of 90 to 120 feet, has a comparatively low throat-to-nozzle ratio (1.6 to 1); in this case about 2.5 gallons of water must be circulated for every gallon delivered. Note an interesting relationship: the deeper the well, the

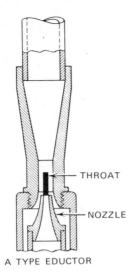

A TYPE EDUCTOR

A $\frac{\text{AREA OF THROAT}}{\text{AREA OF NOZZLE}} = 2.75$ B $\frac{\text{AREA OF THROAT}}{\text{AREA OF NOZZLE}} = 2.1$ C $\frac{\text{AREA OF THROAT}}{\text{AREA OF NOZZLE}} = 1.6$

example 5–30. The throat-to-nozzle ratio is extremely important in jet pumps. The text describes differences for ratios of 2.75, 2.1, and 1.6 to 1.

greater the water circulation for every gallon delivered.

The centrifugal eductor or ejector pump meets the conditions of the "weak" well better than any other type of deep-well pump, when properly installed.

As mentioned earlier, the characteristics of the straight centrifugal pump and the eductor pump do vary in several ways. With the centrifugal pump, as capacity increases with decreasing pressure, the horsepower requirement increases accordingly. With the *eductor* pump, the capacity is always the same. As the discharge pressure or well lift increases, the percentage of water that is circulated increases, and the water discharged by the pump decreases accordingly.

These characteristics must be considered when selecting the correct pump for various types of installations. For an installation where there is considerable elevation between the pump location and the places where the water is to be used, the necessary pressure must be checked; if too high for efficient eductor pump operation, a reciprocating, a dual-impeller eductor, or a submersible pump should be selected. Pipe-friction loss must also be figured since excessive head loss will result in reduced and unsatisfactory delivered pressures.

Regardless of the make of the jet pump, the pressure-switch settings should not be raised higher than shown in the manufacturer's catalog. When the switch setting is raised higher than recommended, the capacity will be inadequate.

Two-Stage Centrifugal Eductor Pump

Two impellers develop more pressure than is possible with a single impeller of the same horsepower. This higher pressure makes it ideally suited for deeper wells or for installations where a higher discharge pressure is necessary because of elevations or excessive pipe friction loss.

Although single-impeller eductor pumps are usually limited to lifts of not more than 120 feet, there seems to be no theoretical limit for which a multi-impeller pump could be used. But on the basis of cost and efficiency, there is a point at which it loses its appeal because the same work or more can be done for less money by other types of deep-well pumps.

Submersible Pumps

Although the submersible pump as we know it today was not in general use before 1945, it actually had its conception before World War I. Early development work was done in foreign countries where the submersible was designed for pumping oil wells and mine dewatering. These were very large capacity pumps and some were available for lifts of several thousand feet. Also, due to the nature of their use, they have heavy ornate castings and were hand-made units.

With the introduction of submersibles for small-diameter wells on a production basis, the quality of the equipment has gone up and prices have come down. Submersibles are most efficient in securing water from deep wells; they are also used for all kinds of shallow-well service.

Submersibles in fractional- and integral (whole number)-horsepower sizes are designed for 4-inch (inside diameter) and larger wells. This accounts for the cylindrical construction shown in example 5–31.

example 5–31. Submersible-pump unit and associated control box. (Courtesy F. E. Myers Company.)

In operation, the entire pump-and-motor assembly is submerged in the water supply. The motor drives the pump impellers directly, so there is no rpm difference between the impeller and the pump motor. Water enters at the inlet to the pump and is forced upward through a single drop pipe to the required point of discharge, which can be a pressure tank or other place where water is needed. This is truly a simple pump, yet one that has required considerable experimental work and development because it is designed to "live" under water, out of sight.

The performance of the submersible pump is the same as the multistage centrifugal in that the pressure increases as the capacity and also the horsepower requirement decreases. In the range of operation from low to high pressure or low to high capacity, there can be no overloading of the motor on the pump.

A submersible may be designed with emphasis on pressure (for deeper settings) or on capacity, or a combination of both. But one is obtained only at the expense of the other unless the horsepower is increased. Thus a $\frac{1}{2}$-horsepower pump with 10 impellers (designed primarily for capacity) would deliver greater capacity at 80 feet than a $\frac{1}{2}$-horsepower pump with 12 impellers (designed for pressure). But the latter might raise water from an extra 100 feet of lift. Horsepower alone would not determine the capacity of the submersible, but horsepower plus the number of impellers would indicate the relative capacity and pumping depth.

Liquid End

As with the centrifugal pump, in general the *pressure* for the submersible varies with the number of impellers, and capacity varies with the design of the impellers. The submersible operates like any centrifugal multistage pump; each stage added increases the head or pressure by a set amount. For a submersible that will deliver 10 gallons per minute per stage, each stage added will increase the head pressure by 10 pounds per square inch. There will be no increase in gallons per minute.

A *stage* is an impeller, diffuser, and housing. The *diffuser* consists of stationary vanes that have tapered sections similar to venturi tubes. The high-velocity water from the impeller is slowed down in these passages and converted into useful pressure. The diffuser also guides the water into the next impeller inlet.

There are several types of diffusers used on submersible pumps; each has certain features for specific applications. Example 5–32 shows a *pancake* impeller with mating parts to make up a stage for submersibles

example 5–32. Pancake-type submersible stage. (Courtesy F. E. Myers Company.)

when good capacities and high heads are desirable. These are of the direct-diffusion design, which uses a small impeller in relation to the pump shell and gives high efficiency.

With this construction, water discharged from the impeller enters the diffusion ports directly and then the guide vanes. More impellers are needed to make a given head because the impeller diameters are smaller.

Example 5–33 shows the impeller and diffuser needed when high capacities and medium heads are desired. This is called a *bowl type* and is used on all larger submersibles and deep-well turbines.

With this construction the water leaves the impeller and is directed into tapered passages that spiral upward in a smooth curve, resulting in high efficiency. This construction also permits the use of a relatively large impeller in relation to the outside diameter of the pump.

Power End

Because submersible motors must live their entire lives submerged in water, the principal design objective is to

example 5–33. Bowl-type submersible stage. (Courtesy F. E. Myers Company.)

make them wet-proof. There are two methods for achieving this: (1) *canning* the motor into a sealed container and (2) submerging the motor in a continuous oil bath.

The canned motor is made by taking the stator winding and completely encasing it in a can or a housing. The inner liner is usually stainless steel of from 5 to 10 thousandths of an inch thick. The outer shell and end caps are also stainless steel. The outer and inner tubes are welded to the end plates. After this is completed, the can is filled with a plastic compound. The rotor then operates in water or oil emulsion.

With the oil immersion approach, the windings are open and operate completely surrounded by oil. The rotor, bearings, and seal all operate in clean oil.

The canned motor allows the rotor and bearings to operate in water. This construction has the advantage that the motor is not filled with any liquid other than the well water. However, one disadvantage is that heat dissipation is relatively low due to the plastic filler. The motor will operate satisfactorily at normal current, but if the current increases (due to low voltage or a tight pump) and the overload does not function properly, the plastic swells and causes the inner liner to grab the rotor, stopping the motor. Operation under this condition can cause the liner to fail and allow the windings to become wet. Another disadvantage is that many well waters are corrosive. Canned motor repairs are expensive.

With the oil immersion approach, neither the quantity nor the quality of the oil will vary over years of motor use because the oil does not become contaminated with water or foreign substances.

Sizing a Submersible

Because a submersible is a high-capacity pump, it follows that the well in which it is installed should be worthy of it. The well should be free of sand, straight, and of sufficient capacity to warrant the installation. Before installing the submersible, the well should be pumped clean of sand and any other foreign matter with a test pump.

Where the water supply is a lake, stream, or spring, an extra screen cage is required to keep leaves and other debris from clogging the screen of the pump. Pumps will operate efficiently in a horizontal position in shallow pools.

In sizing a pump, there are six basic items that must be known:

- Size of well
- Depth of well
- Capacity of well
- Pumping level
- Capacity required by user
- Discharge pressure required

Size of Well. This is very important since most submersibles on the market today require a minimum well diameter of 4 inches. Even in a 4-inch well, the well must be almost perfectly straight.

Depth of Well. This is important. The pump must not be lowered into any mud, sand, or gravel that might be in the bottom of the well. There is a possibility of ruining both the pump and motor if this is done.

Capacity of Well. This is usually the most important factor in selecting a pump. If a well produces only 1 gpm, there is no use putting a pump into it that will pump 15 gpm. There are certain applications where this is done; but there must be some type of control to keep from pumping the well dry.

Pumping Level. This should not be mistaken to mean *static water level.* Pumping level is the lowest level of the water in the well with the pump running. In many cases the pumping level is not known until after the pump is installed, but the driller will usually have a good idea where it is. Pumping level will vary with the capacity of the pump installed.

Capacity Required. The average homeowner doesn't realize how much water he or she uses. But you can estimate your needs according to the 100-gallon-per-day-per-person rule. But always allow for *growth,* such as additions to the family, new water-using appliances, and so on.

Discharge Pressure Required. Low pressure is a common complaint among homeowners who have outdated water systems. With the new appliances on the market now, the required discharge pressure is gradually moving up. As we have seen, pipe sizes used in the plumbing of a house affect the required discharge pressure. Normally, the smaller the plumbing pipe size, the higher the required pressure-switch setting.

If these six items are known, the pump can be easily selected from catalog selection charts. Select a water system with enough capacity to meet both present and future requirements. As already noted, a pump of greater capacity than that of the well should not be installed unless some method of controlling pump flow is used. If the well draws down to the pump inlet, air enters the pump and can result in air binding and vibration that often cause damage to the pump.

Installation of Pumps

The key to success in operating pumps is their proper installation. If you do your job right and the equipment is properly installed, it should perform satisfactorily for a long time. Of course, proper care and maintenance are also essential for continued efficient operation. However, even with the most perfect care and maintenance, you will find it difficult or impossible to overcome faulty installation.

Remember that pumps, especially the centrifugal type, are built in many designs and for different purposes. Study the manufacturer's instruction manual for the equipment you are installing; where specific directions or requirements are furnished, follow them.

Above-Ground Installation

In most cases, the pump should be placed as close as possible to the source of the water or other liquid, so that the suction pipe can be short and direct and the suction lift comparatively low.

The pump should also be placed so that it will be accessible for regular inspection during its operation and that head room (a trap or ceiling opening) is available when it is necessary to use a crane, hoist, or tackle. If possible, a dry place should be selected, and the pump should be protected against the elements or weather.

The foundation of a pump must be substantial enough to absorb vibration and serve as a rigid support for the pump's baseplate. A concrete foundation or a solid base is the most desirable.

In the ideal installation, foundation J-bolts are imbedded in the concrete foundation according to a baseplate template. The bolts should be longer than actually needed, to allow for shimming up the pump to make it level and for grouting under the pump base. A pipe sleeve about $2\frac{1}{2}$ times the diameter of the bolt is used to allow for final positioning. If the bolt shown in example 5–34 were 1 inch in diameter, a $2\frac{1}{2}$-inch pipe sleeve would be used.

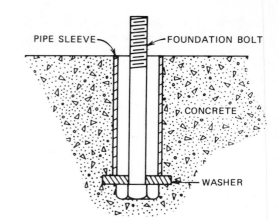

example 5–34. Installing a pump. This illustration shows the pipe sleeve and how it is used to hold down a washer that positions a foundation bolt.

Leveling and Alignment. A small pump is normally aligned and the two major parts bolted together before leaving the factory. The parts will not usually require alignment after the pump has been set on the foundation, but be careful that you don't spring them out of alignment. It is, of course, important that the pump be properly leveled and secured to the foundation. In setting the pump, a spirit level will be required: place the level on the machined surfaces in two directions; it may be necessary to remove the top casing or bearing cover to accomplish this. If a large-sized pump is shipped in sections, you will have to align the water ends with the power ends after they have been placed on the foundation.

In leveling a pump unit, first use small metal wedges (as shown in example 5–35) and then place metal blocks and shims close to the foundation bolts. In each case, space the supports directly under the part carrying the most weight and close enough to give uniform support. Leave a gap of about $\frac{3}{4}$ to 1 inch between the baseplate and the foundation for grouting with cement.

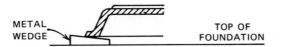

example 5–35. Wedging a baseplate.

Adjust the supports or wedges until the shafts of the pump and the driver are level. By means of a level, check the coupling faces and suction and discharge flanges to see that they are plumb and level. If necessary, correct the position by adjusting the supports or wedges as required.

In addition to checking for parallel alignment, check the angular alignment between the pump shaft and the drive shaft. You can do this by inserting a taper gage or feeler at four points between the coupling faces, as shown in example 5–36. The points should be spaced at 90-degree intervals around the coupling. When the measurements are all alike and the coupling faces are the same distance apart at the four points, the unit will be in angular alignment. Correct any misalignment by adjusting the wedges or shims under the baseplate. Always remember that an adjustment in one direction may disturb adjustments made in another direction.

In all cases, after you have correctly aligned the pump, tighten the foundation bolts evenly, but not too firmly. Then completely fill the baseplace with grout. It is desirable to grout the leveling pieces, shims, or

example 5–36. Angular alignment is very important when attaching an electric motor to a pump.

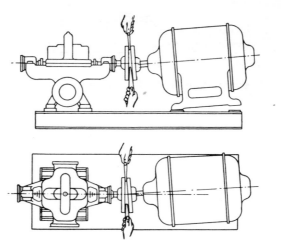

wedges in place. Foundation bolts should not be fully tightened until the grout has hardened, usually about 48 hours after pouring.

After the grout has set and the foundation bolts have been properly tightened, the pump should be checked again for parallel and angular alignment.

Installing the Piping. You are now ready to connect the piping. The pipes must line up naturally. Don't force them into place as this might draw the pump out of alignment. The pipes should also be supported independently so that they will not put any strain on the pump. One or more of these supports should be as near the pump as possible.

The size of the suction or discharge pipe leaving the pump is sometimes increased to reduce the loss of head from friction in the piping. Unnecessary bends should be eliminated from the piping to prevent loss of head. Whenever possible, bends should have a long radius. In order to make removal of the pump as easy as possible, flanged fittings or unions in all connecting lines should be placed close to the pump.

After the piping has been connected, check the pump alignment again. Reconnect the coupling halves and start the pump. Let it warm up thoroughly and operate under normal conditions for a short time; then shut the pump down and immediately check for alignment of the couplings. All alignment checks must be made both with the coupling halves disconnected and again after they are connected.

It is important to check alignment in all directions after making any adjustment. You may find it necessary to readjust the alignment slightly from time to time while the unit and foundation are new.

Submersible Installations

Galvanized steel pipe is generally used for installing submersible pumps; however, if rigid plastic or flexible plastic pipe is used, some method must be used to keep the pump from twisting with starting torque. The starting torque of the motor tends to give a considerable twist, causing the motor shell to rub against the casing. If the well is crooked or rough on the inside, it is possible this rubbing will eventually wear a hole in the motor shell or cause the power cable to rub against the casing, causing a short.

When connecting the first length of pipe and placing the pump in the casing, care should be exercised to avoid misaligning the pump. A wrench should never be applied to the shell of the pump or to the motor casing, but used instead on the top discharge casting. Some installers find it easier to handle the pump if a short piece of pipe is installed first rather than a whole length (example 5–37). In lifting the first length of pipe, use care not to pry on the pump, as the discharge casting may be damaged.

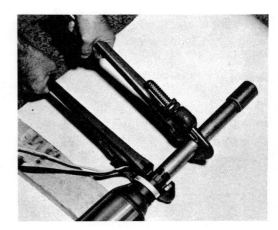

example 5–37. Screw a short pipe and coupling into the discharge casting while holding the casting securely with a wrench. The nipple provides a place for gripping in a clamp vise and allows handling the unit without fear of damage. (Courtesy F. E. Myers Company.)

Use a pump setting rig or tripod to suspend the pump over the well while mounting the drop pipe (example 5–38).

The cable should be taped to the pipe just above the pump and at 10-foot intervals thereafter. This will keep the cable from becoming tangled. Don't let the cable drag over the edge of the well casing, since this may cause the insulation to become cut or chafed. Any rough spots on the casing should be smoothed out with hammer or file. Never pull or lift on the power cable at any time during installation.

Lower the pump into the well slowly without forcing. Use a vise or foot clamp to hold the pipe while connecting the next length. (On deep settings, it is recommended that a check valve be installed in the drop pipe 200 feet above the pump and every 200 feet thereafter to prevent water shock from traveling back to the pump.) Lower the pump to a minimum of 10 feet below the maximum drawdown of the water, if possible, and never closer than 5 feet from the well bottom.

When the pump is at the desired depth, install the throttle valve in the discharge line near the well for a preliminary test run. Wire the pump through the control box following packaged instructions.

Throttle the pump and run until water is clear of sand or any other impurities. Gradually open the throttle valve. If the pump lowers the water in the well to a point at which the pump loses its prime, either lower the pump in the well or throttle it to the capacity of the well. A fluid level control, flow valve, or automatic back-pressure regulator can also be used on low-capacity wells.

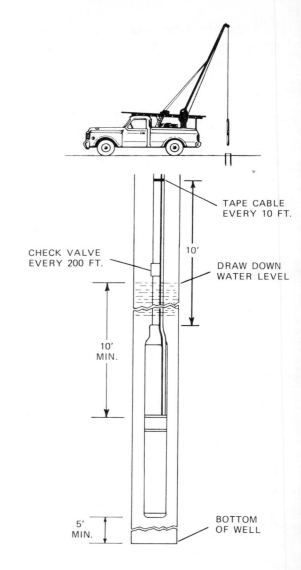

TAPE CABLE EVERY 10 FT.

CHECK VALVE EVERY 200 FT.

10'

DRAW DOWN WATER LEVEL

10' MIN.

5' MIN.

BOTTOM OF WELL

example 5–38. The drop cable should be taped in a straight line to the pipe—never use spiral windings. Tape the cable adjacent to the pump and at intervals of 10 feet or so.

Water-System Servicing Information

Today every pump installer and service worker should have at least three reliable electrical instruments: an ammeter, a voltmeter, and an ohmmeter.

It is not necessary to have high-grade, delicate instruments intended for laboratory use. About the most versatile and convenient of all instruments devised for field work is the clip-on or tong-type combination ammeter-voltmeter. This device permits quick measurement of current in all lines without the need to break connections. Voltage between lines can also be measured at any uninsulated point without breaking connections.

An ohmmeter is also a must. It should have several scales capable of measuring relatively low winding resistance as well as high ground or insulation resistance. Resistance must *never* be measured with the pump power on. All wires of the device being measured must be disconnected from the power source to prevent damage to the ohmmeter and possible shock to the user.

The modern water system of today is driven by an electric motor. It is therefore important to understand some of the basic principles of the electrical circuit. If you are to be the pump installer, you should also know how to use these instruments and interpret their readings.

Electrical Instruments

Each instrument is designed for a specific purpose and each is used in a slightly different way. However, the basic operating principle is the same for all. A knowledge of this principle is helpful in using and understanding these devices.

Example 5–39 shows basic meter construction. The dial is calibrated in the divisions of the electrical quantities it is to indicate—volts, amps, or ohms. An indicating needle is attached to a magnet, called an *armature,* and suspended between two pivot points and held in a normal *zero* position by a spring. A stationary magnet is mounted so that its magnetic field passes through the magnetic field of the armature. As current from the electrical circuit flows through the coil, the magnetic field set up as a result in the armature tends to deflect the armature and needle from the ends of the permanent magnet in proportion to the amount of current flowing.

In order to obtain a range of scale readings, various amounts of resistance can be switched in or out of

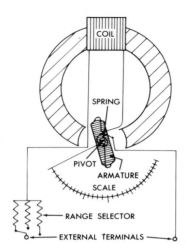

COIL

SPRING

PIVOT

ARMATURE

SCALE

RANGE SELECTOR

EXTERNAL TERMINALS

example 5–39. Arrangement of components in a typical electrical servicing instrument. Current flowing in the armature causes it to deflect in the field of a permanent magnet, and the needle pivots to indicate a value on the scale.

the meter circuit. The ohmmeter has an internal source of voltage in the form of a battery and an adjusting knob to compensate for the battery as it weakens.

Basic System Testing

Understanding an electrical circuit will help you understand the instruments and what they indicate. An electrical circuit may be compared with a hydraulic or water-system circuit. Voltage is electrical "pressure." It is measured between any two wires or between one wire and ground in the same manner that pressure in a pipe is measured between two places in a piping system or from inside the pipe to outside the pipe. It is this electrical pressure that causes current to flow.

Electrical current may be likened to water flowing in a piping system. It is normally measured in *amperes;* an ampere is a quantity of electricity (electrons) that flows past a given point in a given period of time just as gallons per minute represent the amount of water flowing through a piping system in a given period of time.

The *resistance* of an electrical system is measured in *ohms,* and it may be compared to pipe friction in a pipe of a given size.

The relationship of voltage (pressure), current (electrical flow), and resistance (friction) is expressed in *Ohm's law,* which is shown in several ways:

voltage $\quad$ = current X resistance

current $\quad$ = voltage ÷ resistance

resistance = voltage ÷ current

Voltmeter Tests

The *voltmeter* is a device for measuring *electrical pressure* between wires or from wire to ground. Example 5–40 shows a reading being taken at the terminals of a typical submersible motor control box. Note that it is not necessary to have the motor connected or running in order to obtain a reading. In fact, readings should be made both with the motor

example 5–40. Voltmeter reading being taken at the control box of a typical submersible pump. The power can be on or off for this check.

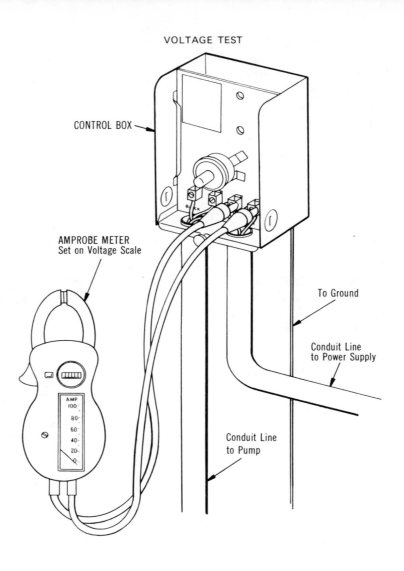

VOLTAGE TEST

CONTROL BOX

AMPROBE METER
Set on Voltage Scale

To Ground

Conduit Line
to Power Supply

Conduit Line
to Pump

AMP
100
80
60
40
20
0

off and with the motor running. A significant difference would indicate an undersized power supply transformer or undersized input wiring.

In making voltage tests, always start with the highest voltage scale setting and work down to a scale that will give a reading between midscale and full scale for best accuracy. Never touch bare leads, as an electric shock could result.

Ammeter Tests

Since the *ammeter* measures *current flow,* the motor must be running in order to get a meter reading. Example 5–41 shows a typical use of tong-type ammeter. For maximum accuracy, follow these steps:

1 Pull the wire being measured away from all other wires.

2 Set the ammeter to the highest scale. (If starting a motor, leave on this scale until the current settles down.)

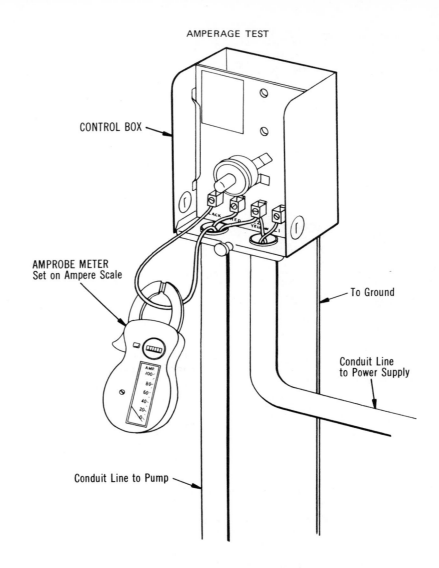

AMPERAGE TEST

CONTROL BOX

AMPROBE METER
Set on Ampere Scale

To Ground

Conduit Line
to Power Supply

Conduit Line to Pump

example 5–41. Ammeter reading taken at control box. The current must be on for this check.

3 Place the tongs of the meter around the wire.

4 Change the meter scale to one that gives the best accuracy. This would be a reading between midscale and full scale.

5 Compare your reading with the manufacturer's data.

Ohmmeter Tests

Ohmmeter tests are always made with power *off* and all wires disconnected from the control box or starter. Always follow these general instructions carefully:

1 Turn off the power and disconnect items to be tested from all other circuits.

2 Turn the selector knob to a multiplier that will give a reading as close to midscale as possible.

3 Clip leads together and adjust the zero-ohms knob until the needle is over the zero. Always ''zero'' the meter before every use and every time the selector switch is changed.

4 Unclip the leads and clip to the wires to be tested.

5 Never allow the bare meter leads or any bare wires of the circuit being tested to touch the ground, water, or any part of the body, especially the fingers. Otherwise, a false reading will result.

Motor Continuity Test. The motor continuity test, for circuit resistance, is used to determine the condition of the motor windings and circuit. Refer to example 5–42 and proceed as follows:

1 Set the selector knob on the lowest scale (RX1) and follow the general instructions listed above.

2 Attach the ohmmeter leads to two of the three motor leads or cable leads at a time and compare the ohmmeter readings with the readings provided by the pump or motor manufacturer.

3 A reading significantly higher than that specified indicates a possible burned (open) winding, loose connection, or wrong motor (different horsepower or voltage rating than that being checked for).

4 A considerably lower reading than the manufacturer's data indicates a possible shorted (burned together) winding, or wrong motor (different horsepower or voltage rating than that being checked).

5 An unbalanced reading for a three-phase motor indicates a burned winding or faulty connection.

6 A correct reading but for a wrong wire color combination indicates improper matching at splice. To correct at the surface, take the following steps:
   a. Ignore the color of the wires and locate the two wires that give the highest ohmmeter reading. Mark the remaining wire ''yellow.''
   b Mark the wire ''black'' that in combination with the ''yellow'' wire (as determined in step a) gives the lowest reading.
   c Mark the remaining wire ''red.''
   (Note: Colors are in line with NEMA standards for submersible pump cable.)

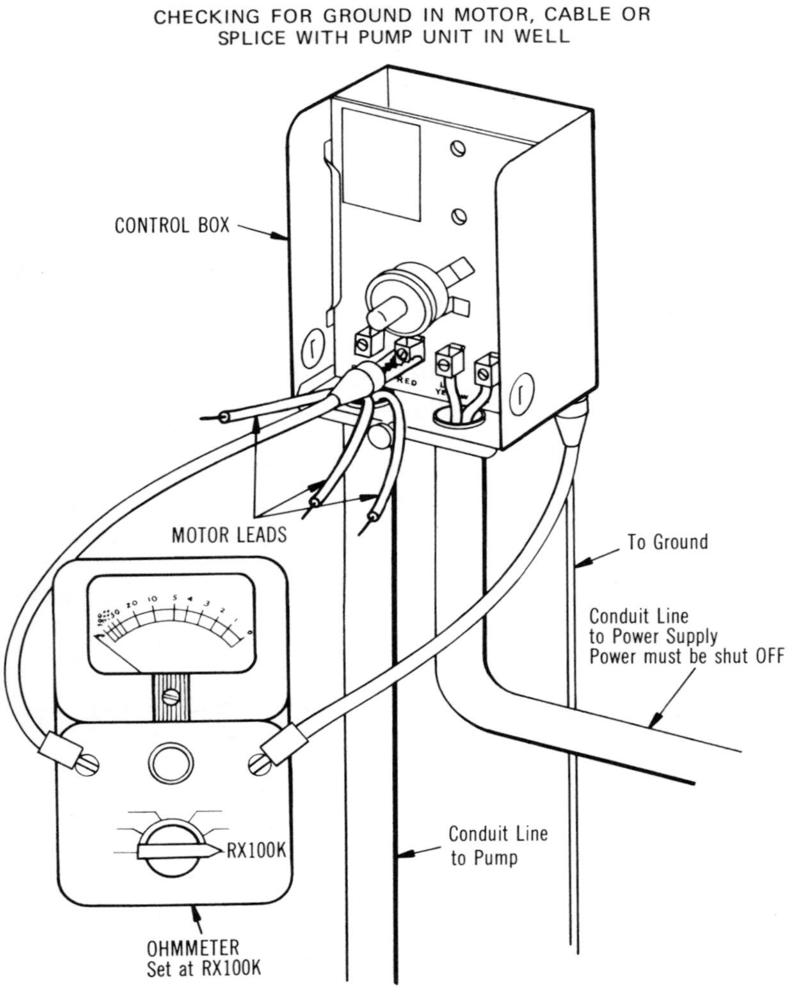

CHECKING FOR GROUND IN MOTOR, CABLE OR
SPLICE WITH PUMP UNIT IN WELL

CONTROL BOX

MOTOR LEADS

To Ground

Conduit Line
to Power Supply
Power must be shut OFF

RX100K

Conduit Line
to Pump

OHMMETER
Set at RX100K

Insulation Test. To obtain accurate readings, the supply must be turned off and the motor leads disconnected from control box. Set the ohmmeter selector knob on RX100K and adjust the needle to zero after clipping the leads together. After the ohmmeter is adjusted, clip one lead to the well casing or discharge pipe and the other lead to each motor lead wire individually, as shown in example 5–43. If, after touching any one of the three motor leads, the needle deflects at all, the pump should be pulled to determine which is causing the connection—motor, cable, or splice.

example 5–42. Motor continuity testing. Set the scale at R X 1 and connect leads to any two cable wires.

After the pump-motor unit is pulled out, cut the cable off on the motor side of the splice and check the motor separately. The check is the same as in the preceding list except that one ohmmeter lead is clipped to some part of the pump-motor unit and the other to each lead separately. If the needle deflects all the way to zero, the unit must be replaced. If the motor is not grounded, check the cable and splice.

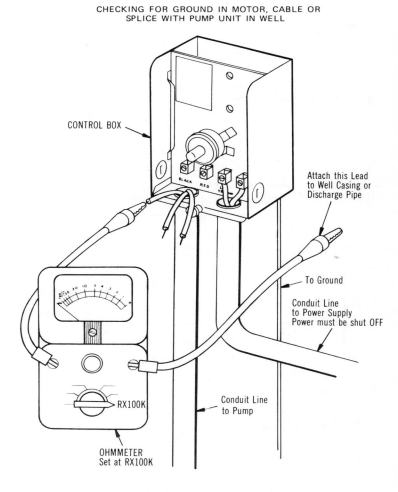

CHECKING FOR GROUND IN MOTOR, CABLE OR SPLICE WITH PUMP UNIT IN WELL

CONTROL BOX

BLACK RED

Attach this Lead to Well Casing or Discharge Pipe

To Ground

Conduit Line to Power Supply
Power must be shut OFF

Conduit Line to Pump

RX100K

OHMMETER
Set at RX100K

example 5–43. You can test insulation with an ohmmeter, but the results aren't always fully reliable, particularly if the insulation has high leakage without direct shorting. Set the meter for R X 100K (or higher, if the range permits). Connect one lead to any cable wire and the other to the metal-wall casing (at ground potential). Check each wire in the cable similarly; none of them should cause the meter to indicate at all.

Ground Leakage. Submerge the cable and splice in the steel barrel of water with both ends out of the water (example 5–44).

Set the ohmmeter selector knob on RX100K and adjust the needle to zero by clipping the ohmmeter leads together.

If the needle deflects to zero on any of the cable leads, pull the splice up out of the water. If the needle falls back to "no reading," the leak is in the splice. If leak is not in the splice, pull the cable out of the water slowly until needle falls back to no reading (indicating "infinity"). When the needle falls back, the leak is at that point.

If the cable or splice is bad, it should be repaired with a cable splice kit.

Vacuum Gages

A centrifugal or jet pump in good operating condition should be capable of pulling a vacuum up to 25 inches of mercury (Hg), which is equivalent to 28 feet of water. But because capacity decreases quite rapidly with an increase in lift, these pumps are normally not operated beyond 22 in. Hg, or 25 feet of water. Conversely,

example 5–44. To locate an existing cable leak, ground one ohmmeter lead and clip the other to a cable wire. Then gradually withdraw the cable from the water and watch the meter. When the resistance jumps to infinity, the leak is immediately above the water level. (This, like other tests, is not a 100-percent reliable indicator; it depends to a large extent on the conductivity of the water.)

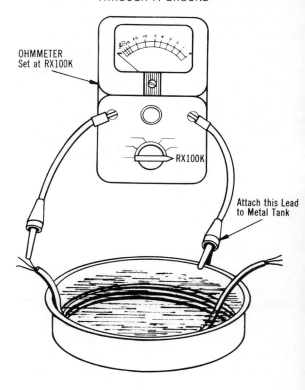

CABLE AND SPLICE TEST FOR LEAKS THROUGH A GROUND

OHMMETER Set at RX100K

RX100K

Attach this Lead to Metal Tank

the more water the pump is handling, the lower the vacuum reading will normally be in the suction pipe.

Because there are always conditions where the suction line will be subject to pressure, a combination pressure and vacuum gage (called a *compound gage*) should always be used to avoid gage damage.

Most vacuum gages are calibrated in inches of mercury. This can be converted to feet of water by multiplying by 1.13, or for rough estimates, reading the gage directly as feet of water.

Vacuum-Gage Installation. When installing a vacuum gage, make certain that it is located ahead of the ejector (example 5–45). Some units have a tapped opening for attaching the air-volume control, located between the ejector and the impeller. Installed here, the gage does not show total suction lift. A vacuum gage located at the pump suction will register the total suction lift at which the pump is operating. This includes the vertical distance to the water as well as friction loss in the foot valve, strainer, pipe, and fittings. When the

example 5–45. In a shallow-well installation, locate the vacuum gage ahead of the ejector, as shown here.

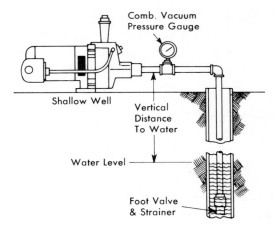

Comb. Vacuum Pressure Gauge

Shallow Well

Vertical Distance To Water

Water Level

Foot Valve & Strainer

pump is stopped, the gage will record storage tank pressure if the line is equipped with a foot valve. If not, the gage will record the vertical distance to the water level.

Vacuum-Gage Readings. Most shallow-well performance problems center around a loss of capacity. In general, low capacity coupled with a higher vacuum-gage reading than is warranted by the estimated water level indicates a restriction in the suction pipe. This may result from a plugged strainer or foot valve or excessive friction loss. It could also indicate a water level lower than estimated.

Low capacity accompanied by a lower vacuum reading than warranted would indicate either an air leak in the suction line or excessive impeller wear-ring clearance. A compound vacuum gage installed in a deep-well system as shown in example 5–46 will indicate some of the pumping problems encountered with deep-well jet pumps. Because part of the pump capacity is recirculated through the pressure pipe back to the ejector installed in the well, it is not possible to determine the water level with the gage reading. In fact, it is possible to

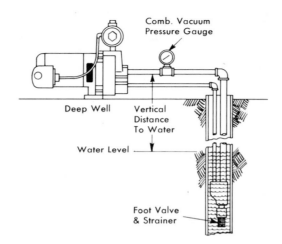

example 5–46. Compound vacuum gage installed on deep-well jet pump is located on the suction line, as shown here.

develop a positive pressure by changing the setting of the regulating valve. Some manufacturers indicate the most efficient point of operation of their jet pumps at a given suction-line value, and the regulating valve can be set using the gage and this information. In general, this value would range from 18 to 25 feet of suction at the pump as indicated by the gage.

The compound vacuum gage can also be used to determine the amount of suction in the line available for operating the air-volume control. Most vacuum-operated AVCs require a minimum of 5 to 6 inches of mercury during pump operation and a zero or slight pressure reading when the pump is stopped. Problems are often encountered with shallow-well installations where the water level is near the surface. A gage connected to the AVC tapping in the pump case will determine if conditions permit operation of the control.

Pump Maintenance

There is no pump existing that is designed to be put into service and forgotten forever. Every pump requires periodic care. But some pumps are inherently more reliable than others. Let's examine some of the more common types of pumps to see what special problems each presents when employed in an operational water supply system. Bear in mind, however, that each pump manufacturer will make specific recommendations regarding the upkeep of its own products. These manufacturer-originated maintenance procedures should take precedence over the general guidelines given here.

Rotary Pumps

The "weak spot" in a rotary pump is the clearance space between rotating parts; therefore, pump makers usually recommend periodic inspection to check for pump slippage. Any time you have to dismantle a pump of this class, be sure to check all clearances against the maker's specifications.

In operation, the rotary pump is subject to "hydraulicking," or operation against the head buildup generated by water turbulence. The discharge valve must be in the open position before the pump is started. And even though the rotary pump is considered to be self-priming, it's a good idea to prime it before use—particularly when the pump has been standing idle for a long period or after being put back into service following a repair. Remember that the rotating elements of a rotary pump are lubricated by the water being pumped; it follows that if the pump end is filled with water before starting, then unnecessary friction and wear of the rotating parts are minimized.

Centrifugal Pumps

Centrifugal pumps are typically fitted with stuffing boxes and various types of bearings that will require occasional inspection. Unlike positive displacement pumps, in the centrifugal pump, the discharge stop valve must be in the closed position before starting. This allows the pump to work against the sealed discharge and build up an effective pressure head before distributing the water downstream. After the pump is up to its rated speed and the discharge valve is open, the pump will continue to maintain that pressure head unless the operating conditions are changed.

There is no problem of hydraulicking with this type of pump; in fact, if the pump were to continue operation with the discharge sealed, it would simply build up to its maximum discharge pressure and then start churning the water. The discharge pressure would overcome the suction pressure and the water would then slip back to the suction side of the pump. Eventually the pump would begin to build up heat because the water would not be able to carry it away; but the pump would not "self-destruct" in the process.

When other pumps are operating in parallel with the centrifugal pump, the common discharge of the other pumps will provide sufficient head to work against in starting. And of course the turbine well pump has a built-in pressure head to work against by virtue of the static pressure of the water above the impellers. So these pumps can be started satisfactorily with the discharge valve in the open position.

Like many other types of pumps, centrifugals require occasional *packing,* or the sealing of moving machinery parts, such as joints, pistons, shafts, valve stems, and the like, against water encroachment. You can think of *packing* as a close-fitting bearing that has to prevent leakage without binding or causing excessive frictional wear on any moving part. Packing generally has lubricating qualities of its own, but this self-lubricating quality is typically enhanced by some of the water being permitted, in controlled quantities, to seep past the seal.

Example 5–47 is an illustrated 14-step packing routine that should help guide you through the process with any pump where it's required. The packing material takes the form of coils, rings, spirals, and the like. What you have to do is insert the packing into a stuffing box that is usually fitted around the sliding or rotating joint. The compressions of the packing around the joint (and the leakage rate) are controlled by adjusting gland nuts.

Unfortunately, there isn't one universal packing applicable to all pumps in all situations. The packing that you'll have to use will depend on the type of joint (sliding or rotating), plus any special requirements imposed by the manufacturer of the pump you're servicing.

Adjustment of the gland nuts should be done with considerable care to insure that compression is even all the way around the joint. If the gland pressure is greater at one point than it is at another, excessive friction will occur at one spot and the packing will wear out prematurely; even worse, the rotating or sliding shaft will likely become scored. And when this happens, you'll have to replace or refinish the shaft or install a new pump, neither of which is an attractive alternative.

It's a good idea to run the pump initially (after stuffing) where you can touch the packing box every few minutes or so during initial operation to check for heat buildup. When the pump first starts, lubrication of the packing might be relatively poor. This initial friction could make the joint heat up excessively; the resulting expansion of the compressible material could cut off the controlled back-seepage of water, thus further compromising the lubrication. You can't always solve the problem by backing off the gland nut, either, since the liquid pressure in the pump can force the packing to move outward as a unit. It's better to shut down the pump if packing-box overheating occurs; after cooldown, you can restart the pump. You might have to start the pump and let the packing box cool several times before operation is satisfactory.

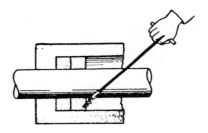

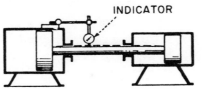

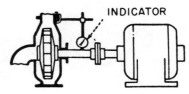

1 Remove *all* old packing. Aim packing hook at bore of the box to keep from scratching the shaft. Clean box thoroughly so the new packing won't hang up

2 Check for bent rod, grooves or shoulders. If the neck bushing clearance in bottom of box is great, use stiffer bottom ring or replace the neck bushing

3 Revolve rotary shaft. If the indicator runs out over 0.003-in., straighten shaft, or check bearings, or balance rotor. Gyrating shaft beats out packing

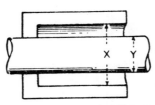

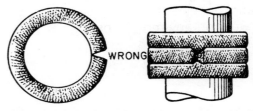

PIPE

4 To find the right size of packing to install, measure stuffing-box bore and subtract rod diameter, divide by 2. Packing is too critical for guesswork.

5 Wind packing, needed for filling stuffing box, snugly around rod (for some size shaft held in vise) and cut through each turn while coiled, as shown. If the packing is slightly too large, never flatten with a hammer. Place each turn on a clean newspaper and then roll out with pipe as you would with a rolling pin

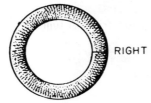

RIGHT

WRONG

example 5–47. Installing packing in a centrifugal pump. The process is similar for other pump types.

6 Cutting off rings while packing is wrapped around shaft will give you rings with parallel ends. This is very important if packing is to do job

7 If you cut packing while stretched out straight, the ends will be at an angle. With gap at angle, packing on either side squeezes into top of gap and ring, cannot close. This brings up the question about gap for expansion. Most packings need none. Channel-type packing with lead core may need slight gap for expansion

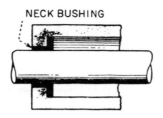

NECK BUSHING

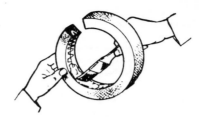

8 Install foil-wrapped packing so edges on inside will face direction of shaft rotation. This is a must; otherwise, thin edges flake off, reduce packing life

9 Neck bushing slides into stuffing box. Quick way to make it is to pour soft bearing metal into tin can, turn and bore for sliding fit into place

10 Swabbing new metallic packings with lubricant supplied by packing maker is OK. These include foil types, lead-core, etc. If the rod is oily, don't swab it

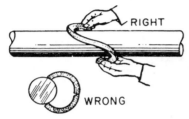

RIGHT

WRONG

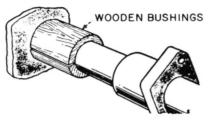

WOODEN BUSHINGS

11 Open ring joint sidewise, especially lead-filled and metallic types. This prevents distorting molded circumference—breaking the ring opposite gap

12 Use split wooden bushing. Install first turn of packing, then force into bottom of box by tightening gland against bushing. Seat each turn this way

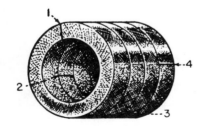

LANTERN RING   GLAND

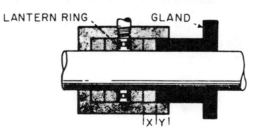

X Y

13 Stagger joints 180 degrees if only two rings are in stuffing box. Space at 120 degrees for three rings, or 90 degrees if four rings or more are in set

14 Install packing so lantern ring lines up with cooling-liquid opening. Also, remember that this ring moves back into box as packing is compressed. Leave space for gland to enter as shown. Tighten gland with wrench—back off finger-tight. Allow the packing to leak until it seats itself, then allow a slight operating leakage.

Airlift Pumps

The inherent simplicity of the airlift pump is reflected in the fact that the only maintenance involved is with the associated air compressor. The most important consideration as far as the pump is concerned is the regulation of compressed air. The amount of air used should be the minimum required to produce a continuous flow of water. Too little air results in water being discharged either not at all or in spurts. Too much air causes an increase in the volume of discharge, but it decreases the output pressure. Even the discharge volume drops off if the amount of air is increased further.

Jet Pumps

The jet pump is very probably the most popular type in current use for domestic water systems. Understanding the functions of all components of this system will enable you to deal with troubles in a logical and systematic way; if you do not fully understand the operation of this type of pump, reread the appropriate subsection before beginning any service. The water system that uses a jet pump can be divided into five main component groups: the centrifugal pump, the jet (ejector or eductor), the air controls, the tank, and the electrical group (including the motor, related controls, and pressure switch). Functional details of these components are given in the preceding text.

The five charts of Table 5–2 should provide you with some basic troubleshooting guidelines that will aid in solving most of the problems you're apt to run into with jet pumps. Chart A lists probable causes and correctional steps to take when the pump won't start or run; chart B lists common sources of trouble resulting in motor overheating; chart C will help you determine the cause of a problem that results in too-frequent pump starting and stopping; chart D tells how to handle the situation when the pump won't shut off when it should; and chart E covers most of the difficulties that could result in insufficient water delivery.

TABLE 5–2. JET-PUMP TROUBLESHOOTING

Chart A—Jet Pump Won't Start or Run.

CAUSE	CHECKOUT PROCEDURE	CORRECTION
Blown fuse.	Check condition of fuse.	Replace with proper fuse.
Low line voltage.	Use voltmeter to check pressure switch or terminals nearest pump.	Check size of wires from main switch on property; if okay, contact power company.
Loose, broken, or incorrect wiring.	Check wiring against diagram. Make sure all connections are tight and no short circuits exist (due to worn insulation, etc.).	Rewire incorrect circuits, tighten connections, replace defective wires.
Defective motor.	Make sure switch is closed.	Perform motor repair.
Defective pressure switch.	Check switch setting; examine contacts for dirt or wear.	Adjust switch settings; clean contacts with emery cloth.
Plugged pressure switch tubing.	Remove and blow out tubing.	Clean or replace.
Binding impeller.	Turn off power; use screwdriver to turn motor or impeller.	Remove housing to locate source of binding.
Defective starting capacitor.	Check resistance across capacitor. Infinite resistance means open capacitor; no resistance means shorted capacitor.	Replace capacitor or take motor to service station.
Motor shorted.	If fuse blows when pump is started—and external wiring is okay —motor is shorted.	Replace motor.

Chart B—Motor Overheats and Overload Breaker Trips.

CAUSE	CHECKOUT PROCEDURE	CORRECTION
Incorrect line voltage.	Check voltage at pressure switch or terminals nearest pump.	If voltage is low, check wiring size from main switch on property; if okay, contact power company.
Motor wired incorrectly.	Check wiring diagram.	Reconnect for proper voltage according to diagram.
Inadequate ventilation.	Check air temperature in pump location.	Ventilate or move pump.
Prolonged low pressure delivery.	Continuous operation at at very low pressure can cause overload breaker to trip.	Install globe valve on discharge line and throttle to increase pressure.

Chart C—Pump Starts and Stops too often.

CAUSE	CHECKOUT PROCEDURE	CORRECTION
Leak in pressure tank.	Apply soapy water to entire surface above waterline; bubbles indicate a leak.	Stop leaks or replace tank.
Defective air volume control.	(This will lead to a waterlogged tank.) Make sure control works properly; if it doesn't, remove and examine for obstruction.	Clean or replace control.

Chart C—Pump Starts and Stops too often. (continued)

CAUSE	CHECKOUT PROCEDURE	CORRECTION
Faulty pressure switch.	Check switch setting. Examine contacts for dirt or wear.	Adjust switch settings. Clean dirty contacts with emery cloth.
Leak on suction side of system.	On shallow-well units, install pressure gage on suction side. On deep-well systems, at attach pressure gage to pump. Close discharge line valve and, using bicycle pump or compressor, apply 30 psi pressure to system; if system will not maintain pressure with compressor off, there is a leak.	Make sure above-ground connections are tight. Repeat test; if necessary, pull piping and repair leak.
Leak in foot valve.	Pull piping and examine foot valve.	Repair or replace valve.
Loss of precharge (in float tanks).	Check water level in tank by sweat line or feel the tank (the part is water). If waterline is above $\frac{2}{3}$ tank height, recharge with air.	Add air to tank according manufacturer's instructions.

Chart D—Pump Won't Shut Off.

CAUSE	CHECKOUT PROCEDURE	CORRECTION
Incorrect pressure-switch setting or setting has drifted.	Lower switch setting; if pump shuts off, setting was incorrect.	Adjust switch to proper setting.
Defective pressure switch.	Examine points and other switch parts for defects (arcing may have caused switch contacts to fuse).	Replace defective switch.
Plugged pressure switch tubing.	Remove and blow out tubing.	Clean or replace.
Loss of prime.	If no water is delivered, check pump prime and well piping.	Reprime as necessary.
Low well level.	Check well depth against pump performance table to make sure pump and ejector are the right size.	Replace pump or ejector with proper-size components.
Plugged ejector.	Remove and inspect ejector.	Clean and reinstall.

Chart E—Pump Operates with Inadequate Water Delivery.

CAUSE	CHECKOUT PROCEDURE	CORRECTION
System incompletely primed.	When no water is delivered, check pump prime and well piping.	Reprime as necessary.
Air lock in suction line.	Check horizontal piping between well and pump. If it doesn't pitch upward from well to pump, an air lock may form.	Rearrange piping to eliminate air lock.
Undersized piping.	If system delivery is low, the piping or plumbing lines may be too small. Recompute friction loss.	Replace piping or install larger pump.
Leak in air volume control or tubing.	Disconnect air-volume control tubing at pump and plug hole. If capacity increases, there is a leak in the control tubing.	Tighten fittings; replace control as necessary.
Pressure regulating valve stuck or set incorrectly (in deep wells).	Inspect valve for defects and check setting.	Reset, clean, or replace.
Leak on suction side of system.	On shallow-well units, install pressure gage on suction side. On deep-well systems, attach pressure gage to pump. Close discharge line valve and, using bicycle pump or compressor, apply 30 psi pressure to system; if system will not maintain pressure with compressor off, there is a leak.	Make sure above-ground connections are tight. Repeat test; if necessary, pull piping and repair leak.

Chart E—Pump Operates with Inadequate Water Delivery. (continued)

CAUSE	CHECKOUT PROCEDURE	CORRECTION
Low well level.	Check well depth against pump-performance table to make sure pump and ejector are the right size.	Replace pump or ejector with proper sized components.
Wrong pump/ejector combination.	Check pump and ejector models against manufacturer's performance tables.	Replace ejector with correct model.
Low well capacity.	Shut off pump and allow well to recover. Restart pump and note if delivery drops after continuous operation.	If well is weak, lower ejector (deep well pumps), use a tail pipe ((deep well pumps), or switch from shallow to deep well equipment.
Plugged ejector.	Remove and inspect injector.	Clean and reinstall.
Defective or plugged foot valve or strainer.	Pull and inspect foot valve. Partial clogging will result in lack of water flow. A defective foot valve may cause pump to lose prime, resulting in no delivery.	Clean, repair, or replace.
Worn or defective pump parts or plugged impeller.	Low delivery may result from worn impeller or other worn pump parts. Disassemble and inspect.	Replace worn parts or entire pump; clean parts as necessary.

Reciprocating Pumps

Pumps of this class have been steadily losing popularity in recent years, being supplanted with simpler, more reliable units of other classes. There are so many variations of the reciprocating pump that maintenance suggestions must be considered as being only general in nature. The most common service problems are listed in seven troubleshooting charts, designated alphabetically in Table 5–3. Chart A describes the possible sources of failure when the pump won't start; chart B suggests ways to cure the problem of an operating pump that delivers no water; chart C describes various causes of low capacity; chart D lists possible causes and corrective steps to take when the pump repeatedly loses its prime; chart E is a symptom-and-cure listing for too-frequent pump cycling; chart F should be consulted when the pump refuses to shut off; and chart G lists some probable causes for excessive noise and vibration.

TABLE 5–3. RECIPROCATING-PUMP TROUBLESHOOTING
Chart A—Pump Won't Start or Run.

CAUSE	CHECKOUT PROCEDURE	CORRECTION
Blown fuse.	Check fuse.	Replace as necessary.
Low line voltage.	With motor energized, check voltage at pressure switch terminals nearest pump.	If voltage is low, check wiring size from main switch on property; if okay, contact power company.
Loose, broken, or incorrect wiring.	Check wiring against circuit diagram. Make sure there are no loose connections or short circuits due to worn insulation, crossed wires, and so on.	Rewire incorrect circuits, tighten connections, replace defective wires.
Defective pressure switch.	Check switch setting; examine contacts for dirt or wear.	Adjust settings; clean contacts with emery cloth.
Plugged pressure-switch tubing.	Remove and blow out tubing.	Clean or replace.
Pump binding.	Turn off power, turn pump by hand.	Locate source of binding and repair.
Defective starting capacitor.	Disconnect capacitor. Check resistance across capacitor; Infinite resistance indicates open capacitor; no resistance indicates shorted capacitor.	Replace capacitor or take motor to service station.
Motor shorted.	If fuse blows when pump is started—and external wiring is okay—motor is shorted.	Replace motor.
Overload protection cut out.	Check manual reset.	Correct cause of overload; reset breaker.

Chart B—Pump Runs, but Doesn't Deliver Water.

CAUSE	CHECKOUT PROCEDURE	CORRECTION
Low line voltage.	With pump motor energized, check voltage at pressure-switch terminals nearest pump.	If voltage is low, check size of wiring from main switch on property; if okay, contact power company.
Loss of prime (piston pumps).	Check pump prime and well piping.	Reprime as necessary.
Broken rod (working heads).	If rod can be lifted easily after being disconnected, it is broken.	If rod is broken, remove upper part and fish for lower part, or pull drop pipe; repair break.
Low well level.	On piston pumps, make sure water level is no more than 25 feet below pump (less if at high altitude). On working heads, lower well cylinder by adding more drop pipe and rod; if this results in water delivery, cylinder was above water level.	If water level is below 25 feet, piston pump won't work. (Replace with submersible or deep-well jet pump.) Leave cylinder at lower level.
Air lock in suction line (piston pumps).	Check horizontal piping between well and pump. If it doesn't pitch upward, an air lock may form.	Rearrange piping to eliminate air lock.
Suction valves stuck open (piston pumps).	Remove cover from water end of pump and inspect valves.	Clean out foreing matter between valves and valve plate; insure a watertight closure.

Chart B—Pump Runs, but Doesn't Deliver Water. (continued)

CAUSE	CHECKOUT PROCEDURE	CORRECTION
Leak in suction line (piston-pumps or drop-pipe working heads).	On piston pumps, install a vacuum gage on the suction side and start pump; low vacuum means leaky suction line.	Repair leaks.
	On working heads, attach a pressure gage to the discharge line, upstream of the main valve; close the valve. Using a bicycle pump or compressor, apply 30 psi pressure to the system; loss of pressure indicates leak in drop pipe.	Repair leaks.
Open foot valve (piston pumps) or cylinder check valve (working heads).	Fill drop pipe or suction line with water. If water level drops, the lower valve may be defective. Pull drop pipe and inspect foot valve (piston pumps) or cylinder check valve (working heads).	Clean, repair, or replace.
Clogged drive point.	If drive point was used in well, pull suction line and inspect.	Clean and reinstall drive point.

Chart C—Low Capacity.

CAUSE	CHECKOUT PROCEDURE	CORRECTION
Undersized piping.	If system discharge is low, discharge piping or plumbing lines may be too small; recompute friction loss.	Replace piping or install larger pump.
Low well capacity.	Stop pump and allow well to recover. Restart pump and note if delivery drops after continuous operation. (On working head units, low well capacity is indicated by a violent jarring of the drop pipe after continuous operation, meaning air is entering the well cylinder.)	Lower suction line or well cylinder to permit greater drawdown; if this can't be done, throttle discharge line so delivery matches well recovery time.
Leaky relief valve (piston pumps).	Examine internal relief valve for defects.	Repair or replace.
Worn parts.	On piston pumps, examine valves, valve plates, plunger leathers, and gaskets. On working-head units, pull rod and examine leathers.	Replace worn or defective parts.

Chart D—Pump Loses Prime (Piston Types).

CAUSE	CHECKOUT PROCEDURE	CORRECTION
Suction valve stuck open.	Remove cover from water end of pump and inspect valves.	Clean out foreign matter between valves and valve plate; insure watertight closure.
Leak in suction line.	On piston pumps, install a vacuum gage on the suction side and start pump. Low vacuum means a leaky suction line.	Repair leaks.
	On working heads, attach a pressure gage to the discharge line, upstream of the main valve. Using a bicycle pump or compressor, apply 30 psi pressure to the system; loss of pressure indicates a leak in the drop line.	Repair leaks.
Defective relief valve.	Examine internal relief valve for defects.	Repair or replace leaky valve.

Chart E—Pump Starts and Stops too often.

CAUSE	CHECKOUT PROCEDURE	CORRECTION
Leak in pressure tank.	Apply soap solution to entire surface above waterline; bubbles indicate leak in tank.	Repair leaks or replace tank.
Defective air-volume control.	This will lead to a waterlogged tank. Make sure control is operating properly; if it isn't, examine for obstruction.	Clean or replace defective control.
Fault in pressure switch.	Remove tubing. Determine if switch is jet or reciprocating type and inspect opening for restrictor. Check pressure differential.	Insert restrictor (or replace switch if jet type). Adjust pressure differential to 20 psi.
Leak on discharge side of system.	Make sure all plumbing fixtures are shut off, then check all units (especially ballcocks) for leaks; listen for running water.	Repair leaks.
Leak in suction line (piston pumps) or drop pipe (working heads).	On piston pumps, install a vacuum gage on the suction side and start pump; low vacuum means a leaky suction line. On working heads, attach a pressure gage to the discharge line. Using a bicycle pump or compressor, apply 30 psi pressure to the system; loss of pressure indicates a leak in the drop pipe.	Repair leaks. Repair leaks.
Leak in foot valve (piston pumps).	Fill drop pipe or suction line with water. If the water level drops, the lower valve may be defective. Pull drop pipe and inspect foot valve (piston pumps) or cylinder check valve (working heads).	Clean, repair, or replace.

Chart F—Pump Won't Shut Off.

CAUSE	CHECKOUT PROCEDURE	CORRECTION
Wrong pressure-switch setting or setting has drifted.	Lower switch setting. If pump shuts off, this was the problem.	Adjust switch to proper setting.
Defective pressure switch.	Examine points and other switch parts for defects (arcing may have caused switch contacts to fuse).	Replace defective switch.
Plugged pressure switch tubing.	Remove and blow out tubing.	Clean or replace.
Loss of prime (piston pumps).	When no water is delivered, check pump prime and well piping.	Reprime as necessary.
Low well level.	On piston pumps, make sure water level is no more than 25 feet below pump (less if at high altitude). On working heads, lower well cylinder by adding more drop pipe and rod. If this results in water delivery, cylinder was above water level.	If water level is below 25 feet, piston pump won't work. (Replace with submersible or deep-well jet pump.) Leave cylinder at lower level.

Chart G—Excessive Noise.

CAUSE	CHECKOUT PROCEDURE	CORRECTION
Waterlogged tank or air chamber.	Remove and examine control for plugging that might be causing improper operation.	Clean or replace defective control.
Undersized suction line (piston pumps).	Check manufacturer's recommendations for suction line size.	Replace with larger line as necessary.
Suction valves sticking (piston pumps).	Remove cover from water end of pump and inspect valves.	Clean out foreign matter between valves and valve plate.
Rod slapping against drop pipe (working heads).	Check rod for play (especially if made of steel).	Install rod guides at 10-foot intervals or replace steel rod with wood rod.
Low well level.	On piston pumps, make sure water level is no more than 25 feet below pump (less if at high altitude). On working heads, lower well cylinder by adding more drop pipe and rod. If this results in water delivery, cylinder was above water level.	If water level is below 25 feet, piston pump won't work. (Replace with submersible or deep-well jet pump.) Leave cylinder at lower level.
Cylinder valves sticking or noisy (working heads).	Pull cylinder and examine valves.	If valves operate sluggishly, install cylinder with spring-loaded valves. (If noise is the problem, use cylinder rubber-faced valves.)

Submersible Pumps

The submersible pump has, in the past few years, gained in acceptance. Simplicity of operation, ease of installation, and improved electrical and mechanical design have contributed much to its popularity. In order to gain this acceptance, the manufacturers have had to do much educational work. It was necessary to teach installers how to handle and service the electrical parts of this system. Installers were also required to purchase and learn the use of such electrical instruments as the ohmmeter, ammeter, and voltmeter, which are necessary to help locate sources of trouble and prevent the unnecessary pulling of a satisfactory unit. The proper use of these instruments and an adequate knowledge of the function of electrical components have overcome much of the fear of installing and servicing this equipment.

Once having mastered the instruments and the procedure for checking a unit, you have only to follow the procedure outlined in the six charts of Table 5–4, which cover most of the submersible service work. You can do most of the checking without touching the pump itself. It is always desirable to make all the above-ground checks before pulling the pump from the well.

Because much of the checkout work on submersibles involves electrical tests, it is essential that you be thoroughly familiar with the electrical components of the water system. But the tests are not complicated, and manufacturers' service literature generally includes explicit instructions for making electrical tests.

Each of the six charts in Table 5–4 outlines the procedure for tracing down the sources of major operating trouble. Together, the six charts cover 95 percent or more of all submersible service work. And, as noted, you can do most of this checking without touching the pump.

TABLE 5-4. SUBMERSIBLE-PUMP TROUBLESHOOTING

Chart A—Fuses Blow or Circuit Breaker Trips When Motor Is Started.

CAUSE	CHECKOUT PROCEDURE	CORRECTION
Incorrect line voltage.	Check line voltage terminals in the control box (or connection box in 2-wire models); make sure voltage is within range specified by manufacturer.	If line voltage is incorrect, contact power company.
Defective control-box wiring.	Check all motor and power-line wiring in control box, using attached wiring diagram. Look for loose connections, worn insulation, crossed wires, and so on.	Rewire incorrect circuits; tighten loose connections, replace worn wires.
Incorrect control-box components.	Check all control box components; they should be the size and type specified by the pump manufacturer.	Use correct control components as specified by the pump manufacturer.
Defective control-box starting capacitor. (Not applicable to 2-wire models.)	Check resistance across capacitor. Infinite resistance means open capacitor or defective relay points; no resistance means capacitor is shorted.	Replace defective capacitor.
Defective control box relay. (Not applicable to 2-wire models.)	Check relay coil for continuity; its resistance should be that specified by the manufacturer.	Replace relay if coil resistance is incorrect or points are defective.

Chart A—Fuses Blow or Circuit Breaker Trips When Motor Is Started.
(continued)

CAUSE	CHECKOUT PROCEDURE	CORRECTION
Defective pressure switch.	Check voltage across pressure-switch points; if less than line voltage determined as first entry above, the switch points are causing low voltage due to improper contact.	Clean points with an emery cloth or replace switch.
Pump in crooked well.	If wedged in a crooked well, the motor and pump may become misaligned, resulting in a locked rotor.	If pump doesn't move freely, pull the pump and straighten the well or install a jet pump).
Shorted or open motor winding.	Check continuity of motor winding using manufacturer's wiring diagram. If the resistance is below that specified, the motor winding may be shorted; if infinite resistance is shown, motor winding is open.	If motor winding is defective, pull and repair motor.
Grounded motor cable or winding.	Ground one lead of an ohmmeter to the drop pipe or or shell (casing), then touch the other lead to each motor wire terminal. If high resistance is shown, the cable or winding is grounded.	Pull the pump and inspect the cable for damage; replace damaged cable. If cable is okay, winding is grounded.

Chart A—Fuses Blow or Circuit Breaker Trips When Motor Is Started.
(continued)

CAUSE	CHECKOUT PROCEDURE	CORRECTION
Tight motor.	Check current drawn by motor. The reading should not be higher than that recommended.	Pull pump; repair or replace motor.
Sand-locked pump.	Test as above. The reading should not be higher than that recommended.	Pull, disassemble, and clean pump. Allow sand to settle in well before replacing. (A submersible pump should not be used in an extremely sandy well.)

Chart B—Pump Operates with Inadequate Water Delivery.

CAUSE	CHECKOUT PROCEDURE	CORRECTION
Pump air-locked.	Stop and start pump several times, waiting a minute between cycles. If pump resumes normal delivery, it was air-locked.	If the test fails to correct problem, go on to next entry.
Well water level too low.	Well production may be too low for pump capacity. Restrict pump output, allow well to recover, and start pump.	If partial restriction corrects problem, leave valve or cock at new setting; otherwise, lower pump in well if depth is sufficient (do not lower if sand clogging can occur).
Discharge-line check valve installed backward.	Examine check valve to make sure flow-direction arrow is pointed correctly.	Reverse valve as necessary.
Leak in drop pipe.	Raise and examine pipe for leaks.	Replace damaged pipe section.
Pump check valve jammed by drop pipe.	After examining drop pipe for leaks, pull pump and examine drop-pipe connection to pump outlet. If drop pipe has been screwed in too far, it may be jamming check valve closed.	Unscrew drop pipe and cut off portion of threads.
Blocked pump-intake screen.	Examine intake screen for blockage by sand or mud.	Clean screen and reinstall at least 10 feet above well bottom.

Chart B—Pump Operates with Inadequate Water Delivery.
(continued)

CAUSE	CHECKOUT PROCEDURE	CORRECTION
Worn pump parts.	Abrasives in water may wear impeller, casing, or other close clearance parts. Before pulling pump, reduce setting of pressure switch; if pump shuts off, parts may be worn.	Pull pump and replace worn components.
Loose motor shaft.	After looking for worn components as outlined above, see if the coupling between motor and pump shaft is loose.	Tighten all connections and fasteners.

Chart C—Pump Starts too Frequently.

CAUSE	CHECKOUT PROCEDURE	CORRECTION
Pressure switch defective or maladjusted.	Check switch for setting and defects.	Reduce setting or replace switch.
Leak in pressure tank above water level.	Apply soap solution to entire surface of tank; bubbles indicate a leak.	Repair or replace tank.
Leak in plumbing system.	Examine house service line and distribution branches for leaks.	Repair leaks.
Leaky discharge line check valve.	Remove and examine valve.	Replace defective valve.
Air volume control plugged.	Remove and inspect control.	Clean or replace.
Snifter valve plugged.	Remove and inspect valve.	Clean or replace.

Chart D—Fuses Blow When Motor Is Running.

CAUSE	CHECKOUT PROCEDURE	CORRECTION
Incorrect voltage.	Check line-voltage terminals in control box (or connection box in 2-wire models models). Make sure the voltage is the range specified by the manufacturer.	If voltage is low, contact power company.
Overheated overload-protection box.	If sunlight or other heat source has made box too hot, circuit breakers may trip or fuses may blow; if box is hot, this may be the problem.	Ventilate or shade box, or move it.
Defective control-box components. (Not applicable to 2-wire models.)	Check the resistance across the running capacitor. Infinite resistance indicates open capacitor or defective relay points; zero resistance indicates shorted capacitor. Ideal response is a low resistance reading that gradually increases to some fixed value.	Replace defective components.

Chart E—Pump Won't Shut Off.

CAUSE	CHECKOUT PROCEDURE	CORRECTION
Defective pressure switch.	Examine points and other switch parts for defects (arcing may have caused switch points to fuse).	Clean points or replace switch.
Water level in well too low.	Well production may be too low for pump capacity. Restrict pump output, allow well to recover, and start pump.	If partial restriction corrects problem, leave valve or cock at new setting; otherwise, lower pump in well if depth is sufficient. (Do not lower if sand clogging can occur.)
Leak in drop pipe.	Raise and examine pipe for leaks.	Replace damaged pipe section.
Worn pump parts.	Abrasives in water may wear impeller, casing, and other close clearance parts. Before pulling pump, reduce setting on pressure switch to see if pump shuts off; if it does, parts are worn.	Pull pump and replace worn components.

Chart F—Motor Doesn't Start, but Fuses Don't Blow.

CAUSE	CHECKOUT PROCEDURE	CORRECTION
No overload protection.	Check condition of fuses or circuit breaker.	Replace blown fuses; reset breaker.
No power.	Check power supply to control box (or overload protection box); voltage should approximate that nominal for line.	If box is without power, contact power company.
Defective control box.	Examine control-box wiring for tight contacts. Check voltage at motor-wire terminals; no voltage indicates defective wiring.	Correct faulty wiring; tighten loose contacts.
Defective pressure switch.	Check voltage across closed switch; if voltage drop equals line voltage, switch is not making contact.	Clean points or replace switch.

Section 6

Sewage Disposal

Section 6

The *house sewer line,* as the term is used in this section, is the pipeline used to carry sewage from the building or house drain to a private sewer, septic tank, or other point of disposal. It usually extends from a point 3 to 8 feet outside the perimeter of the building, where it is connected to the building's drain system.

If you're thinking about constructing an individual sewage-disposal system, you should first consult with officials having jurisdiction in your local area. A number of state and local governments have developed requirements that have been incorporated into their official regulations, all soundly based on conditions peculiar to those areas and adequately representing good practice there. The recommendations and guidelines contained in this section should be considered *supplemental* to such local requirements.

Sewage consists mainly of waterborne waste matter discharged from water closets, showers and tubs, laundry, and kitchen areas of the home. As such, it may contain soap and detergent wastes and floating solids such as grease, paper, matches, and other debris. Sewage is a liquid that usually has animal, vegetable, and mineral matter in suspension and in solution. It also contains a very large number of bacteria, most of which are actually harmless; however, as some of the bacteria may have been discharged by victims or carriers of infectious diseases, sewage as a whole is dangerous. Although sewage has a number of elements in its makeup, the biggest by far is water. By weight, water is about 99.9% of the total.

The term *sewage* is a general inclusive term used to designate the combination of liquidborne waste originating in residences, business buildings, institutions, and industrial establishments with or without such ground, surface, or storm water as may become mixed with the waste when it is admitted into or passes through the sewers. Sewage may further be defined, to indicate the character of the liquid waste carried, as sanitary sewage, industrial sewage (more commonly called industrial waste), and storm sewage.

Sanitary sewage originates in the sanitary conveniences of dwellings, business buildings, factories,

institutions, and the like. *Storm sewage* flows in combined or storm sewers during or following rainfall and is the direct result of the rainfall.

Sewers are often designated as to type of sewage carried. *Sanitary sewers* carry sanitary sewage, which may or may not be mixed with some ground water percolating into the sewers. *Storm sewers* carry roof and surface runoff following rains. A single sewer designed to carry both sanitary sewage and storm water is called a *combined sewer.*

Sewage treatment, or *sewage purification,* designates any artificial process to which sewage is subjected to remove or alter its objectionable constituents and to render it less dangerous.

Occasionally, in designing a household plumbing system, it is necessary to include plans for final disposal of the sewage. Many factors enter into any plan for treatment of sewage; and the proper operation of the treatment plant, no matter how small, depends on the careful consideration of these factors in the design. When it is not possible to place the design in the hands of a competent sanitary engineer, the

advice of government health agencies should be obtained.

The usual method of disposing of sewage where the amount is small is through seepage beds, cesspools, or small septic-tank subsurface disposal systems. Although definite recommendations relative to design of these types of disposal should be obtained from health agencies, certain information may be of value in preparing preliminary plans.

A *leaching cesspool* retains the solids present in the sewage while the liquid leaches or seeps into the surrounding soil. The success of this type of disposal is dependent on the character of the soil and its suitability for rapid dispersal of liquid. Cesspools tend to seal themselves; the time required to do this depends on the character of the surrounding soil. The solids in the sewage entering the cesspool accumulate and at intervals must be removed. In the design, therefore, the character of the soil must be considered and allowances made for the storage of solids.

Where water supplies are developed in the same area, *cesspools should be located at least 100 feet away from wells or other sources of water supply and on ground that slopes away from such sources.* Cesspools should not extend into the ground-water table.

Septic-tank installations generally consist of a watertight tank and a subsurface system for disposal of the effluent from the tank. The subsurface disposal system may consist of open-joint tile, filter trench, or subsurface sand filter.

A septic tank is a settling tank that retains the sewage solids (sludge) from the sewage flowing through the tank for a sufficient period to secure a satisfactory decomposition of organic solids by anaerobic bacterial action. Septic tanks should not be confused with other types of tanks used in connection with sewage treatment.

The solids (sludge) retained in a septic tank must be removed whenever there is a sufficient accumulation to reduce materially the liquid capacity of the tank. The effluent from the septic tank is only partially treated sewage and is in a septic condition.

This effluent should not be discharged to a small stream; nor can it be discharged to the surface of the ground without nuisance.

A septic tank should be located as close as practical to the subsurface disposal field. It should be located at least 100 feet from wells, springs, or underground water-storage basins; at a lower elevation; and where there will be the least danger of sewage overflowing or leaking in such a way as to contaminate a water supply.

Grease interceptors or traps normally can be omitted on small septic-tank installations.

Septic tanks should be designed with a minimum liquid capacity equal to 24 hours' flow of sewage. The minimum-size tank should not be at less than 500 gallons liquid capacity. When the effluent of a septic tank having a capacity in excess of 1000 gallons is discharged to a subsurface tile disposal field, a dosing tank and an automatic sewage siphon are advisable. Tanks of less than 1000 gallons liquid capacity do not require automatic siphons.

Subsurface disposal of the effluent from septic tanks is possible where the character of the soil will permit its absorption. The percolation test is the most logical and practical method of determining the suitability of the soil for this method of disposal.

Suitability of Soil

The first steps in the design of subsurface sewage-disposal systems are to determine whether the soil is suitable for the absorption of septic tank effluent and, if so, how much area is required. The soil must have an acceptable percolation rate, without interference from ground water or impervious strata below the level of the absorption system. In general, three conditions must be met:

1 The percolation time is within the range specified in Table 6–1.
2 The maximum seasonal elevation of the ground water table is at least 4 feet below the bottom of the trench or seepage pit.
3 Rock formations or other impervious strata are at a depth greater than 4 feet below the bottom of trench or seepage pit.

Unless these conditions can be satisfied, the site is unsuitable for a conventional subsurface sewage-disposal system.

Percolation Tests

Subsurface explorations are necessary to determine subsurface formations in a given area. An auger with an extension handle, shown in example 6–1, is often used for making the investigation. In some cases, an examination of road cuts, stream embankments, or building excavations will give useful information. Wells and well drillers' logs can also be used to obtain information on ground water and subsurface conditions. In some areas, subsoil strata vary widely in short distances, and borings must be made at the site of the system. If the subsoil appears suitable, percolation tests should be made at points and elevations selected as typical of the area in which the disposal field will be located.

The percolation tests help to determine the acceptability of the site and establish the design size of the subsurface disposal system. The time required for percolation tests will vary in different types of soil. The safest method is to make tests in holes that have been kept filled with water for at least 4 hours, and preferably overnight. This is particularly desirable if the tests are to be made by an inexperienced person; and in some soils it is necessary even if the individual has had considerable experience (as in soils that swell upon wetting).

Percolation rates should be figured on the basis of the test data obtained after the soil has had an opportunity to become wetted or saturated and allowed to swell for at least 24 hours. Enough tests should be made in separate holes to assure that the results are valid.

The percolation test developed at the Robert A. Taft Sanitary Engineering Center incorporates these principles. Its use is particularly recommended when knowledge of soil types and soil structure is limited.

1 *Number and location of tests.* Six or more tests should be made in separate holes spaced uniformly over the proposed absorption field site.
2 *Type of test hole.* Dig or bore a hole with horizontal dimensions of from 4 to 12 inches and vertical sides to the depth of the proposed absorption trench.

TABLE 6–1. ABSORPTION-AREA REQUIREMENTS FOR INDIVIDUAL RESIDENCES[a]. (WITH PROVISION FOR GARBAGE GRINDER AND AUTOMATIC CLOTHES WASHING MACHINES.)

PERCOLATION RATE (TIME REQUIRED FOR WATER TO FALL 1 INCH)	REQUIRED ABSORPTION AREA PER BEDROOM [b], STANDARD TRENCH [c], SEEPAGE BEDS [c], AND SEEPAGE PITS [d]	PERCOLATION RATE (TIME REQUIRED FOR WATER TO FALL 1 INCH)	REQUIRED ABSORPTION AREA PER BEDROOM [b], STANDARD TRENCH [c], SEEPAGE BEDS [c], AND SEEPAGE PITS [d]
1 minutes or less·	70 sq ft	10 minutes or less	165 sq ft
2	85	15	190
3	100	30 [e]	250
4	115	45 [e]	300
5	125	60 [e,f]	330

[a] It is desirable to provide sufficient land area for entire new absorption system if needed in future. *

[b] In every case, sufficient land should be provided for the number of bedrooms (minimum of two) that can be reasonably anticipated, including the unfinished space available for conversion as additional bedrooms.

[c] Absorption area is figured as trench-bottom area and includes a statistics allowance for vertical side-wall area.

[d] Absorption area for seepage pits is figured as effective side-wall area beneath the inlet.

[e] Unsuitable for seepage pits if over 30 sq ft.

[f] Unsuitable for absorption systems if over 60 sq ft.

* Section 5.1(b)(2)(A), page 20 of Recommended State Legislation and Regulations: Urban Water Supply and Sewerage Systems Act and Regulations, Water Well Construction and Pump Installation Act and Regulations, Individual Sewerage Disposal Systems Act and Regulations. *U.S.D.H.E.W., Public Health Service, July 1965.*

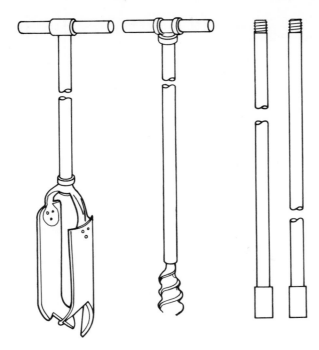

example 6-1. Auger and extension handle for test borings.

3 *Preparation of test hole.* Carefully scratch the bottom and sides of the hole with a knife blade or other sharp-pointed instrument to remove any smeared soil surfaces and to provide a natural soil interface into which water may percolate. Remove all loose material from the hole. Add 2 inches of coarse sand or fine gravel to protect the bottom from scouring and sediment.

4 *Saturation and swelling of the soil.* It is important to distinguish between saturation and swelling. *Saturation* means that the void spaces *between* soil particles are full of water. This can be accomplished in a short period of time. *Swelling* is caused by the intrusion of water *into* the individual soil particles. This is a slow process, especially in a soil of clay composition, therefore requiring a prolonged soaking period.

Carefully fill the hole with clear water to a minimum of 12 inches over the gravel. In most soils, it is necessary to refill the hole, possibly by means of an automatic siphon, to keep water in the hole for 4 hours. This procedure insures that the soil is given ample opportunity to swell and to approach its condition during the wettest season of the year. Thus, the test will give comparable results whether made in a dry or wet season. In sandy soils containing little or no clay, the swelling procedure is not essential, and the test may be made after the water from one filling of the hole has completely seeped away.

5 *Percolation rate measurement.* Except for sandy soils, percolation rate measurements must be made on the day following the procedure described under item 4.

If water remains in the test hole after the overnight swelling period, adjust the depth to approximately 6 inches over the gravel. From a fixed reference point, measure the drop in water level for 30 minutes. This drop is used to calculate the percolation rate.

If no water remains in the hole after the overnight swelling period, add clear water to approximately 6 inches over the gravel. From a fixed reference point, measure the drop in water level at approximately 30-minute intervals for 4 hours, refilling 6 inches over the gravel as necessary. The drop occurring during the final 30-minute period is used to calculate the percolation rate. The drops during prior periods provide information for modification of the procedure to suit local circumstances.

With soils that are sandy or in which the first 6 inches of water seeps away in less than 30 minutes, after the overnight swelling period, the interval between measurements should be reduced to 10 minutes and the test allowed to run for one hour. The drop occurring during the final 10 minutes is used to calculate the percolation rate.

Determining Absorption Area

Where the percolation rates and soil characteristics are good, the next step after making the percolation tests is to determine the required absorption area from Table 6–1 or example 6–2 and to select the soil absorption system that will be satisfactory for the area in question. As noted in Table 6–1, soil in which the percolation rate is slower than 1 inch in 30 minutes is unsuitable for seepage pits, and a percolation rate slower than 1 inch in 60 minutes is unsuitable for any type of soil absorption system.

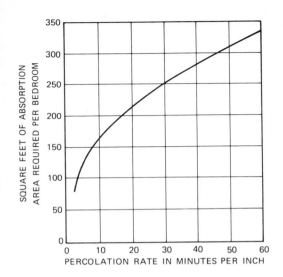

example 6–2. Absorption area requirements for homes.

The Sewer Line

The installation of an underground sewer line for transferring domestic sewage from the source to the disposal point includes (1) trenching and grading, (2) measuring and cutting pipe, (3) laying pipe, (4) joining pipe, (5) testing, and (6) backfilling and tamping. Let's first discuss trenching and grading; information on the other phases of sewer installation is given later in this section.

Trenching and Grading

A trench may be excavated manually or with heavy equipment, depending on the size of the job and the type of soil to be removed. When machines are impractical, you will have to do the job with a pick and shovel. Whichever method is used, the trench must be dug wide enough to allow ample working room to join pipe sections. It is also important that the bottom of the trench be sloped in the direction of flow so that sewage traveling through the pipeline laid in the trench will not be restricted. If local codes are not available, a rule of thumb is to slope the trench $\frac{1}{4}$ inch per foot, the grade at which sewage will flow freely through a pipe. According to a government-published set of plumbing recommendations, horizontal drainage piping should be run in practical alignment, supported at intervals of 10 feet or less. The minimum slopes are the following:

- Not less than $\frac{1}{4}$ inch of fall for each foot of travel for $1\frac{1}{4}$- to 2-inch-diameter pipe

- Not less than $\frac{1}{8}$ inch of fall for every foot of travel for $2\frac{1}{2}$- to 4-inch-diameter pipe

- Not less than $\frac{1}{16}$ inch of fall per foot for 5- to 8-inch diameter pipe

When a pipeline is to be laid in a stable soil such as hard clay or shale the trench should be excavated below the pipe grade. If bell-and-spigot pipe is to be used, excavation must be made for the bells. Be sure that enough undisturbed earth remains at the bottom of the trench to fully support the pipe, both joints and barrel. In areas where the temperature drops below freezing, the trench has to be excavated deep enough to put the pipeline below the frostline. Pipes that cross under roads or driveways must be buried in trenches at least 4 feet or more in depth; when the soil is so unstable that it is not considered safe at greater depths, the trenches should be supported by substantial sheeting, sheet piling, bracing, or shoring. Surface areas adjacent to the sides should be well drained. Trenches in partly saturated, filled, or unstable soils must be suitably braced.

Outside Sewage Piping

Various types of pipe materials are used in sewage systems. However, concrete and vitrified clay are the most common materials used for sewer lines. These pipes are joined together by hub (bell) and spigot ends.

Vitrified clay pipe is made of moistened powdered clay. It is available in *laying* lengths of 2, $2\frac{1}{2}$, and 3 feet and in diameters ranging from 4 inches up. It has a bell end and a spigot end to facilitate joining. After the pipe is taken from the casting, it is glazed and fired in large kilns to create a moisture-proof baked finish. Clay pipe is used for house sewer lines, sanitary sewer mains, and storm drains. The types of fittings for clay pipe are few: bends, tees, and wye branches primarily.

Plain precast concrete pipe is not reinforced with steel. This concrete pipe is similar to vitrified clay pipe in measuring, cutting, and handling.

Vitrified clay and concrete pipe, since both are available in such short lengths, seldom need cutting except at terminals and inlets. If, after

measurement, it is necessary to cut vitrified clay or concrete pipe, score it with a chisel, deepening the cut gradually until the pipe breaks cleanly at the desired point. Vitrified clay and concrete pipes may be cut with *chain* cutters, but these aren't likely to be in your tool inventory.

Example 6–3 shows some common fittings used with vitrified clay and concrete pipes.

Joints on vitrified clay and concrete pipe may be made of cement or bituminous compounds. Cement joints may be made of grout, a mixture of cement, sand, and water. The following procedure may be used as a guide in joining with grout (the procedure is similar when joining with bituminous compounds):

1 Insert the spigot of one length of pipe into the bell or hub of another and align the two pieces to the desired position.
2 Calk a gasket of oakum about $\frac{3}{4}$ inch thick into the bell to prevent the grout from running into the pipe.
3 Mix grout, using one part portland cement, two parts clean sharp washed sand, and sufficient water to thoroughly dampen.

example 6–3. Clay or concrete fittings shown in cross section.

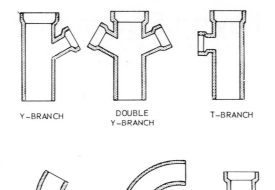

Y–BRANCH DOUBLE Y–BRANCH T–BRANCH

SHORT–RADIUS 1/8 BEND LONG–RADIUS 1/4 BEND REDUCER

INCREASER

4 Fill the joint with grout, using a packing tool.
5 Recalk the joint after 30 minutes. This is necessary to close shrinkage cracks that occur after the initial set of the grout.
6 Smooth and level the grout with a trowel. In hot weather, cover the joint with a wet burlap sack.
7 Remove excess mortar with a scraper.

The use of speed-seal joints (rubber rings) in joining vitrified clay pipe has become widespread. Speed-seal joints eliminate the use of oakum and mortar joints for sewer mains. The speed seal is made part of the vitrified pipe joint at the time of manufacture; it is composed of permanent polyvinyl chloride and called a *plastisol-joint connection*. This type of joint helps to insure tight joints that are root proof, flexible, and so on.

The speed-seal joint can be installed quickly and easily by one person. To make the joint, first insert the spigot end into the hub; then give the pipe a strong push so that the spigot locks into the hub seal. A solution of liquid soap may be spread on the joint to help it slip into place easily. You will find that other types of mechanical seal joints also are available, all installed in about the same way.

Special mechanical seal adapters are available for joining vitrified clay pipe to cast-iron soil pipe or soil pipe to vitrified pipe; this is necessary where the building disposal outlet connects to the sewer line.

Laying Sewer Pipe

Small pipes can be assembled and joined in sections on top of the ground and laid in the trench by hand. Large, heavy pipes are usually laid in the trench and then joined. These pipes may be lowered into the trench by rope, cable, or chain.

Sewer pipes should be laid on a compacted bed of sand, gravel, or material taken from the trench excavation, if suitable, in order to provide a slightly yielding and uniform bearing. This will assure safe support for the pipe, fill, and surface loads. When pipes are laid on sand, gravel, or similar material, the weight of the pipe will usually provide a suitable equalizing bed.

Pipelines should be placed carefully so that they do not settle at any point. Settling causes suspended matter to collect in the lower portion of the pipe, restricting the flow and reducing the capacity of the line.

Checking Grade of Sewer Line

After you have laid the pipes, the next step is to check the grade and align the pipeline. This is important in installing a below-ground sewer system. Remember that sewage will not flow uphill. The pipe is laid with the hub end upstream, and the laying proceeds upgrade. The spigot is inserted into the hub of the previously laid length. Each length is checked as to its grade and alignment before the next length is placed.

When grading for the proper pitch per foot the method illustrated in example 6–4 may be used as a guide. This illustration shows a ditch with batter boards used in transferring line and grade to trench; a stick for checking grade is shown in position.

Place batter boards across the trench at about 25-foot intervals. Put a mark on the stakes at some even foot distance above the invert (the lowest point on the inside of the pipe). Drive a nail in the top of the batter boards and stretch a cord from board to board as shown. You can transfer the center line for the pipe from the cord to the bottom of the trench by means of a plumb bob. Grade is transferred by means of a stick marked off by the foot, with a short piece fastened at a right angle to its lower end. Check the grade by placing the short piece on

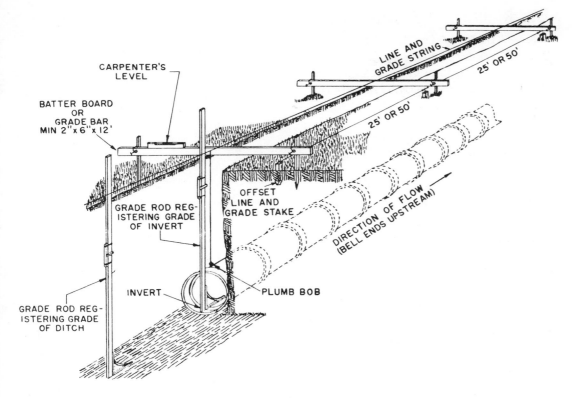

the invert of each length of sewer pipe and aligning the proper mark on the grade rod to the cord.

Testing

The purpose of testing any pipeline is to make sure the joints are tight enough to withstand working pressure. Before pipe is covered with dirt, it must be tested for leakage. There are several methods of effecting this test. The most widely used test is the *water test*, although an air test or odor test may be used.

Water Test. Here are the main steps to follow in making a water test. At the lowest point of the section to be tested, insert a test plug in the open end of the pipe or at a test tee; plug other openings. Fill the pipe to its highest level with water; about a 10 foot head is all that is necessary. Leave the water in the pipe for at least 15 minutes prior to starting any test; this will allow the oakum to soak up some water before you look for leaks. Refill the pipe and check each joint for leaks.

CARPENTER'S LEVEL

BATTER BOARD OR GRADE BAR MIN 2"x 6"x 12'

LINE AND GRADE STRING

25' OR 50'

25' OR 50'

GRADE ROD REGISTERING GRADE OF INVERT

OFFSET LINE AND GRADE STAKE

DIRECTION OF FLOW (BELL ENDS UPSTREAM)

GRADE ROD REGISTERING GRADE OF DITCH

INVERT

PLUMB BOB

example 6–4. Laying sewer pipe to line and grade.

Air Test. Before making an air test, fill the system with water and allow it to stand until the oakum expands at the joints. Drain the water from the lines and reinsert the test plug. Close all openings and apply air pressure of at least 5 pounds per square inch (psi). In a satisfactory test, the line should hold 5 psi for 15 minutes. If it does not, cover the joints with a soapy water solution and check for bubbles at the leak.

Backfilling and Tamping

After all pipe has been laid and tested, the system is ready to be covered up; this process is known as *backfilling and tamping*. The following method should give satisfactory results for most sewer lines.

Fine material, free from stones and other debris, is tamped in uniform layers with a small hand- or air-operated tamper under, around, and over the pipe. Use a hand shovel to backfill the ditch until the pipe has a 2-foot covering. This fill should be placed in the ditch and tamped in 4-inch layers or less, and it should proceed evenly on each side of the pipe so that injurious side pressure cannot occur. Make sure you do not walk on the pipe until you have at least 1 foot of soil tamped over the pipe. Until 2 feet of fill has been placed over the pipe, the filling should be done carefully with hand shovels; after that, machinery may be used for faster backfilling. Don't let the machinery run over the line!

Puddling or flooding with water to consolidate the backfill should not be done for a sewer line. The sections of pipe used are in short lengths and will tend to settle very rapidly to form pockets or low spots in the line.

Soil Absorption System

After a soil absorption system is determined to be usable, three types of design may be considered: absorption trenches, seepage beds, and seepage pits. The selection of the absorption system will depend to some extent on the location of the system in the area under consideration.

A safe distance should be maintained between the site and any source of water supply. Since the distance that pollution will travel underground depends on numerous factors, including the characteristics of the subsoil formations and the quantity of sewage discharged, no specified distance would be absolutely safe in all localities. Ordinarily, of course, the greater the distance, the greater will be the safety provided. In general, the location of the components of sewage disposal systems should be as shown in Table 6–2.

You can't use seepage pits in areas where domestic water supplies are obtained from shallow wells, or where there are limestone formations and sinkholes connected to underground channels, through which pollution may travel to water sources.

Details pertaining to local water wells, such as depth, type of construction, vertical zone of influence, and other parameters, together with data on the geological formations and porosity of subsoil strata should be considered in determining the safe allowable distance between wells and subsurface disposal systems.

Absorption Trenches

Soil absorption fields contain 12-inch lengths of 4-inch agricultural drain tile, 2- to 3-foot lengths of vitrified clay sewer pipe, or perforated nonmetallic pipe. In areas having unusual soil or water characteristics, local experience should be reviewed before selecting piping materials. The individual laterals preferably should not be over 100 feet long, and the trench bottom and tile distribution lines should be level. Use of more and shorter laterals is preferred because if something should happen to disturb one line, most of the field will still be serviceable. From a moisture-flow viewpoint, a spacing of twice the depth of gravel would prevent taxing the percolative capacity of the adjacent soil.

A variety of designs may be used in laying out subsurface disposal fields. The choice will depend on the size and shape of the available disposal area, the capacity required, and the topography of the disposal area.

Typical layouts of absorption trenches are shown in examples 6–5 and 6–6.

TABLE 6–2. MINIMUM DISTANCE BETWEEN COMPONENTS OF SEWAGE DISPOSAL SYSTEM.

COMPONENT OF SYSTEM	HORIZONTAL DISTANCE(FT)				
	WELL OR SUCTION LINE	WATER-SUPPLY LINE (PRESSURE)	STREAM	DWELLING	PROPERTY LINE
BUILDING SEWER	50	10 [a]	50	—	—
SEPTIC TANK	50	10	50	5	10
DISPOSAL FIELD AND					
SEEPAGE BED	100	25	50	20	5
SEEPAGE PIT	100	50	50	20	10
CESSPOOL [b]	150	50	50	20	15

[a] *Where the water supply line must cross the sewer line the bottom of the water service within 10 feet of the point of crossing shall be at least 12 inches above the top of the sewer line. The sewer line shall be of cast iron with leaded or mechanical joints at least 10 feet on either side of the crossing.*

[b] *Not recommended as a substitute for a septic tank. To be used only when found necessary and approved by the health authority.*

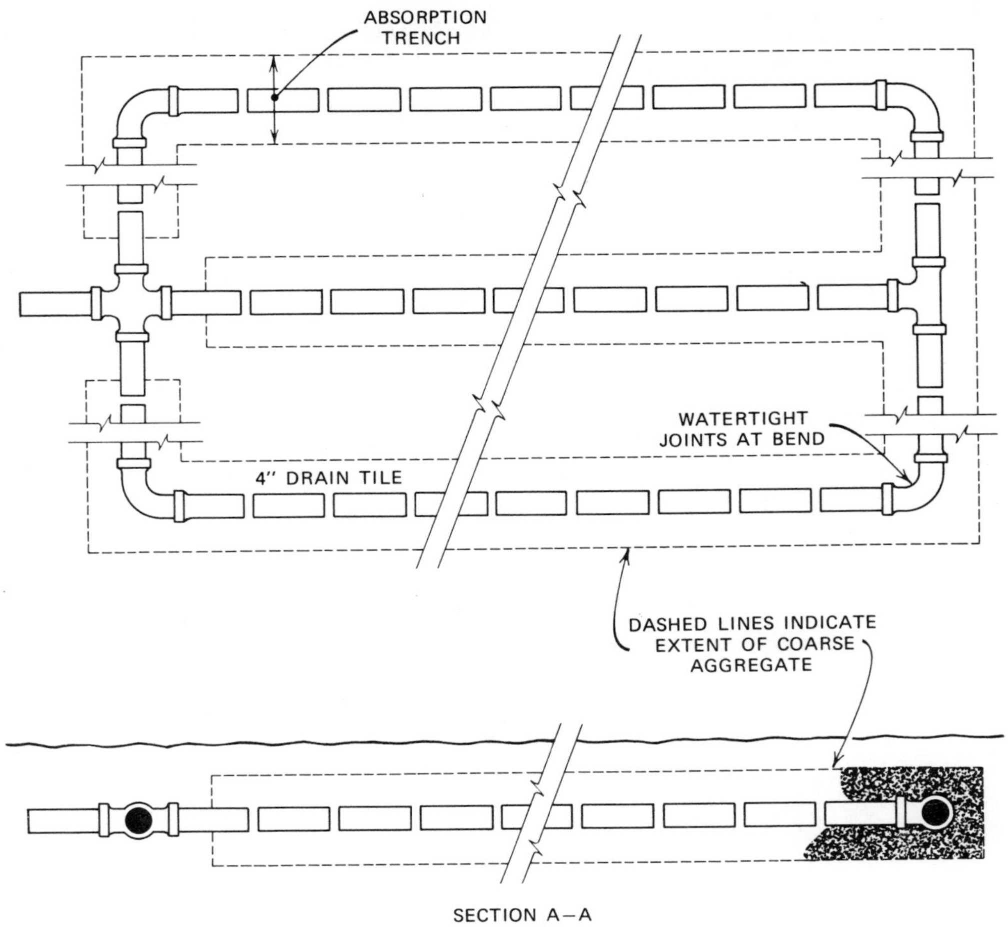

ABSORPTION TRENCH

example 6-5. Typical absorption trench layout.

WATERTIGHT JOINTS AT BEND

4" DRAIN TILE

DASHED LINES INDICATE EXTENT OF COARSE AGGREGATE

SECTION A—A

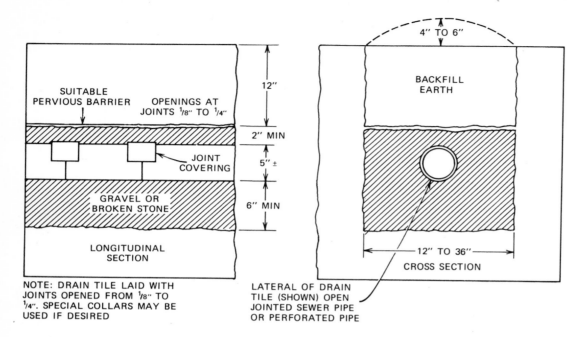

SUITABLE
PERVIOUS BARRIER   OPENINGS AT
JOINTS 1/8" TO 1/4"

JOINT
COVERING

GRAVEL OR
BROKEN STONE

LONGITUDINAL
SECTION

12"

2" MIN

5" ±

6" MIN

NOTE: DRAIN TILE LAID WITH
JOINTS OPENED FROM 1/8" TO
1/4". SPECIAL COLLARS MAY BE
USED IF DESIRED

4" TO 6"

BACKFILL
EARTH

12" TO 36"

CROSS SECTION

LATERAL OF DRAIN
TILE (SHOWN) OPEN
JOINTED SEWER PIPE
OR PERFORATED PIPE

example 6–6. Absorption trench and lateral.

Dimensional Considerations

To provide the minimum required gravel depth and earth cover, the depth of the absorption trenches should be at least 24 inches. Additional depth may be needed for contour adjustment, extra aggregate under the tile, or other design purposes. The maintenance of a 4-foot separation between the bottom of the trench and the water table is required to minimize ground-water contamination.

In considering the depth of the absorption field trenches, the possibility of tile lines freezing during prolonged cold periods is raised. Freezing rarely occurs in a carefully constructed system that is *kept in continuous operation*. It is important during construction to assure that the tile lines are surrounded by gravel. Pipes under driveways or other surfaces usually cleared of snow should be insulated.

The required absorption area is predicated on the results of the soil percolation test, and may be obtained from column 2 or 4 of Table 6–1 or example 6–2. Note especially that the area requirements are *per bedroom.*

The area of the lot on which the system is to be developed should be large enough to allow room for an additional system if the first one fails. For a 3-bedroom house on a lot where the minimum percolation rate is 1 inch in 15 minutes, the necessary absorption area for one system will be 3 bedrooms X 190 square feet per bedroom, or 570 square feet.

For trenches 2 feet wide with 6 inches of gravel below the drain pipe, the required total length of trench would be 570 ÷ 2, or 285 feet. If this were divided into 5 portions, or *laterals,* the length of each line would be 285 ÷ 5, or 57 feet. The spacing of trenches is generally governed by practical construction considerations—type of equipment, safety, and so on.

For serial distribution on sloping ground, trenches should be separated by 6 feet to prevent short-circuiting. Table 6–2 gives the various distances the system has to be kept away from wells, dwellings, property lines, and so on; in the example cited, each is 2 feet wide times 5 trenches = 10 feet;

adding the 6 feet between and multiplying by 4 spaces gives 24 feet. The total width of 34 feet times 57 feet in length = 1938 square feet, plus the additional land required to keep the field away from wells, property lines, and dwellings.

Construction Considerations

Careful construction is important in obtaining a satisfactory soil absorption system. Attention should be given to the protection of the natural absorption properties of the soil, care must be taken to prevent sealing of the surface on the bottom and sides of the trench, and trenches should not be excavated when the soil is wet enough to smear or compact easily. *Soil moisture is right for safe working only when a handful will mold with considerable pressure.* Open trenches should be protected from surface runoff to prevent the entrance of silt and debris.

If it is necessary to walk in the trench, a temporary board laid on the bottom will reduce the damage. Some smearing and damage are bound to occur, however. Smeared or

compacted surfaces should be raked to a depth of 1 inch and loose material removed before the gravel is placed in the trench.

The pipe, laid in a trench of sufficient width and depth, should be surrounded by clean graded gravel or rock, broken hard-burned clay brick, or similar aggregate. The material may range in size from $\frac{1}{2}$ inch to $2\frac{1}{2}$ inches. Cinders, broken shell, and similar materials are not recommended because they are usually too fine and may lead to premature clogging.

The material should extend from at least 2 inches above the top of the pipe to at least 6 inches below the bottom. If tile is used, the upper half of the joint openings should be covered, as shown in example 6–4. The top of the stone should be covered with either untreated building paper or a 2-inch layer of hay, straw, or similar pervious material to prevent the stone from becoming clogged by the earth backfill. An impervious covering should not be used, as this interferes with *evapotranspiration* at the surface.

Although generally not figured in the calculations, evapotranspiration is often an important factor in the operation of horizontal absorption systems. Evapotranspiration is a giving up of soil water through evaporation, and through absorption into roots, and expulsion from leaves of vegetation.

Drain-tile connectors, collars, clips, or other spacers with covers for the upper half of the joints are of value in obtaining uniform spacing, proper alignment, and protection of tile joints, but use of such aides is optional. They are made of galvanized iron, copper, and plastic.

Root problems may be prevented by using a liberal amount of gravel or stone around the tile. Clogging due to roots occurs mostly in lines with insufficient gravel under the tile. Furthermore, roots seek locations where moisture conditions are most favorable for growth; in the small percentage of cases where they become troublesome in well-designed installations, there is usually some explanation involving moisture conditions.

At a residence used only during the summer, for example, roots are most likely to penetrate drain tile when the house is uninhabited, or when moisture immediately below or around the gravel becomes less plentiful than when the system is in use. In general, trenches constructed within 10 feet of large trees or dense shrubbery should have at least 12 inches of gravel or crushed stone beneath the tile.

The top of a new absorption trench should be hand tamped and overfilled with about 4 to 6 inches of earth. Unless this is done, the top of the trench may settle to a point lower than the surface of the adjacent ground. This will cause the collection of storm water in the trench, which can lead to premature saturation of the absorption field and possibly to complete washout of the trench. Machine tamping or hydraulic backfilling of the trench should be avoided.

When you use sloping ground for the disposal area, you'll also have to construct a small temporary dike or surface water-diversion ditch above the field to prevent the disposal area from being washed out by rain. The dike should be maintained or the ditch kept free of obstructions until the field becomes well covered with vegetation.

A heavy vehicle would readily crush the tile in a shallow absorption field. For this reason, heavy machinery should be excluded from the disposal area unless special provision is made to support the weight. All machine grading should be completed before the field is laid. The use of the field area must be restricted to activities that will not contribute to the compaction of the soil with the consequent reduction in soil aeration.

Seepage or Leach Beds

Common design practice for soil absorption systems for private residences provides for trench widths up to 36 inches, but variations of design utilizing even more width are being used in many areas. Absorption systems having trenches wider than 3 feet are referred to as *seepage beds*. The design of trenches is based on an empirical relationship between the percolation test and the bottom area of the trenches. The use of seepage beds has been limited by the lack of experience with their performance and the absence of design criteria comparable to that for trenches.

Advantages

Studies sponsored by the Federal Housing Administration have demonstrated that the seepage bed is a satisfactory device for disposing of effluent in soils that are acceptable for soil absorption systems. The studies have further demonstrated that the empirical relationship between the percolation test and the bottom area

required for trenches is applicable for seepage beds.

There are three main elements of a seepage bed: absorption surface, rockfill or packing material, and the distribution system. The design of the seepage bed should be such that the total intended absorption area is preserved, sufficient packing material is provided in the proper place to allow for further treatment and storage of excess liquid, and a means for distributing the effluent is protected against siltation of earth backfill and mechanical damage.

An outline of the construction details for a conventional seepage bed follows. Tabulation of construction details for the conventional seepage bed is not intended to preclude other designs, which may provide the essential features in a more economical or otherwise desirable manner. Specifically, there may be equally acceptable or even superior methods developed for desirable distribution of the liquid than by tile or perforated pipe covered with gravel.

The use of seepage beds results in the following advantages:

1 A wide bed makes more efficient use of land available for absorption systems than a series of long narrow trenches with wasted land between the trenches.
2 Efficient use may be made of a variety of modern earthmoving equipment employed at housing projects.

Construction Considerations

When seepage beds are used, the following design and construction procedures should be observed:

1 The amount of bottom absorption area required shall be the same as shown in Table 6–1.
2 Percolation tests should be conducted.
3 The bed should have a minimum depth of 24 inches below natural ground level to provide a minimum earth backfill cover of 12 inches.
4 The bed should have a minimum depth of 12 inches of rockfill or packing material extending at least 2 inches above and 6 inches below the distribution pipe.

5 The bottom of the bed and distribution tile or perforated pipe should be level.

6 Lines for distributing effluent shall be spaced not more than 6 feet apart and not more than 3 feet from the bed sidewall.

7 When more than one bed is used, there should be a minimum of 6 feet of undisturbed earth between adjacent beds, and the beds should be connected in series.

Serial Distribution

Distribution boxes can be eliminated from septic tank soil absorption systems in favor of some other method of distribution without inducing increased failure of disposal fields. In fact, evidence indicates that distribution boxes as such may be harmful to the system.

On sloping ground, of course, a method of distribution is needed to prevent failure of any one trench before the capacity of the entire system is utilized.

Serial distribution is achieved by arranging individual trenches of the absorption system so that each trench is forced to form into a pond to the full depth of the gravel fill before liquid flows into the succeeding trench.

We can see that serial distribution minimizes the importance of variable absorption rates by forcing each trench to absorb effluent until its ultimate capacity is utilized. The variability of soils even in the small area of an individual absorption field raises doubt about the desirability of uniform distribution. Any one or a combination of factors may lead to nonuniform absorptive capacity of the several trenches in a system; some factors are the varying physical and chemical characteristics of soil, construction damage such as soil interface smearing or excessive compaction, poor surface drainage, and variation in depth of trenches.

Since serial distribution causes successive trenches in the absorption system to be used to full capacity, it has a distinct advantage on sloping terrain. With imperfect division of flow in a parallel system, one trench could become overloaded, resulting in a surcharged condition. If the slope of the ground and elevation of the distribution box were such that a surcharged trench continued to receive more effluent than it could absorb, local failure would occur before the full capacity of the system could be utilized.

The cost of the distribution box is eliminated in serial distribution. And long runs of closed pipe connecting the box to each trench are unnecessary.

Fields in Flat Areas

Where the slope of the ground surface does not exceed 6 inches in any direction within the area used for the absorption field, the septic tank effluent may be applied to the absorption field through a system of interconnected tile lines and trenches in a continuous system. The following specific criteria should be followed:

1 A minimum of 12 inches of earth cover is provided over the gravel fill in all trenches of the system.
2 The bottom of the trenches and the distribution lines are level.

One type of a satisfactory absorption system layout for "level" ground is shown in example 6–7.

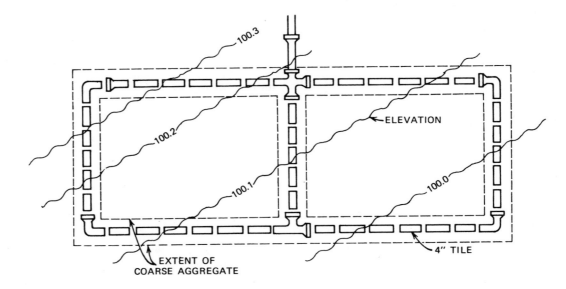

example 6–7. Absorption-field system for level ground.

Fields in Sloping Ground

Serial distribution may be used in all situations where a soil absorption system is permitted and should be used where the fall of the ground surface exceeds approximately 6 inches in any direction within the area utilized for the absorption field. The maximum ground slope suitable for serial distribution systems should be governed by local factors affecting the erosion of the ground used for the absorption field. Excessive slopes not protected from surface water runoff or without adequate vegetation cover to prevent erosion should be avoided.

Generally, ground having a slope greater than one vertical unit of measure for each two horizontal units of measure should be investigated carefully for the possibility of erosion. Also, the horizontal distance from one side of the trench to the ground surface should be adequate to prevent lateral flow of effluent and breakout on the surface (in no case less than 2 feet).

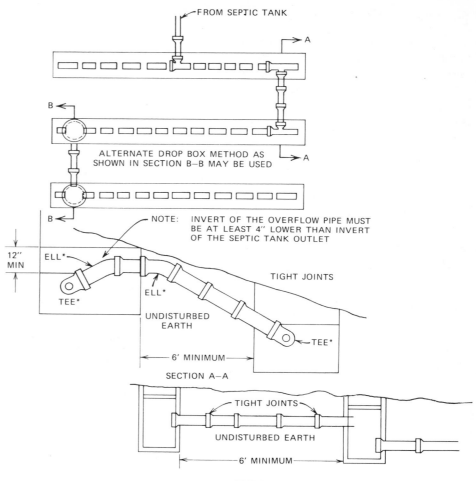

example 6–8. Relief-line configuration for serial distribution.

In serial distribution, each adjacent trench (or pair of trenches) is connected to the next by a closed pipe laid on an undisturbed section of ground, as shown in example 6–8. The arrangement is such that all effluent is discharged to the first trench until it is filled. Excess liquid is then carried by means of a closed line to the next succeeding or lower trench. In that manner, each portion of the subsurface system is used in succession. When serial distribution is used, the following design and construction procedures should be followed:

1 The bottom of each trench and its distribution line should be level.

2 There should be a minimum of 12 inches of ground cover over the gravel fill in the trenches.

3 The absorption trenches should follow approximately the ground-surface contours so that variations in trench depth will be minimized.

4 There should be a minimum of 6 feet of undisturbed earth between adjacent trenches and between the septic tank and the nearest trench.

5 Adjacent trenches may be connected with the relief line or a drop-box arrangement (example 6–6) in such a manner that each trench is completely filled with septic tank effluent to the full depth of the gravel before effluent flows to succeeding trenches. Trench-connecting lines should be 4-inch tight-joint sewers with direct connections to the distribution lines in adjacent trenches or to a drop-box arrangement.

Care must be exercised in constructing relief lines to insure an undisturbed block of earth between trenches. The trench for the relief pipe where it connects with the preceding absorption trench should be no deeper than the top of the gravel. The relief line should rest on undisturbed earth and backfill should be carefully tamped.

The relief lines connecting individual trenches should be as far from each other as practical to prevent short-circuiting.

6 The invert of the overflow pipe in the first relief line must be at least 4 inches lower than the invert of the septic-tank outlet (example 6–8).

Deep Trenches and Beds

When the depth of filter material below the tile exceeds the standard 6-inch depth, the added absorption area provided in deeper trenches permits a resultant decrease in length of trench. Such ''credit'' is given in accordance with Table 6–3, which gives the percentage of length of standard absorption trench (as computed from Table 6–1), based on 6-inch increments of increase in depth of filter material.

For trenches or beds having width not shown in Table 6–3, the percent of length of standard absorption trench may be computed. Percent of length, standard trench is:

$$\frac{w+2}{w+1+2d} \times 100$$

where

w = width of trench in feet

d = depth of gravel below pipe in feet

To use Table 6–3, consider the example given under *Absorption Trenches* (Dimensional Considerations). Using a trench 2 feet wide with 6 inches of gravel under tile, 285 feet is required. If the depth of gravel is increased to 18 inches, keeping trench width at 2 feet, only 66% of 285 feet is required, or 188 feet. If 4 laterals are used, the length would be $188 \div 4 = 47$ feet.

The space between lines for serial distribution on sloping ground is 6 feet × 3 spaces = 18 feet, plus 4 lines × 2 feet = 8 feet. Total land required is 26 feet in width × 47 feet in length, or 1222 square feet, plus additional area required to keep the field away from wells, property lines, and so on.

Cesspools

As with all soil absorption systems, cesspools, or *seepage pits,* should never be used where there is a likelihood of contaminating underground waters, nor where adequate seepage beds or trenches can be provided. When seepage pits are to be used, the pit excavation should terminate 4 feet above the ground water table.

In some states, seepage pits are permitted as an alternative when absorption fields are impractical, and where the top 3 or 4 feet of soil is underlaid with porous sand or fine gravel and the subsurface conditions are otherwise suitable for pit installations. Where circumstances permit, cesspools may be either supplemental or alternative to the more shallow absorption fields. When seepage pits are used in combination with absorption fields, the absorption areas in each system should be prorated, or based on the weight average of the results of the percolation tests.

It is important that the capacity of a seepage pit be computed on the basis of percolation tests made in

TABLE 6–3. PERCENTAGE OF LENGTH OF STANDARD TRENCH.

DEPTH OF GRAVEL BELOW PIPE (INCHES)	TRENCH WIDTH 12" (PERCENT)	TRENCH WIDTH 18" (PERCENT)	TRENCH WIDTH 24" (PERCENT)	TRENCH WIDTH 36" (PERCENT)	TRENCH WIDTH 48" (PERCENT)	TRENCH WIDTH 60" (PERCENT)
12	75	78	80	83	86	87
18	60	64	66	71	75	78
24	50	54	57	62	66	70
30	43	47	50	55	60	64
36	37	41	44	50	54	58
42	33	37	40	45	50	54

The standard absorption trench is one in which the filter material extends 2 inches above and 6 inches below the pipe.

each vertical stratum penetrated. The weighted average of the results should be computed to obtain a design figure. Soil strata in which the percolation rates are in excess of 30 minutes per inch should not be included in computing the absorption area. Adequate tests for deep pits are somewhat difficult to make, time-consuming, and expensive. Although few data have been collected comparing percolation test results with deep pit performance, the results of such percolation.tests, combined with competent engineering judgment, are the best means of arriving at design data for seepage pits.

Table 6–1 and example 6–2 give the absorption area requirements per bedroom for the percolation rate obtained. The effective area of the seepage pit is the *vertical wall area* (based on dug diameter) of the pervious strata below the inlet. No allowance should be made for impervious strata or bottom area. With this in mind, Table 6–4 may be used for determining the effective side-wall area of circular, or cylindrical, cesspools.

Sample Calculations

Assume that a seepage pit absorption system is to be designed for a 3-bedroom home on a lot where the minimum percolation rate of 1 inch in 15 minutes prevails. According to Table 6–1, 3 X 190 (or 570) square feet of absorption area would be needed. Assume that the water table does not rise above 27 feet below the ground surface, that seepage pits with an effective depth of 20 feet can be provided, and that the house is in a locality where it is common practice to install seepage pits of 5 feet in diameter (4 feet to the outside walls, surrounded by about 6 inches of gravel). Design of the system is as follows:

$$\text{Let } d = \text{depth of pit in feet}$$
$$D = \text{pit diameter in feet}$$
$$Dd = 570 \text{ sq ft}$$
$$3.14 \text{ X } 5 \text{ X } d = 570 \text{ sq ft}$$

Solving for d, the depth of the pit is about 36 feet.

In other words, one 5-foot-diameter pit 36 feet deep would be needed, but since the maximum

effective depth is 20 feet in this location, it will be necessary to increase the diameter of the pit or the number of pits, or both.

$$2 \text{ X } 3.14 \text{ X } 10 \text{ X } d = 570 \text{ sq ft}$$
$$d = 9.1 \text{ ft deep}$$

To design for 2 pits with a 5 ft diameter:

$$2 \text{ X } 3.14 \text{ X } 5 \text{ X } d = 570 \text{ sq ft}$$
$$d = 18 \text{ ft (approximately)}$$

Multiple cesspools should be separated by a distance equal to 3 times the diameter of the largest pit. For pits over 20 feet in depth, the minimum space between pits should be 20 feet (example 6–9). The area of the lot should be large enough to maintain this distance between the pits while still allowing room for additional pits if the first ones should fail. If this can be done, such an absorption system may be approved; if not, other suitable sewerage facilities would be required.

TABLE 6–4. VERTICAL WALL AREAS OF CIRCULAR SEEPAGE PITS.

DIAMETER OF SEEPAGE PIT (FT)	EFFECTIVE STRATA DEPTH BELOW FLOW LINE, SQ FT									
	1 ft	2 ft	3 ft	4 ft	5 ft	6 ft	7 ft	8 ft	9 ft	10 ft
3	9.4	19	28	38	47	57	66	75	85	94
4	12.6	25	38	50	63	75	88	101	113	126
5	15.7	31	47	63	79	94	110	126	141	157
6	18.8	38	57	75	94	113	132	151	170	188
7	22.0	44	66	88	110	132	154	176	198	220
8	25.1	50	75	101	126	151	176	201	226	251
9	28.3	57	85	113	141	170	198	226	254	283
10	31.4	63	94	126	157	188	220	251	283	314
11	34.6	69	104	138	173	207	242	276	311	346
12	37.7	75	113	151	188	226	264	302	339	377

Example: A 5-ft diameter pit of 6-ft depth (below the inlet) has an effective area of 94 sq ft. A pit 5 ft in diameter and 16-ft depth has an area of 94 + 157, or 251 sq ft.

example 6–9. Dual-seepage-pit disposal system.

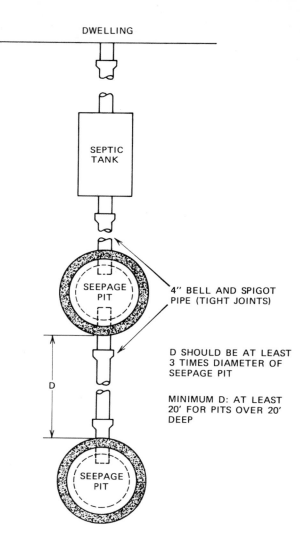

DWELLING

SEPTIC TANK

SEEPAGE PIT

4" BELL AND SPIGOT PIPE (TIGHT JOINTS)

D SHOULD BE AT LEAST 3 TIMES DIAMETER OF SEEPAGE PIT

MINIMUM D: AT LEAST 20' FOR PITS OVER 20' DEEP

D

SEEPAGE PIT

Construction Considerations

Soil is susceptible to damage during excavation. Digging in wet soils should be avoided as much as possible, cutting teeth on mechanical equipment kept sharp, bucket-augered pits reamed to a larger diameter than the bucket, and all loose material removed from the excavation. (See example 6–10.)

Pits should be backfilled with clean gravel to a depth of 1 foot above the pit bottom or 1 foot above the reamed ledge to provide a sound foundation for the lining. Preferred lining materials are clay or concrete brick, block, or rings. Rings should have seep holes or notches to provide for seepage. Brick and block should be laid dry with staggered joints. Standard brick should be laid flat to form a 4-inch wall. The outside diameter of the lining should be at least 6 inches less than the least excavation diameter. The annular space formed should be filled with clean, coarse gravel to the top of the lining, as shown in example 6–10.

Either brick dome or flat concrete covers are satisfactory. They should

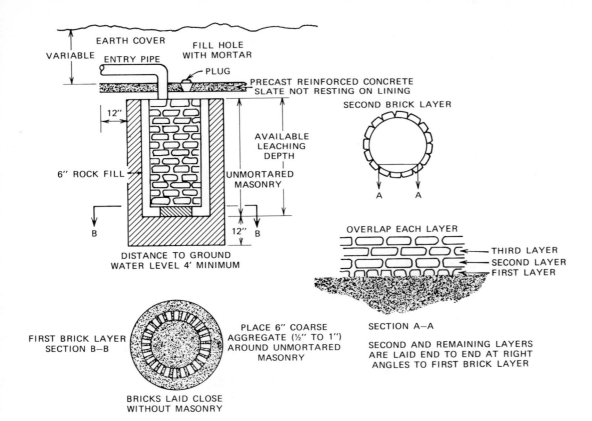

EARTH COVER

VARIABLE

ENTRY PIPE

FILL HOLE
WITH MORTAR

PLUG

PRECAST REINFORCED CONCRETE
SLATE NOT RESTING ON LINING

12"

6" ROCK FILL

AVAILABLE
LEACHING
DEPTH

UNMORTARED
MASONRY

SECOND BRICK LAYER

A   A

B   12"   B

DISTANCE TO GROUND
WATER LEVEL 4' MINIMUM

OVERLAP EACH LAYER

THIRD LAYER
SECOND LAYER
FIRST LAYER

FIRST BRICK LAYER
SECTION B–B

PLACE 6" COARSE
AGGREGATE (½" TO 1")
AROUND UNMORTARED
MASONRY

SECTION A–A

SECOND AND REMAINING LAYERS
ARE LAID END TO END AT RIGHT
ANGLES TO FIRST BRICK LAYER

BRICKS LAID CLOSE
WITHOUT MASONRY

example 6–10. Seepage pit.

be based on undisturbed earth and extend at least 12 inches beyond the excavation and should not bear on the lining for structural support. Bricks should be either laid in cement mortar or have a 2-inch covering of concrete. If flat covers are used, a prefabricated type is preferred, and they should be reinforced to be equivalent in strength to an approved septic-tank cover. A 9-inch capped opening in the pit cover is convenient for pit inspection. All concrete surfaces should be located with a protective bitumastic or similar compound to minimize corrosion.

Connecting lines should be of a sound durable material, the same as used for the house-to-septic-tank connection. All connecting lines should be laid on a firm bed of undisturbed soil throughout their length. The grade of a connecting line should be at least 2%. The pit inlet pipe should extend horizontally at least 1 foot into the pit with a tee or ell to divert flow downward to prevent washing and eroding of the side walls. If multiple pits are used, or in the event repair pits are added to an existing system, they should be connected in series.

Septic Tanks

Assuming that the lot will be large enough to accommodate one of the types of absorption systems, and that construction of the system is permitted by local authority, the next step is the selection of a suitable septic tank.

Function

Untreated liquid household wastes (sewage) will quickly clog all but the most porous gravel formations. The tank conditions sewage so that it may be more readily percolated onto the subsoil of the ground. Thus, the most important function of a septic tank is to provide protection for the absorption ability of the subsoil. Three functions take place within the tank to provide this protection:

1 *Removal of Solids.* Clogging of soil with tank effluent varies directly with the amount of suspended solids in the liquid. As sewage from a building sewer enters a septic tank, its rate of flow is reduced so that larger solids sink to the bottom or rise to the surface. These solids are retained in the tank, and the clarified effluent is discharged.

2 *Biological Treatment.* Solids and liquid in the tank are subjected to decomposition by bacterial and natural processes. Bacteria present are of a variety called *anaerobic,* which thrive in the absence of free oxygen. Sewage that has been subjected to such treatment causes less clogging than untreated sewage containing the same amount of suspended solids.

3 *Sludge and Scum Storage. Sludge* is an accumulation of solids at the bottom of the tank, while *scum* is a partially submerged mat of floating solids that may form at the surface of the fluid in the tank. Sludge, and scum to a lesser degree, will be digested and compacted into a smaller volume. However, no matter how efficient the process, a residue of inert solid material will remain. Space must be provided in the tank to store this residue during the interval between cleanings; otherwise, sludge and scum will eventually be scoured from the tank and may clog the disposal field.

If adequately designed, constructed, maintained, and operated, septic tanks are effective in accomplishing their purpose.

The relative position of a septic tank in a typical subsurface disposal system is shown in example 6–11. The liquid contents of the house sewer (A) are discharged first into the septic tank (B), and finally into the subsurface absorption field (C).

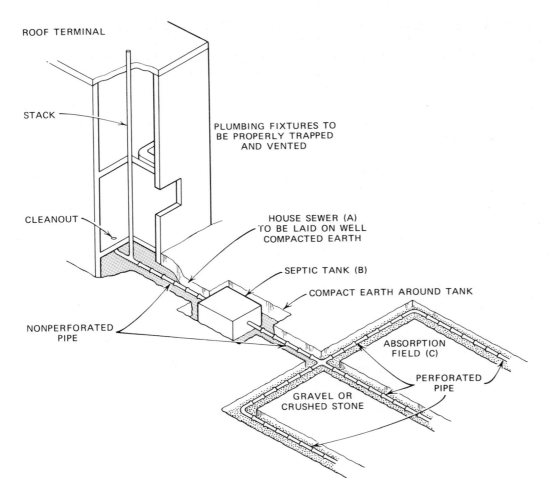

ROOF TERMINAL

STACK

CLEANOUT

NONPERFORATED
PIPE

PLUMBING FIXTURES TO
BE PROPERLY TRAPPED
AND VENTED

HOUSE SEWER (A)
TO BE LAID ON WELL
COMPACTED EARTH

SEPTIC TANK (B)

COMPACT EARTH AROUND TANK

ABSORPTION
FIELD (C)

PERFORATED
PIPE

GRAVEL OR
CRUSHED STONE

example 6–11. Septic-tank sewage-disposal
system.

The heavier sewage solids settle to the bottom of the tank, forming a blanket of sludge. The lighter solids, including fats and greases, rise to the surface and form a layer of scum. A considerable portion of the sludge and scum are liquefied through decomposition or digestion. During this process, gas is liberated from the sludge, carrying a portion of the solids to the surface, where they accumulate with the scum. Ordinarily, they undergo further digestion in the scum layer, and a portion settles again to the sludge blanket on the bottom. This action is retarded if there is much grease in the scum layer. The settling is also retarded because of gasification in the sludge blanket. Further, there are relatively wider fluctuations of flow in small tanks than in the large units. This effect has been recognized in Table 6–5, which shows the recommended minimum liquid capacities of household septic tanks.

TABLE 6–5. LIQUID CAPACITY OF TANK (GALLONS).

NUMBER OF BEDROOMS	RECOMMENDED MINIMUM TANK CAPACITY	EQUIVALENT CAPACITY PER BEDROOM
2 or less	750	375
3	900	300
4	1000	250

For each additional bedroom, add 250 gallons.

Location

Septic tanks should be located where they cannot cause contamination of any well, spring, or other source of water supply. Underground contamination may travel in any direction and for considerable distances unless filtered effectively. Underground pollution usually moves in the same general direction as the normal movement of the ground water in the locality; ground water moves in the direction of the slope or gradient of the water table, from the area of higher water table to areas of lower water table.

In general, the water table follows the general contour of the ground surface. For this reason, septic tanks should be located *downhill* from wells or springs. Sewage from disposal systems occasionally contaminates wells having higher surface elevations. Obviously, the elevations of disposal systems are almost always higher than the *level of water* in such wells as may be located nearby; hence pollution from a disposal system on a lower surface elevation may still travel downward to the water-bearing stratum (see example 6–12). So rely

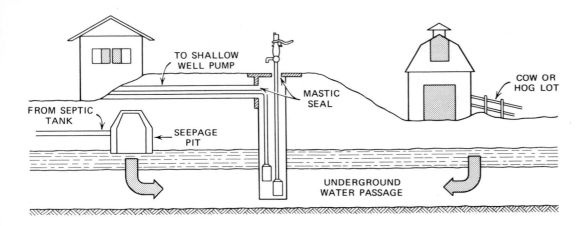

example 6–12. Well pollution from sources with lower surfaces.

TO SHALLOW WELL PUMP

MASTIC SEAL

FROM SEPTIC TANK

SEEPAGE PIT

COW OR HOG LOT

UNDERGROUND WATER PASSAGE

on horizontal as well as vertical distances for protection. Tanks should never be closer than 50 feet from any source of water supply; and greater distances are preferred where possible.

The septic tank should not be located within 5 feet of any building because structural damage may result during construction or seepage may enter the basement. The tank should not be located in swampy areas, nor in areas subject to flooding. In general, the tank should be located where the largest possible area will be available for the disposal field. Consideration should also be given to the location for easiest cleaning and maintenance. Where public sewers may be installed at a future date, provision should be made in the household plumbing system for connection to such sewer.

Effluent

Contrary to popular belief, septic tanks do not accomplish a high degree of bacteria removal. Although the sewage undergoes treatment in passing through the tank, this does not mean that infectious agents will be removed; hence, septic tank effluents cannot be considered safe. The liquid that is discharged from a tank is, in some respects, more objectionable than that which goes in; it is dirty and smelly. This, however, does not detract from the value of the tank. As previously explained, its primary purpose is to condition the sewage so that it will cause less clogging of the disposal field.

Further treatment of the effluent, including removal of harmful bacteria, is effected by percolation through the soil. Disease-producing bacteria will, in time, die out in the unfavorable environment afforded by soil. In addition, bacteria are also removed by certain physical forces during filtration. This combination of factors results in the eventual purification of the sewage effluent.

Capacity

Capacity is one of the most important considerations in septic tank design. Studies have proved that liberal tank capacity is not only important from a functional standpoint, but good economy. The liquid capacities recommended in Table 6–5 allow for the use of all household appliances, including food-waste disposal units.

Specifications

Septic tanks should be watertight and constructed of materials not subject to excessive corrosion or decay; good construction materials are concrete, coated metal, vitrified clay, heavyweight concrete blocks, and hard-burned bricks. Properly cured precast and cast-in-place reinforced concrete tanks are acceptable everywhere. Steel tanks meeting Commercial Standard 177–62 of the U. S. Department of Commerce are also generally acceptable. Heavyweight concrete blocks should be laid on a solid foundation and mortar joints should be well filled. The interior of the tank should be surfaced with two $\frac{1}{4}$-inch coats of portland cement and sand plaster.

Precast tanks should have a minimum wall thickness of 3 inches, and should be adequately reinforced to facilitate handling. When precast slabs are used as covers, they should be watertight, have a thickness of at least 3 inches, and be adequately reinforced. All concrete surfaces should be coated with a bitumastic or similar compound to minimize corrosion.

Backfill around septic tanks should be made in thin layers thoroughly tamped in a manner that will not produce undue strain on the tank. Settlement of backfill may be done with water, provided the material is thoroughly wetted from the bottom upwards and the tank is first filled with water to prevent floating.

Inlet and Outlet Devices

Adequate access must be provided to each compartment of the tank for inspection and cleaning. Both the inlet and outlet devices should be accessible. Access should be provided to each compartment by means of either a removable cover or a 20-inch manhole in least dimension. Where the top of the tank is located more than 18 inches below the finished grade, manholes and inspection holes should extend to approximately 8 inches below the finished grade (see example 6–13), or can be extended to the finished grade if a seal is provided to keep odors from escaping. In most instances, the extension can be made using clay or concrete pipe, but proper attention must be given to the accident hazard involved when manholes are extended close to the ground surface.

The inlet invert should enter the tank at least 3 inches above the liquid level in the tank to allow for momentary rise in liquid level during discharges to the tank. This free drop prevents backwater and stranding of solid material in the house sewer leading to the tank.

A vented inlet tee or baffle should be provided to divert the incoming sewage downward. It should penetrate at least 6 inches below the liquid level, but in no case should the penetration be greater than that allowed for the outlet device. A number of arrangements commonly used for inlet and outlet devices are shown in example 6–14.

It is important that the outlet device penetrate just far enough below the liquid level of the septic tank to provide a balance between sludge- and scum-storage volume; otherwise, part of the advantage of capacity is lost. A vertical section of a properly operating tank would show it divided into three distinct layers: scum at the top, a middle zone free of solids (called *clear space*), and a bottom layer of sludge. The outlet device retains scum in the tank, but at the same time it limits the amount of sludge that can be accommodated without scouring, which results in sludge discharging in the effluent from the tank. The outlet device should generally extend to a distance below the liquid surface equal to 40% of the liquid depth; for horizontal cylindrical tanks, this should be reduced to 35%

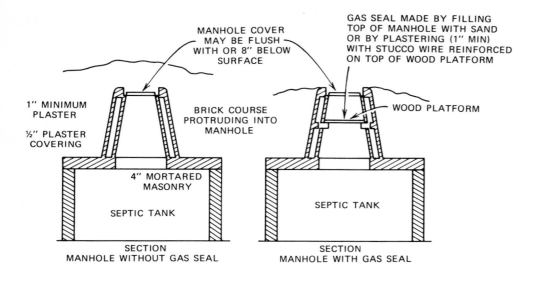

MANHOLE COVER
MAY BE FLUSH
WITH OR 8" BELOW
SURFACE

1" MINIMUM
PLASTER

½" PLASTER
COVERING

BRICK COURSE
PROTRUDING INTO
MANHOLE

4" MORTARED
MASONRY

SEPTIC TANK

SECTION
MANHOLE WITHOUT GAS SEAL

GAS SEAL MADE BY FILLING
TOP OF MANHOLE WITH SAND
OR BY PLASTERING (1" MIN)
WITH STUCCO WIRE REINFORCED
ON TOP OF WOOD PLATFORM

WOOD PLATFORM

SEPTIC TANK

SECTION
MANHOLE WITH GAS SEAL

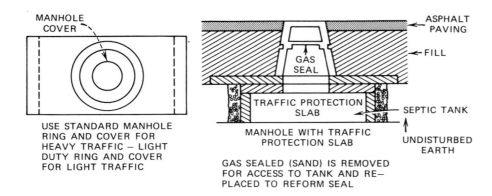

MANHOLE
COVER

USE STANDARD MANHOLE
RING AND COVER FOR
HEAVY TRAFFIC — LIGHT
DUTY RING AND COVER
FOR LIGHT TRAFFIC

ASPHALT
PAVING

FILL

GAS
SEAL

TRAFFIC PROTECTION
SLAB

SEPTIC TANK

UNDISTURBED
EARTH

MANHOLE WITH TRAFFIC
PROTECTION SLAB

GAS SEALED (SAND) IS REMOVED
FOR ACCESS TO TANK AND RE-
PLACED TO REFORM SEAL

example 6–13. Manhole designs.

(see example 6–14). For example, in a horizontal cylindrical tank having a liquid depth of 42 inches, the outlet device should penetrate 42 X 0.35 = 14.7 inches below the liquid level.

The outlet device should extend above the liquid line to approximately 1 inch from the top of the tank (see example 6–14). The space between the top of the tank and the baffle allows gas to pass off through the tank into the house vent.

Tank Proportions

For tanks of a given capacity, shallow tanks function as well as deep ones. Also, for tanks of a given capacity and depth, the shape of a septic tank is unimportant. However, it is recommended that the smallest dimension be at least 2 feet. Liquid depth may range between 30 and 60 inches.

Storage Above Liquid Level. Capacity is required above the liquid line to provide for the portion of the scum that floats above the liquid. Although some variation is to be expected, some 30% of the total scum will accumulate above the liquid line. In

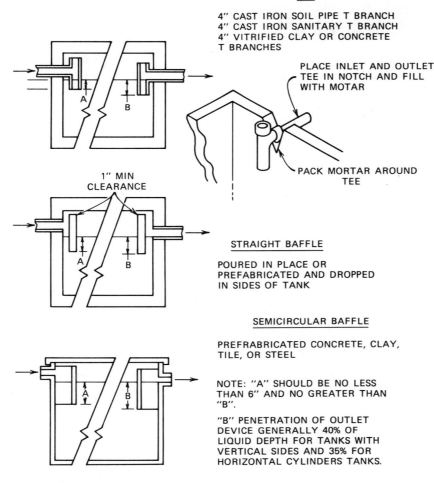

TEE

4" CAST IRON SOIL PIPE T BRANCH
4" CAST IRON SANITARY T BRANCH
4" VITRIFIED CLAY OR CONCRETE
T BRANCHES

PLACE INLET AND OUTLET
TEE IN NOTCH AND FILL
WITH MOTAR

PACK MORTAR AROUND
TEE

1" MIN
CLEARANCE

STRAIGHT BAFFLE

POURED IN PLACE OR
PREFABRICATED AND DROPPED
IN SIDES OF TANK

SEMICIRCULAR BAFFLE

PREFRABRICATED CONCRETE, CLAY,
TILE, OR STEEL

NOTE: "A" SHOULD BE NO LESS
THAN 6" AND NO GREATER THAN
"B".

"B" PENETRATION OF OUTLET
DEVICE GENERALLY 40% OF
LIQUID DEPTH FOR TANKS WITH
VERTICAL SIDES AND 35% FOR
HORIZONTAL CYLINDERS TANKS.

example 6–14. Inlet and outlet devices.

addition to the provision for scum storage, 1 inch is usually provided at the top of the tank to permit free passage of gas back to the inlet and house vent pipe.

For tanks having straight vertical sides, the distances between the top of the tank and the liquid line should be approximately 20% of the liquid depth. In horizontal cylindrical tanks, area equal to approximately 15% of the total circle should be provided above the liquid level. This condition is met if the liquid depth (distance from outlet invert to bottom of tank) is equal to 70% of the diameter of the tank.

Use of Compartments. Although a number of arrangements are possible, compartments, as used here, refer to a number of units in series. These can be either separate units linked together, or sections enclosed in one continuous shell, with watertight partitions separating the individual compartments.

A single-compartment tank will give acceptable performance. A two-compartment tank, with the first compartment equal to 50 to 65% of the total volume, provides better suspended-solids removal, which may be especially valuable for protection of the soil absorption system. Tanks with three or more equal compartments give performance at least as good as single-compartment tanks of the same overall capacity. Each compartment should have a minimum-plan dimension of 2 ft with a liquid depth ranging from 30 to 60 inches.

An access hole should be provided to each compartment. Venting between compartments should be provided to allow free passage of gas. Inlet and outlet fittings in the compartmented tank should be proportioned as for a single tank (example 6–12.) The same allowance should be made for storage above the liquid line as in a single tank.

General Information

Septic tanks should be cleaned before too much sludge or scum is allowed to accumulate. If either the sludge or scum approaches too closely to the bottom of the outlet device, particles will be scoured into the disposal field and will clog the system. Eventually, liquid may break through to the ground surface and the seepage may back up in the plumbing fixtures. When a disposal field is clogged in this manner, it is not only necessary to clean the tank, but it may be necessary to construct a new disposal field.

The tank capacities given in Table 6–5 will give a reasonable period of good operation before cleaning becomes necessary, but there are wide differences in the rate that sludge and scum will accumulate from one tank to the next. Tanks should be inspected at least once a year and cleaned when necessary.

Inspection

Although it is difficult for most homeowners, actual inspection of sludge and scum accumulations is the only way to determine definitely when a given tank needs to be pumped. When a tank is inspected, the depth of sludge and scum should be measured in the vicinity of the outlet baffle. The tank should be cleaned when the bottom of the scum mat reaches within approximately 3 inches of the bottom of the outlet device, or when sludge comes within the limits specified in Table 6–6. Scum can be measured with a stick to which a weighted flap has been hinged, or with any device that can be used to feel out the bottom of the scum mat. The stick is forced through the mat, the hinged flap falls into a horizontal position, and the stick is raised until resistance from the bottom of the scum is felt. With the same tool, the distance to the bottom of the outlet device can be found (see example 6–13).

A long stick wrapped with rough white toweling and lowered to the bottom of the tank will show the depth of the sludge and the liquid depth of the tank. The stick should be lowered

TABLE 6–6. ALLOWABLE SLUDGE ACCUMULATION.

LIQUID CAPACITY OF TANK	LIQUID DEPTH			
	$2\frac{1}{2}$ ft	3 ft	4 ft	5 ft
	DISTANCE FROM BOTTOM OF OUTLET DEVICE TO TOP OF SLUDGE			
750 gallons	5 in.	6 in.	10 in	13 in.
900	4	4	7	10
1000	4	4	6	8

Tanks smaller than the capacities listed will require more frequent cleaning.

behind the outlet device to avoid scum particles. After several minutes, if the stick is carefully removed, the sludge line can be distinguished by sludge particles clinging to the toweling.

In most communities where septic tanks are used, there are firms that specialize in cleaning them. The local health department can make suggestions on how to obtain this service. Cleaning is usually accomplished by pumping the contents of the tank into a tank truck. Tanks should not be washed or disinfected after pumping. A small residual of sludge should be left in the tank for seeding purposes. The material removed may be buried in uninhabited places or, with permission of the proper authority, emptied into a sanitary sewer system. *It should never be emptied into storm drains or discharged directly into any stream or water course.* Methods of disposal must be approved by health authorities.

Chemicals

The functional operation of septic tanks is not improved by the addition of disinfectants or other chemicals. In general, the addition of chemicals to a septic tank is *not* recommended. Several proprietary products that are claimed to clean septic tanks contain sodium hydroxide or potassium hydroxide as the active agent. Such compounds may result in sludge bulking and a large increase in alkalinity, and interfere with digestion. The resulting effluent may severely damage soil structure and cause accelerated clogging, even though some temporary relief may be experienced immediately after application of the product.

Frequently, however, the harmful effects of ordinary household chemicals are overemphasized. Small amounts of chlorine bleaches, added ahead of the tank, may be used for odor control and will have no adverse effects. Small quantities of lye or caustics added to plumbing fixtures are not objectionable as far as operation of the tank is concerned. Dilution of the lye or caustics in the tank will be enough to overcome any harmful effects that might otherwise occur.

According to the *Joint Committee on Rural Sanitation,* some 1200 products, many containing enzymes, have been placed on the market for use in septic tanks, and extravagant claims have been made for some of them. As far as is known, however, none has been proved advantageous in properly controlled tests.

Soaps, detergents, bleaches, drain cleaners, or other material normally used in the household, will have no appreciable adverse effect on the system. However, because both the soil and essential organisms might be susceptible to large doses of chemicals and disinfectants, moderation should be the rule. Advice of responsible officials should be sought before chemicals arising from a hobby or home industry are discharged into the system.

Guidelines on Use

It is generally advisable to have all sanitary wastes from a household discharge to a single septic tank and disposal system. For household installations, it is usually more economical to provide a single disposal system than two or more with the same total capacity; normal household waste, including that from the laundry, bath, and kitchen, should pass into a single system.

But roof drains, foundation drains, and drainage from other sources producing large intermittent or constant volumes of clear water should *not* be piped into the septic tank or absorption area. Such large volumes of water will stir up the contents of the tank and carry some of the solids into the outlet line; the disposal system following the tank will likewise become flooded or clogged, and may fail. Drainage from garage floors or other sources of oily waste should also be excluded from the tank.

Toilet-paper substitutes should not be flushed into a septic tank. Paper towels, newspaper, wrapping paper, rags, and sticks may not decompose in the tank, and are likely to lead to clogging of the plumbing and disposal system.

Waste brines from household water softener units will have no adverse effect on the action of the septic tank, but they may cause a slight shortening of the life of a disposal field installed in a structured clay soil.

Adequate venting is obtained through the building plumbing if the tank and the plumbing are designed and installed properly. A separate vent on a septic tank is not necessary.

Inspection

After a sewer line and soil absorption arrangement have been installed and before they are used, the entire system should be tested and inspected. The septic tank should be filled with water and allowed to stand overnight to check for leaks. If leaks occur, they should be repaired. The soil absorption system should be promptly inspected before it is covered to be sure that the disposal system is installed properly. Prompt inspection before backfilling should be required by local regulations, even where approval of plans for the subsurface sewage disposal system has been required before issuance of a building permit. Backfill material should be free of large stones and other deleterious material and trenches and excavations should be overfilled a few inches to allow for settling.

Section 7

Water Conditioning

Section 7

Throughout this book we have talked about the problems that develop in a fresh-water piping system when the water corrodes the inside surfaces or leaves deposits of minerals along the conduit. Generally, the water that causes buildup in pipes is "hard"; that is, it contains certain salts that solidify when we don't want them to. The problem is almost always more severe in the hot-water pipes than in the cold. When hard water is heated, the calcium and magnesium minerals in the water are converted to a less soluble form that gradually builds up as a layer of hard scaly material in the pipes. Example 7–1 shows this buildup in the outlet piping of a hot water heater.

As little as an eighth inch of hard-water scale can boost your heating bill as much as 28 cents for every dollar and reduce heating efficiency by more than 20%. The scale does more than slow the flow and lower the water pressure; it acts as an insulator that develops "hot spots," which prevent the retention of heat by the water and destroy the pipes. All it takes to prevent this is a means of *softening the water.*

In this section, we discuss the relative merits and methods of conditioning water to make it safer to drink, more economical to use in cooking and laundering, and more palatable to the taste. Our jumping-off point is water as it comes to us in nature.

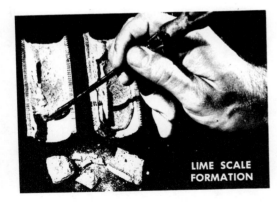

example 7-1. Loose scale, which is mainly calcium and magnesium carbonated in water-heating piping.

LIME SCALE FORMATION

Unconditioned Water

Most people tend to think of rain water as being pure water. Unfortunately, this isn't so. Water always contains impurities, whether it's obtained as rain, from streams, or from wells deep beneath the surface of the earth. It has been estimated that some 16 million tons of water are being precipitated each second and a like amount is being soaked up by the atmosphere in the process we call *evaporation*. But none of it is pure.

Water high up in the atmosphere is pure—but only so long as it's there. The instant it begins to precipitate, it becomes impure. Precipitation is the gathering of water into liquid form around a particle that may be microscopic in size. As the particle and its water buildup begin to fall, the mass accumulates other impurities. It absorbs oxygen, nitrogen, and carbon dioxide from the air, and with them, dust, smoke fumes, and whatever else happens to be airborne and small—even bacteria in the spores of microorganisms.

But running water is purified—right? Perhaps, but not necessarily. The water in streams may be turbid (contain visible materials) with the presence of clay and silt. Animal wastes in agricultural land can pollute waters. Surface waters generally are exposed to an unending torrent of pollution by the sewage of cities and the wastes of industry. No stream known in this country is totally free of human pollution, not even those in the comparatively virgin lands of upper New England.

How about ground water, the water lying deep down below the

reach of our pollutants? Here nature does some "polluting" of its own. As rain waters percolate through the soil, they absorb carbon dioxide, which in turn forms carbonic acid. The acid then dissolves a certain amount of minerals in the soil and rocks through which the water flows; the most common minerals dissolved are limestone, calcium, and magnesium.

Water hardness depends on the quantity of these minerals in the water.

The water obtained from deep wells is usually clearer than that of shallow wells and more often totally colorless by virtue of the filtration taking place in the soil as the waters descend through it. But filtration like this doesn't remove the hardness minerals; it only eliminates the turbidity and certain other impurities. Deep-well water is usually safer to drink than shallow-well water—but it's also generally harder.

Water Hardness

The chemicals of hardness are dissolved calcium and magnesium salts such as lime, epsom, and gypsum. Of course, not all water supplies are undesirably hard—but there are almost no waters anywhere that are entirely free of hardness. Example 7–2 is a water-hardness map that shows the average hardness of waters found in various parts of the U. S. The white areas—note how few there are—indicate regions of relatively little hardness; the dark areas show extreme hardness. It is important to point out that this map illustrates hardness only; it doesn't indicate the degree to which waters may be polluted with such chemicals as iron and corrosive acids.

Most of us are already familiar with the disadvantages of hard water. It:

- increases the quantity of water and soap needed to make clothes clean in the laundry,

- makes a dirty residue on the white surfaces of bathtub and lavatories,

- leaves a lusterless film on dishes and dulls otherwise bright fabrics,

- leaves deposits that cause premature destruction of pipes,

- lowers water pressure,

- raises the cost of operating a water heater, radiator, or water heating system,

and it doesn't feel good. But it's not necessarily unhealthy. As a matter of fact many consider *controlled* hardness of water to be advantageous to the human body.

In this book, we won't take a stand for or against hardness except in its extreme form, which undeniably is objectionable. Of course, our definition of *extreme* has to be arbitrary, so we'll consider anything over 8 grains per gallon as being worthy of softening.

The unit of measure—grains per gallon—for hardness is more or less universal, but it can be translated directly to "parts per million." To convert grains per gallon (as indicated on the hardness map) to parts per million, simply multiply the number of grains per gallon by 17.1. For example, 10.5 grains per gallon translates into 179.55 parts per million (ppm).

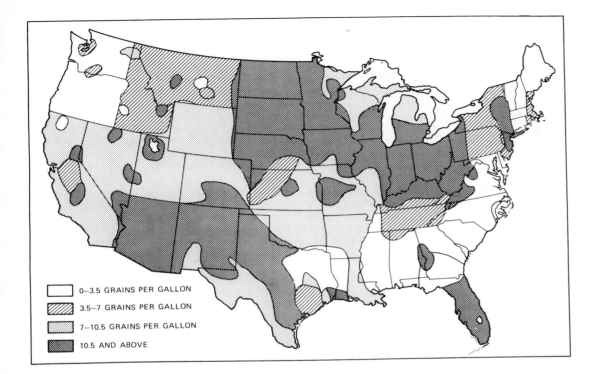

example 7-2. Map showing water hardness areas in recognized variations.

0–3.5 GRAINS PER GALLON

3.5–7 GRAINS PER GALLON

7–10.5 GRAINS PER GALLON

10.5 AND ABOVE

Iron Content

Iron, one of the most common elements in nature, discolors everything it touches when it's suspended in water. As such, it's an undesirable element when it occurs as ferric oxide. It also occurs as ferrous oxide, but in this condition it is completely soluble, tasteless, colorless, odorless; you wouldn't know it's there except for the benefits of the minerals in the water. It takes a set of "favorable" conditions to convert iron into its soluble form; otherwise it occurs as an insoluble oxide, painting everything with that familiar rust stain.

Sometimes iron oxides will convert from one state to the other; deep-well waters, for example, generally carry iron in its colorless form; but after being drawn from the well the water is exposed to air, and the subsequent release of carbon dioxide converts ferrous oxide to ferric oxide.

There's more. Certain living organisms, called *iron bacteria,* accompany the wet oxides of iron. They feed on the pump of a well, the tank, the piping, and the iron fixtures. And they build a thick coating of slime on tanks, water closets, and other surfaces that stay in more or less constant contact with the water.

What is the limit of what's considered acceptable as far as iron content is concerned? According to the U.S. HEW standards, if the iron content is greater than three-tenths of the one part per million (0.3 ppm), it's objectionable from an appearance and taste point of view. Most iron-bearing well waters range from 1 to 5 ppm, although some are in the 5 to 15 ppm range, and a few contain in excess of even these ranges.

Sulfur Content

So-called sulfur water is actually a hydrogen sulfide solution. If you've ever tasted sulfur water, you probably remember it well; hydrogen sulfide makes the water taste and smell like rotten eggs. And it combines with whatever impurities may be present in the water to form *black water,* or water with a disagreeably dark coloration.

Sulfur water occurs in nature, but more often than not the process of its formation is a result of certain types of bacteria that happen to be present in the water table (from nearby mining operations, for example, or as a result of any activity that expels sulfur in almost any form).

Naturally, people who have sulfur water in their homes are usually anxious to get rid of it. It turns liquors black when it's used to mix certain drinks, and it makes coffee, tea, and ade drinks taste "terrible." It's hopeless for canning.

It's corrosive, too. Hydrogen sulfide eats away pump parts, pipes, tanks, water heaters, fixtures, and almost anything made of an alloy of steel, iron, or copper. It can ruin paint and wall paper and turn silverware black almost instantly. Even if we enjoyed the rotten-egg taste and odor, most of us couldn't afford to live with the costly wastes that accompany water laden with hydrogen sulfide.

So how much is allowable before the disadvantages start getting heavy? Waters containing 1 part per million are considered most definitely objectionable. Those that contain less than half a part per million are not, usually. It is not uncommon to find waters with as many as 75 parts per million.

Acidity

All waters have a value of acidity or alkalinity according to a scale numbered from 1 to 14. The midscale position, 7.0, is exactly neutral. Numbers greater than 7 are alkaline—the higher the alkalinity, the greater the number; numbers below 7 are acid—the lower the number, the more intense the acidity. The scale, shown in example 7–3, is referred to as a pH table; and the number that identifies a solution's acidity or alkalinity is called its *pH value.*

It is important to remember that the pH scale is much like Richter's earthquake rating system; that is, each number represents a value 10 times greater than the number immediately preceding it. A water that has a pH value of 2.0 is ten times more acidic than vinegar, with a pH of 3. And a liquid with a pH of 12 is 10 times more alkaline than ammonia, with a pH of 11. Ammonia is a thousand times more alkaline than baking soda. Lye is at the upper (alkaline) extreme of the scale; muriatic acid is at the other extreme.

The pH value of water typically is around 7, but within a range of 6.0 to

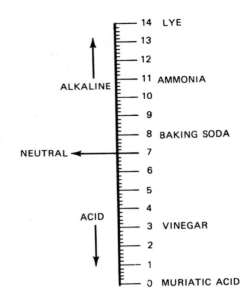

example 7–3. pH table.

8.0. Those areas indicated as nonhard-water concentrations in the water hardness map of example 7–2 will generally be acidic; experience has shown that where water hardness drops below 3.5 grains per gallon the waters tend to be acidic.

Waters tending to the acid side of the pH scale have a knack for eating away metal and corroding pipes, tanks, and fixtures. They cause premature leaks and equipment failures. Where the plumbing pipework is copper, acid waters pick up the copper and deposit it in the form of a bluish green stain on white fixtures; where the pipework is galvanized iron, the fixtures get a deposit of rust.

Turbidity and Color

Turbidity is a measurement of the amount of light passing through a liquid; it should not be confused with color. A liquid that is clear but colored may not be turbid; a liquid that is colorless but cloudy is turbid. Turbidity encompasses muddy water, cloudy water, and water that contains sediment and other particulate matter, which usually occurs in the form of clay or sand, rust (iron oxides), or organic matter, both visible and microscopic.

Pond water may be clear until someone jumps in, then it becomes turbid. Water taken from a fast-moving stream or a river is normally turbid, with the particulate matter and cloudiness a legacy of the stirring action of the water along its path. Springs and wells are normally not turbid; the ground through which the waters flow to reach the point of access usually filters out particles with high efficiency.

Color in water has an almost unlimited number of causes, some of which we have already discussed. It may stem from coal or peat sources, from swamps, and various iron-bearing strata. Very often color in water can be attributed to the presence of tannin in solution, a substance of organic origin that combines readily with water to give it a tea-like coloration.

What's the turbidity and color limit for potable water? Governmental recommendations say that turbidity should be below 10 ppm, and that color should be less than 20 standard units. (Color can be measured on a scale that reads crystal-clear as zero, with increasing value as the color intensifies.)

Taste and Odor

When it comes to measurement of taste and odor, the scales must become subjective—unless impurities are involved, the taste of water and its odor are a matter of personal preference. Water can be quite safe to drink even though completely disagreeable to the nose and palate. Still, you won't know that "bad" tasting water is palatable unless you subject it to a bacteriological examination by individuals trained and equipped to make such tests. Your local board of health will help you in this regard.

Actually, taste is not so reliable an indicator of a water's quality as odor. There are only four basic tastes: sweet, sour, bitter, and salty. Most of what we perceive as taste is in reality a combination of the four in concert with a specific odor, of which there are considerably more than only four. We stated earlier that sulfur water "tastes terrible"; in truth, however, hydrogen sulfide in water is virtually tasteless. It's the *odor* of rotten eggs that makes us think the water tastes bad.

Bitter tastes in water may be caused by iron, manganese, sulfates, or lime concentrations. Waters with excessive salts are said to taste "brackish." But there is no method for determining the taste of water from a quantitative viewpoint, so we rely on our sense of smell for "testing" a water supply's potability; one of the standard tests we can make is the *threshold odor test*.

The threshold odor test is the most widely used method of evaluating odor levels. It consists of comparing different dilutions of water samples (diluted with pure, odor-free water) to a control sample containing only pure, odor-free water. The dilution at which the odor can be barely detected is called the *threshold point,* and the odor at that point is expressed quantitatively as a *threshold number.* The threshold number is simply the number of times the odor-bearing sample is diluted with a like quantity of odor-free water to reach the threshold point.

Let's take an example. Suppose an odor-bearing water sample requires dilution to ten times its volume with odor-free water in order to make the odor just barely perceptible in the sample. Its threshold number will be 10. A more concentrated odor-bearing water may require dilution to 100 times its volume; its threshold number would be 100. The corrective steps to take to render household water totally odor-free will be determined by the threshold number. A number from 1 to 5 may indicate the problem can be overcome with simple filtration; water with a number between 5 and 10 may require special filters or chlorination; higher numbers indicate the need for combinations of techniques.

Contamination

Those of us who live in cities can rest assured that our water is safe to drink regardless of what it tastes like (and it may taste *very* disagreeable, depending on the fluorides, chlorination, and other treatment the water is subjected to). But private water systems are another story entirely. A well that has been used for years—that has been tested perhaps several times in years past—is safe to drink, right? Unfortunately, this isn't necessarily true; nor is it even a likelihood. Wells can become infected in many ways, and contamination can take place overnight. Water that is not continuously treated has no safeguard against disease organisms that enter the system.

When we speak of contaminated water we mean water *containing sewage waste*. Sewage wastes enter the well from contaminated ground water. When people are informed that their water is contaminated, it is often difficult for them to understand, because they may have been drinking the water for some time. We can drink water contaminated with sewage and be exposed to only the diseases of the people whose sewage is contaminating our water.

Most untreated water supplies have periods when they are free from waterborne disease, but other periods when they show sewage contamination. It is this lack of dependable safe water in an untreated water system that has resulted in over 98% of all municipalities in our country chlorinating their water. In other words, for complete protection it can be assumed that the continuous sanitary quality of each water system is not assured unless a proper amount of purifying agent is kept in it continuously.

You can expect to find impurities in practically all water. Some of these impurities do not amount to much—others are extremely dangerous to health. Impurities in water can be broken down into two major categories: *dissolved* and *suspended*.

Dissolved impurities are organic or inorganic materials or chemicals that may cause an unpleasant taste, color, or odor.

Waterborne Diseases

Suspended impurities include organisms as well as organic and inorganic materials that usually make the water turbid or muddy looking. Waterborne diseases caused by dangerous organisms include typhoid, paratyphoid, cholera, amebic dysentery, schistosomiasis, poliomyelitis, infectious hepatitis, and diarrhea. A brief discussion of these waterborne diseases, which will spread if proper treatment is not given to water supply, will help to stress the importance of continual care. *Typhoid fever* is an intestinal disease caused by the bacterium known as *bacillus typhosus.* Symptoms of this disease are rose-colored eruption of the skin accompanied by a high fever (lasting about 4 weeks) and frequent bowel movements. Typhoid-fever organisms are readily destroyed by field chlorination and iodine-dosing methods. Most waterborne diseases do not appear immediately after contaminated water has been used because they need time to grow after entering a person's system. This period is known as the *incubation period.*

Paratyphoid is similar to typhoid in sources of infection and in symptoms; the organisms are, like the typhoid bacillus, readily destroyed by field chlorination and iodine treatment methods.

The incubation period varies from 4 to 10 days. An attack gives a person immunity from a second attack of paratyphoid, but it doesn't give immunity from typhoid.

Cholera germs are discharged from the body in feces, where they live for several days. When water in any form contacts this germ, it is carried along and multiplies.

Amebic dysentery is an infectious intestinal disease. Symptoms are eruptions of the skin and frequent bowel movements. This disease is caused by a very small animal rather than bacteria and resists ordinary chlorination. It is carried by amebic cysts that foray in the intestines and are then discharged in the feces.

Cysts are crustaceans somewhat like tiny shellfish. An ameba forms a hard shell, called a cyst, around itself when it is in unfavorable surroundings. When it is taken into a warm body, the ameba comes out of its shell and enters the human system. Cysts are much larger than bacteria and can be filtered by certain kinds of fine filters. To kill cysts with chlorination, a higher than normal chlorine residue must be maintained; with *iodination,* cysts are destroyed within 30 minutes at normal dosage levels.

Schistosomiasis is caused by a small worm that may enter the body when contaminated water is consumed. Or it may enter through the skin while a person is bathing or swimming in contaminated water. Eggs of this parasite (commonly called *blood flukes*) are discharged from an infected person through the urine or feces. In fresh water, these eggs hatch into very small, free-swimming larvae that are not infectious to humans. However, if these larvae find fresh-water snails into which they can enter; they develop into the next form, *cercariae,* and become highly infectious to human beings, In water, larvae can live for only 24 hours, and cercariae for only 36 hours. The effective remedy, therefore, is to kill all snails at the water source. Once the snails are destroyed, the cycle is broken and the disease ceases.

Infectious hepatitis and poliomyelitis are diseases commonly carried by viruses in a water supply. A *virus* is smaller than a bacterium and usually quite a bit harder to kill. Current research indicates that viruses can survive in chlorine dosages up to 5 times that normally used to purify water, but they don't fare so well with iodination, which can kill a polio virus in 9 minutes at standard dosage levels.

Diarrhea is a name given to several intestinal diseases that are characterized by cramps and frequent bowel movements, with watery feces. Inadequate sanitary protection of food and water can cause diarrhea. Where the disease is caused by food, it is restricted to those who consume the contaminated food; waterborne infection is likely to be widespread. Proper chlorination or iodination measures will control waterborne diarrhea.

Conveyance of Disease Organisms

In cases of both shallow and deep wells, the water originally comes from the earth's surface. It reaches the water table through cracks in the earth, septic-tank tile fields, excavations, rock outcroppings, abandoned wells, sink holes, tree roots, animal borings, and by percolation through the ground. The water comes directly from rain on the earth's surface or from lakes, rivers, ponds, and so on. As water travels these routes to the water table, it carries with it whatever bacteria, viruses, and cysts it picks up. Little is known whether this ground filtration is of any value where viruses are concerned.

Generally shallow wells are the least satisfactory from a sanitary and water quality standpoint. Because water in rock travels for great distances with no ground filtration, wells in rock water are generally poor health risks. Some people say their wells are in solid rock, but if the rock were solid, it would have no water in it. Actually, water-bearing rock has thousands of small cracks and solution channels acting as tiny pipelines that can carry contamination for miles from where it enters the rock water.

Testing for Safety

A water supply can be deemed safe for drinking only if it has been subjected to laboratory analysis. Unfortunately, there is no simple method of testing for bacteria as there is for iron or hardness. To confirm the presence of disease organisms is a long and costly laboratory process and in some cases, such as hepatitis virus, the organism can't be isolated at all! Therefore, a simpler indicator laboratory test providing a reasonable margin of safety has been devised. This test takes from 24 to 48 hours to perform.

The principle of this test is based on the fact that humans and other warm-blooded animals grow in their intestines a family of microorganisms called *coliform bacteria*. These organisms are usually harmless, but are relatively easy and inexpensive to test for. Since coliform bacteria are present in human and animal excrement only, whenever a sample of water shows a presence of coliform, the water is immediately assumed to contain human sewage and is therefore declared unsafe for human consumption.

In water testing, the frequency of tests is of extreme importance. Unfortunately, well-water testing for private systems is very seldom done on a regular basis. In the vast majority of cases, only one sample is taken and if it comes back safe, no tests are ever run again! In many instances, this one test is made on a new well (just chlorinated by the well driller, but not yet having had an opportunity to become developed). Water testing indicates the safety of the submitted sample only and gives no guarantee of the safety of the water 5 minutes, 5 days, or 5 months later! Private water sources should be tested at least once each year.

How To Condition Water

A variety of methods are employed in the treatment and purification of water. Table 7–1 is a quick-reference chart that lists some of the more common impurities of water, the types of problems they cause, and some of the ways they may be corrected. This is not an all-encompassing chart—it does not list those correctional steps typically applied by municipalities and large treatment plants—but it should give a good idea of what can be done to clean up water on an individual basis and what to expect from various forms of treatment.

Controlling Water Hardness

Water softening may be carried out by *ion exchange* or by *lime-soda treatment*. There is another technique, called *threshold treatment,* for controlling scale formations in pipework that prevents the hardness chemicals from building up in a water supply, but this method does not actually "soften" the water in the generally accepted sense of the word.

Ion-Exchange Water Softening

Ion-exchange water softening is a chemical operation in which certain minerals—including calcium and magnesium, the hardness materials in water—are exchanged for other materials that will not form a scale. This action is accomplished by passing the water through a bed of a solid ion-exchange medium such as zeolite, which captures and holds the unwanted mineral ions in the water and releases to the water an equivalent amount of sodium. The ion-exchange materials that are used in the process are insoluble, granular material possessing this unique property of exchange. They may not be natural mineral zeolites but precipitated synthetic substances similar to zeolite in character, or organic or carbonaceous zeolites or synthetic ion-exchange resins.

An *ion* is a charged particle. But that may not mean much to you unless you're in electronics or are a physics or chemistry major. So let's examine the subject a bit.

All things, organic or mineral, whether occurring in the form of a gas, a liquid, or a solid, are made up of minute particles called *ions*. These are attracted to each other electrically to form *atoms*. Atoms differ in the number of particles they contain, and they differ in size and weight. In turn, atoms combine in a vast majority of types and proportions to form *molecules*. The distinctive feature of a molecule is that it consists of the smallest group of atoms by which we can identify the specific substance. A molecule of salt—composed of a sodium atom and a chlorine atom—is bound into a single unit by electrical attraction.

TABLE 7–1. COMMON IMPURITIES IN WATER, WITH REMEDIES

IMPURITIES	PROBLEM	REMEDY
Hardness (calcium and magnesium)	Scale in pipes and water heaters. Causes insoluble soap curd on dishes and fabrics.	Removal by ion-exchange water softener.
Iron	Discolors water, stains plumbing fixtures and fabrics, causes deposits in heaters.	Removal by a softener or by an iron filter when large amounts are present. Also removed by chlorination, then filtration.
Acid in water (low pH)	Corrosion, attacks piping and tanks; red stains from galvanized pipe, blue-green stains from copper.	pH raised by neutralizing filter or by soda ash food.
Hydrogen sulfide	Disagreeable taste and odor, tarnishes silverware.	Removal by oxidizing filters or by chlorination, then filtration.
Mud, clay, silt (dirty water)	Suspended matter in water (water cloudy or dirty).	Sand filter
Taste and odors (organic matter)	Makes water unpalatable.	Carbon filters—in some cases chlorination, then filtration.
Sodium salts	Salty or alkali taste.	Cannot be economically treated.
Bacteria	Source of disease; unfit for human consumption.	Chlorination and filtration, iodination.
Algae	Taste, odor, color.	Chlorination and filtration.

Many substances whose atoms are held together by ionic forces—by electrical attraction—have a characteristic way of reacting (called *ionization*) when they are dissolved in water. The water separates the molecule into its component ions, or charged atoms. Ions that are positively charged are called *cations* (cat-ions), and those that are negatively charged are called *anions* (an-ions).

The fact that some substances break down into their constitutents in water is the key to softening water by ion exchange. The most efficient way of dealing with the problems of calcium and magnesium in water is to remove their troublesome ions before they can do their damage. Ion exchange does it.

How It Works. The ion-exchange process, remember, removes calcium and magnesium ions and substitutes sodium in their stead. During the exchange, two sodium ions—cations—are substituted for each calcium or magnesium ion received; this is because the exchange process has to maintain an ion balance. The calcium and magnesium ions have a charge value of 2, and the sodium ion's value is 1. So when an ion with a value of 2 is removed, it must be replaced with two individual ions that have a value of 1.

The exchange process takes place because the new ions are attracted more strongly to the exchanger than the loosely bound ions of the unconditioned water; so the new ions displace the old and occupy the site the old ones have vacated.

The zeolite substances occur in nature, but it has proved more economical to synthesize them. In the synthesizing process, the exchange value of the "zeolite" is increased substantially. The synthetic zeolite consists of fine porous beads about the size of a pinhead, and they are made from petroleum byproducts; as such, of course, they are organic in character rather than mineral.

A resin can carry out millions of transactions (exchanges) over long periods before its supply of sodium ions must be replenished. When the softener can no longer remove the hardness ions from the incoming water; that is, when its absorption capacity is exhausted, the absorption process is temporarily reversed and a regeneration process is initiated.

Regeneration is accomplished by treating the exchanger with a concentrated solution of salt (sodium chloride), the source of new sodium ions. This regeneration cycle renews the ion absorption power of the softener so that it can be reused for countless numbers of cycles.

It takes about half a pound of salt for every thousand grains of hardness exchange. Softeners are rated in "grain capacity," which is a measure of the number of grains of hardness per gallon of water that can be neutralized by the system. It takes ten pounds of salt, then, to regenerate a water softener with a 20,000-grain capacity.

The Debit Side. As we mentioned, an ion exchange softener turns out water of zero hardness. This water will not form even a thin scale on piping, so the pipework is left unprotected against rust; oxygen in the water can then react with the unprotected inside surfaces of galvanized iron pipes and rust them. Depending on the circumstances, it may be desirable to mix the softened water with untreated water to obtain a small degree of hardness. If the unsoftened water contains a hardness of more than three tenths of a part per million, the unsoftened water can be aerated and filtered, then mixed with the softened water in controlled ratios. The pH of ion-exchanged water can be adjusted to make the water less corrosive by addition of a caustic silicate solution.

Lime-Soda Water Softening

We can soften water by adding lime or lime and soda ash, a process that precipitates the hardness chemicals. After precipitation, the insoluble compounds are removed by sedimentation and filtration. The process is generally not practical for individual installations, because it depends on further treatment downstream.

Lime and soda ash are added to raw water, and the softening "reaction" occurs during mixing and flocculation. (*Flocculation* is a clumping of aluminum sulfate particles added to water to entrap bacteria and absorb water coloration.) The precipitated calcium and magnesium compounds are removed during sedimentation (a settling process used in large treatment plants to eliminate turbidity).

Threshold Treatment

Carbonate scales can be prevented by the addition to water of very low concentrations of polyphosphates, but no true chemical action takes place in the process. The added polyphosphate prevents tiny crystals of scale from growing or accumulating, and the water is stabilized without actual removal of the scale-forming constituents.

The maximum concentration of calcium bicarbonate that can be stabilized by threshold treatment varies with the temperature and the alkalinity of the water. For normal temperatures, 50 to 70°F, water that is quite hard can be stabilized with 2 ppm of glassy phosphates. When polyphosphates are used for scale control, the dosage must be controlled with considerable precision; underdosing won't allow the job to be done, and overdosing will result in a buildup of phosphate scales in the pipes.

Controlling Iron and Manganese Content

Iron is present in many ground waters, and manganese is present occasionally also. The presence of more than 0.3 ppm of iron in a water supply, or of more than 0.5 ppm of manganese, is considered objectionable. Both these compounds stain everything that comes into contact with them when they are in suspension. This is especially true of plumbing fixtures and clothes that are rinsed in such water.

Manganese, when present, is almost invariably associated with iron. Due to its chemical similarity to iron, it is removed by the same process, though the process that removes iron will not generally remove manganese in the same relative weight ratios. Manganese is not oxidized quite as readily as iron.

Iron and manganese affect certain foods by darkening them, some to complete blackness. Iron also favors the growth of *crenothrix,* more commonly known as *iron bacteria.* Crenothrix growths are jellylike and stringy, and at times they become so voluminous in a water system that they actually interfere with the flow of water through the pipes. Iron and manganese may be removed in several ways.

Threshold Treatment

Where the ferrous iron content doesn't exceed 2 or 3 ppm, color and staining can be prevented by the threshold treatment described previously. In this treatment method, 2 ppm or so of polyphosphate has to be added before the water is treated with chlorine and before it comes into contact with air.

For best results in private water systems, the polyphosphate should be fed to the water before it reaches the pressure tank so that the iron is not oxidized by air in the tank. Polyphosphate will not prevent iron from staining after the iron has already been oxidized.

The polyphosphate then should be fed either into the suction side of the pump or into the discharge pipe between pump and tank (example 7–4A). In the case of a submersible pump (example 7–4B), it can be fed into the well at the intake of the pump.

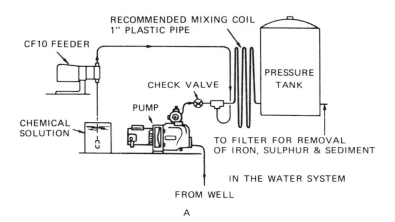

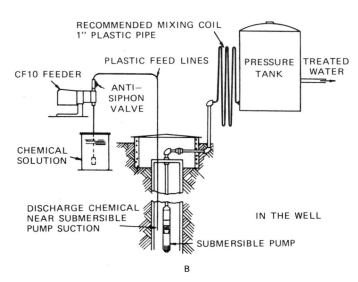

example 7–4. Feeding liquid solutions in jet-pump (A) and submersible-pump (B) installations.

For each ppm of iron present, 4 ppm of polyphosphate is usually fed. To prepare 10 gallons of 4 ppm concentration of polyphosphate for each 100 gallons per hour of pump capacity, 1½ ounces of polyphosphate is required.

Polyphosphates will hold manganese in solution as long as the manganese level does not exceed about 1 ppm; again, the material has to be added before chlorine treatment and air contact.

Water-Softener Iron Removal

Probably the simplest way to remove iron and manganese from clear water is with an ion-exchange water softener. This is an effective approach if the iron content doesn't exceed 3 ppm.

Iron is particularly troublesome on well jobs because the iron is in complete solution as ferrous bicarbonate. When it comes into contact with air that is pressurized in the water tank, part of the iron is oxidized before it reaches any following filter or softener. The softener can suspend iron, but it must be thoroughly backwashed and cleaned often and with regularity. The iron tends to foul the zeolite bed and to bleed through. Iron, of course, is heavier than the softener resin, so the job of backwashing can be quite tedious.

Manganese Zeolite Filtration

A special manganese zeolite filter can be used to remove iron (and manganese) up to a level of 10 ppm. The principle of this filter is to oxidize the iron in order to precipitate it, and then to filter the precipitate within the same unit. The mineral bed in the filter contains *greensand,* or manganese dioxide, which converts ferrous oxide to ferric oxide, thus precipitating the iron. The oxygen in the filter is replenished by adding potassium permanganate.

Chlorination and Filtration

Iron in solution can normally be removed by chlorination and filtration. Chlorine is an excellent oxidizing agent and when fed into the water supply will oxidize and precipitate the iron. This oxidized iron can then be removed by filtration, which can be accomplished by a standard sand or carbon filter.

In addition to oxidizing the iron in solution, chlorine will also destroy the iron bacteria, which cannot be removed by filtration alone.

If water is acidic (pH below 7), the pH must first be raised to a neutral level because iron will not readily oxidize in acid water. Soda ash can be mixed and fed with the chlorine solution from the same container.

A chlorine solution of 1 ppm will oxidize 1 ppm iron. Sufficient chlorine should be fed into the water system to leave a residue of at least 0.2 ppm. Purification as well as iron removal will be accomplished.

To prepare a 1-ppm chlorine solution, mix 6 ounces of household laundry bleach with 10 gallons of water for every 100 gph pump capacity. If residual chlorine is not 0.2

ppm, increase the amount of bleach until the correct residual is maintained. Then use this amount in preparing future solutions.

Iron Removal Summary

Generally, here are the guidelines for selection of the method to employ for removing iron from water:

1 If the iron content is not over 3 ppm—whether ferrous or ferric oxide in structure—a softener alone will eliminate it.

2 If iron is between 3 and 10 ppm, a manganese zeolite filter should handle the exchange effectively.

3 If iron content is above 10 ppm, or if the iron is bacterial, use a chlorinator and filter.

Controlling Sulfur Content

Several methods are used for removal of sulfides from water supplies. These include aeration, passing the water through a manganese zeolite filter, and treatment by chlorination.

Aeration

Aeration is mainly effective for removal of dissolved gases; since it is of little value for removal of algae and other impurities, it is often used in conjunction with other treatment methods. Aeration for removal of sulfides consists of exposing as much water surface as possible to air. The process causes dissolved gases to be released to the atmosphere.

Some types of aerators consist of spraying water over tiered trays or slats. Example 7–5 shows one design; this type is quite efficient and requires little attention.

Aeration isn't recommended with waters that have low alkalinity, because it would tend to increase the corrosiveness of the water.

Manganese Zeolite Filtration

This is the unit mentioned earlier for iron removal. It has an upper limit of 5 ppm of sulfur. In operation, this unit causes oxidation of the sulfur, which renders it tasteless, odorless, colorless, and completely noncorrosive.

Chlorination and Filtration

With small water systems, the most widely used method of sulfur control is chlorination because chlorine accomplishes so many other jobs. The chlorine chemically oxidizes sulfides, but it takes a considerable amount of chlorine to do the job, so the chlorination process is followed by *dechlorination,* which involves putting the water through a carbon filter, which absorbs excess chlorine. The process follows.

Start with a 2-ppm chlorine feed and then increase the feed until the sulfur odor disappears and the water has a free chlorine residual of 0.5 ppm. For 2-ppm chlorine feed, mix 12 ounces of household laundry bleach with 10 gallons of water for each 100

example 7–5. Aerator.

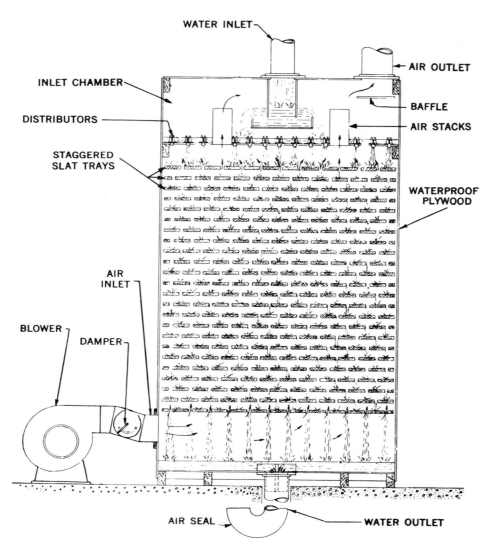

WATER INLET

AIR OUTLET

INLET CHAMBER

BAFFLE

DISTRIBUTORS

AIR STACKS

STAGGERED
SLAT TRAYS

WATERPROOF
PLYWOOD

AIR
INLET

BLOWER

DAMPER

AIR SEAL — WATER OUTLET

gph pump capacity. For example, if pump capacity is 500 gph, use 12 X 5 or 60 ounces of bleach in 10 gallons of water.

Controlling Acidity and Alkalinity

In small water systems, pH control is the most common means of corrosion prevention. In larger systems, various corrosion-inhibiting methods are employed, including a process of neutralizing electrochemical activity by inducing a counter current into the tanks or piping in which corrosion control is to be effected.

Sometimes water is treated to reduce the intensity of corrosion. This treatment involves adding chemicals to the water to adjust the pH and to cause deposition of a thin jellylike film on the inside surfaces of the pipes so that the water and the metal pipe surface have a sort of barrier between them. Some chemicals for treating water in this way are calcium carbonate, phosphates, and sodium silicate.

With calcium carbonate, the pH value of the water is raised slightly above the saturation point of the chemical. Hard alkaline waters may require addition of small amounts of soda ash. Soft waters of low alkalinity may require both lime and soda ash.

When phosphates are used, the rate of application of the phosphate to the metal surfaces is more important than the concentration of the phosphate in the water. A high-zinc-content polyphosphate concentration of about 10 ppm is typically used initially; this forms a thin protective film on the pipes. After this initial dose, the concentration is reduced to 2 or 3 ppm. Polyphosphates aren't much good at pH values below about 5; and sometimes they cause corroded pipes to start releasing clumps of built-up materials into the water. This, coupled with the fact that polyphosphates tend to support bacterial growth, makes polyphosphate application an alternative that isn't the most attractive of those available.

In soft waters, corrosion control may be enhanced by the addition of small amounts of sodium silicate. Like the other chemicals, this causes a thin deposit on the pipes that keeps the metal and water from direct contact. To maintain the film, the chemical must be fed in continuously at a rate of about 30 ppm.

There are two common methods of raising the pH of acid waters: use of neutralizing filters and feeding a soda-ash solution into the water.

Neutralizing Filters

When acidic water passes through a bed of limestone chips (they are calcium carbonate), the chemical combines with carbon dioxide in the water and raises the pH value. But the calcium carbonate tends to dissolve in the water, which raises the hardness level. As long as a water softener is used following the application of this treatment, the hardness-increase problem is overcome.

Soda-Ash Feeding

Perhaps the most satisfactory way of correcting the problem of acid water is to feed a soda-ash solution into the water supply continuously. In a private water system that employs a pump, it is best to feed this solution through a small tube down the well or into the pond at the end of the pump drop pipe; this technique gives protection not only to the water supply piping but to the pipe and fittings, including the tank, between the source and the pump.

Soda ash chemically combines with water and neutralizes it by virtue of the alkalinity of the compound. A soda-ash feeder can be adjusted to raise the pH of water to whatever value is required at a particular site. The biggest advantage of soda-ash feeding is that pH control is effected without raising the hardness of the water. It can be used with an automatic chlorinator to correct conditions of excess sulfur, excess iron, and bacteria, as well as to lower the acidity of the water.

Normally, the soda ash and chlorine bleach solutions are mixed and fed into the water system with a single feed pump and from the same solution container. When it's the initial application, dissolve a 20-ppm soda-ash solution in 10 gallons of water, and then adjust the solution strength so that the pH is 7.5. Then just add the bleach and adjust for the correct chlorine residual. This same ratio should be used to mix future solutions. Remember to dissolve the soda ash first, and then add the chlorine or bleach.

Controlling Turbidity and Color

Under most conditions, turbidity can be controlled by a conventional sand filter. If the suspended matter still won't settle out, applications of certain chemicals will cause flocculation of the particles so they will be large enough to filter completely. The application causes the tiny particles to grow in size so they can be trapped in standard filter elements.

A common floc-forming chemical is aluminum sulfate (filter alum). If sufficient natural alkali is not present in the water to form a good floc, additional alkali (soda ash) must be added. Observe example 7–6, which will give you an idea of how flocculation works. During the growth process, the floc increases in size, absorbs color, enmeshes bacteria particles causing turbidity, and with its increased weight settles to the bottom of a sedimentation basin or tank.

The equipment used to promote the growth of floc is called a *flocculator,* and the basin or chamber in which it is installed is called the flocculating basin. It is in the

example 7–6. The flocculation process.

FLOCCULATION

(COAGULATION)

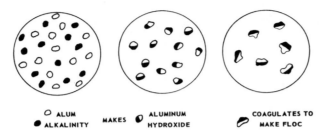

○ ALUM
● ALKALINITY

MAKES

○ ALUMINUM
● HYDROXIDE

COAGULATES TO
MAKE FLOC

processes of coagulation, flocculation, and sedimentation that the greatest removal of suspended matter takes place.

The process of coagulation involves complex chemical and physical reactions. In basic terms, the addition of a coagulant to water produces two actions, one electrical and one chemical. The suspended particles of mud and clay carry minute negative electrical charges. The coagulant floc is a hydrous oxide with a small positive charge. The positive charges attract and neutralize the negative charges, thus causing an agglomeration, or joining together, of the suspended particles and the floc. As this combined particle develops and increases in size, it becomes dense enough to settle out of the water.

A number of different types of chemicals are used as coagulants:

Aluminum Sulfate

Aluminum sulfate (filter alum) may be added to a water containing natural alkalinity in the form of calcium or magnesium bicarbonate.

When water does not contain

sufficient natural alkalinity to react with alum, an alkali such as calcium hydroxide (hydrated lime) is added. Sodium carbonate (soda ash) may also be used with alum to supply the necessary alkalinity.

It is advantageous to use alum alone when possible, since it greatly simplifies control of the treatment process. It does not intensify color that may be present in the water and may actually reduce the color already present to some extent. Ordinarily, the resulting increase in hardness and corrosiveness is of little consequence. The minimum dose should never be less than 5 ppm even with relatively clear water, because smaller doses will not produce the concentration of aluminum hydroxide necessary for floc formation. The usual pH range for use of alum is 5.5 to 6.8.

Ferric Salts

Liquid, crystalline, and anhydrous (without water) ferric chloride are available as coagulants. Ferric chloride is corrosive in the liquid state or as a damp solid. Ferric sulfate is available as a commercial coagulant and has the advantage of being less corrosive than ferric chloride.

Ferric salts are used for the following reasons:

1 Ferric floc is heavier than aluminum floc and settles more readily. It is also more completely precipitated over a wider pH range.

2 Ferric floc does not redissolve at high pH when lime is used for corrosion control.

3 Ferric floc forms more rapidly in cold water than does alum floc.

Coloration is another matter; no one method will remove all coloration, and no single rule applies to all waters. The most common method will remove all coloration, but, again, no single rule applies to all waters. The most common method for color removal is the same as that used for eliminating turbidity: coagulation

(flocculation) with alum, followed by lime or soda ash if necessary. The alum has to be added first and given some reaction time at a pH of 5.5 to 6 before adding the alkali. If the soda ash and lime are added first, they "set" the color to such an extent that removal becomes extremely difficult.

Chlorination is fairly effective in color removal; the chlorine dosage for this is between 1 and 10 ppm, with the correct dose determined experimentally. A contact period of at least 15 minutes is required for this treatment to be effective, and at least 0.1 ppm chlorine residual must be maintained.

Activated carbon (popularly called charcoal) filtration is sometimes effective in controlling turbidity and color in water; but this can be a hit-or-miss situation. If odors accompany coloration, activated carbon filtration is a good gamble.

Controlling Taste and Odor

Taste and odor treatment can be applied as either a preventive or a corrective measure. The chief causes of unpleasant taste or odor are:

- Pollution from industrial wastes
- Algae and the resultant slime deposits on surfaces
- Decomposition of organic materials
- Dissolved gases such as hydrogen sulfide
- Chemicals used to treat water (chlorine, for example)

Depending on the type of problem that exists, chemicals can be added to the water before any subsequent treatment or they may be added immediately before the final filtration process. Some effective methods of taste and odor control are aeration, activated carbon filtration, or the addition of such chemicals as copper sulfate, chlorine, or chlorine dioxide.

Aeration

Aeration is effective when the problem is attributable to dissolved gases in the water. The aeration process involves subjecting as much of the water to air contact as possible so that gases that are dissolved in the water are released to the atmosphere. Water should be filtered following aeration to keep out insects and other foreign matter that tend to mix with the water during the aeration operation.

Activated-Carbon Treatment

Where water is permitted to mix intimately with the activated-carbon filter for a sufficient period, taste and odor control is quite effective. *Activated carbon* is carbon granules that have been specially treated to increase their absorptive capability. This type of filtration provides good insurance against odors and tastes contributed by gases and finely divided solids.

A carbon dose of 3 ppm removes most tastes and odors from water; but where industrial wastes are the cause of the problem, the dosage may have to be significantly larger.

There are several ways to employ activated carbon filtration. The particles can be added to the raw water, but this isn't usually done when the waters are turbid. When sedimentation is used as part of the water treatment, the carbon may be added immediately ahead of the sedimentation tank or basin. Or the treatment can be incorporated as part of a conventional filtration technique, whereby the carbon is used as the filter medium in a gravity- or pressure-filtering system. In this case, operation is the same as with sand filters; the difference would be in the backwash rate, which is reduced to prevent excessive loss of the carbon granules during the flushing operation.

When the activated carbon filter has absorbed all the taste- and odor-causing impurities that it can hold, it has to be replaced. For typical applications, the useful life of such a filter is at least a full year but no more than three years. The actual longevity of the filter depends on the types of impurities being removed from the water and the total quantity of water processed by the filter during its life.

Disinfection

Disinfection is the chemical destruction of bacteria. Because of its economy, dependability, efficiency, and ease of handling, chlorine is universally (though not exclusively) used for this purpose. All new, altered, or repaired water supply facilities should be disinfected before they are placed in service. Water from surface supplies may be disinfected before filtration or before coagulation and sedimentation to prevent the growth of organisms. This procedure is known as *prechlorination*. It may also be disinfected after filtration to kill organisms that still remain and to provide a safeguard against recontamination. This procedure is known as *postchlorination*.

Chlorination

Chlorine is presently the only widely accepted agent that destroys organisms in water and leaves an easily detectable residual that serves as indicator of the completeness of treatment. The sudden disappearance of residual chlorine may signal contamination in the system. Disinfection action is faster at higher temperatures, but is retarded by pH. If the pH is above 8.4, the rate of disinfection decreases sharply!

Chlorine Disinfectants. Chlorine disinfectants are available in a number of different forms as described in the following paragraphs.

Calcium hypochlorite is a relatively stable powder that is readily soluble, forming a chlorine solution. It has 65 to 70% available chlorine by weight. Because of its concentrated form and ease of handling, calcium hypochlorite is preferred over other hypochlorites.

Sodium hypochlorite is generally furnished as a solution that is highly alkaline, and therefore reasonably stable. Federal specifications call for solutions having 5 and 10% available chlorine by weight. Shipping costs limit its use to areas where it is available locally. It is also furnished as powder under various trade names such as *Lobax* and HTH-15; the powder generally consists of calcium hypochlorite and soda ash, which react in water to form sodium hypochlorite.

Ordinary household bleach is a sodium hypochlorite solution containing 2.5% available chlorine and is often used.

When chlorine is introduced into pure water, some of it reacts to form hypochlorous acid, and the rest remains as dissolved chlorine. Both these forms of chlorine are termed *free available chlorine,* because their oxidizing and disinfecting ability is fully available. Because most natural waters contain small amounts of ammonia and nitrogenous organic substances, free available chlorine will react with these substances to form chloramines and other complex chlorine–nitrogen compounds. These forms of chlorine compounds are termed *combined available chlorine;* part of the chlorine oxidizing disinfecting ability is lost.

Both free available chlorine and combined chlorine will react with oxidizable substances in water until their oxidizing and disinfecting ability is depleted. The amount of chlorine consumed in reacting with organic substances in water in a given time (usually 10 minutes) is called the *chlorine demand.* Chlorine remaining in excess of the chlorine demand is the *total chlorine residual,* or *residual chlorine.* Residual chlorine is composed of both free available chlorine and combined available chlorine. The time elapsing between the introduction of chlorine and permitted use of the water is 30 minutes and is termed the *contact period.*

The bactericidal effectiveness of chlorine depends on the chlorine residual, contact period, temperature, and pH.

Chlorine effectiveness increases rapidly with an increase in the residual. However, free available chlorine is 20 to 30 times as effective as combined chlorine under the most favorable conditions of pH (7.0) and water temperature (68 to 77° F). Therefore, the relative amounts of free

and combined available chorine in the total residual is important.

Within normal limits, the higher the chlorine residual, the lower the required contact period. If the residual is halved, the required contact period is doubled.

The effectiveness of free available chlorine at 35 to 40° F is approximately half of what it is at 70 to 75° F.

The effectiveness of free chlorine is highest at pH 7 and below. At pH 8.5, it is one-sixth as effective as at pH 7; and at pH 9.8, it may require 10 to 100 times as long for a 99% bacteria kill as at pH 7.

Simple chlorination is a single application of chlorine as the only treatment before discharge to the distribution system. Prechlorination is the application of chlorine to raw water before coagulation, sedimentation, or filtration. *Postchlorination* is the application of chlorine after filtration, but before the water leaves the treatment plant. *Rechlorination* is the application of chlorine into the distribution system or into a previously chlorinated purchased supply to maintain the chlorine residual.

Other Uses of Chlorine. Chlorine is also used to control tastes and odors in water. It reacts with the substances causing taste and odor, such as hydrogen sulfide, minute organisms, algae, and organic compounds. If the reaction is incomplete, the taste and odor of some substances may be intensified or become more objectionable. Chlorine is also used to a limited extent to oxidize iron and manganese and to remove color.

Iodination

Iodination is a process of introducing controlled quantities of a special form of the chemical iodine into water for purification. Iodine is not effective in clarification or odor-and-taste control, but it has some distinct advantages over chlorine when it comes to the killing of dangerous micro-organisms. Iodination kills bacteria, viruses, fungi, and amebic cysts the latter of which is notoriously resistant to chlorination.

Iodine is very weak chemically and so it reacts quite slowly with organic materials; for this reason the added iodine must be allowed time to interact in solution before the treated water is used. Chlorine reacts rapidly and so tends to be used up at a high and often unpredictable rate; and while temperature, pH balance, and sunlight tend to affect the germicidal properties of chlorine, they have negligible effect on iodine.

In practice, a special tank *(Iodinator)* containing iodine crystals is placed in the water line immediately ahead of a holding tank (in a private water system the holding tank is the pressure tank associated with the well pump). A simple system of valves diverts a small fraction of the total water flow through the Iodinator, after which the flow is returned to the main water line. The iodine then reacts in the holding tank to do its germ-killing job. At the Iodinator (a tradename of the Iodinamics Corporation) the valves are set for a concentration of about 300 ppm; after recombination with the untreated water in the holding tank, the total iodine concentration is diluted to about 0.5 ppm.

The Iodinamics flow-through process offers several practical advantages over the chlorinator: the Iodinator requires no electricity and contains no moving parts. Once an Iodinator is installed, the depth of the iodine bed must be checked only after a couple of years of service—chlorinators must be checked and serviced quite frequently.

Water-Treatment Equipment

The principles we have been discussing in this section apply to water-treatment facilities of all types, from the very small single-faucet add-on to the very large facility designed to serve a good-sized community. But we haven't said much about the equipment that performs the function of making water drinkable.

Because the methods are similar regardless of the water problem so are the equipment items designed to solve the problems. There is no single description that would apply to all makes of equipment, and there is no single procedure for maintaining the equipment in good working order. But we can make generalizations about the equipment and the care of it—and these generalizations can cut across the broad spectrum of design differences and appearance variations.

As a rule, no maintenance procedures are required for filter units the size of fire extinguishers, because they typically employ filter cartridges. All you have to do is shut the water off every so often and unscrew the top or bottom of the unit to remove the used-up filter element and slip in a new one. Installation is simple for this type of filter element, too—it is no more difficult than installing a tee in the water supply pipe. It's always handy to have a valve on each side of a filter, but it isn't required.

Automatic Water Softener

The automatic softener requires no attention except for the occasional task of adding salt to the brine tank. In operation, unconditioned water enters the top of the tank and flows down through the filter bed. Water with a hardness value of zero comes out the bottom. As the water flows down through the bed, it is filtered and softened.

The first step in the process is a *backwash,* which continues for ten minutes or so, depending on the make and model of the softener. This loosens deposits filtered from the water and flushes them down the drain. The next step is *regeneration,* the adding of new salt to the system. The *rinse* operation follows for about an hour, which brings the automatic water softener back into the *service* mode, where it stays until another backwash is needed.

The four-step process is the same as a manual softener: backwash, regeneration, rinse, and service.

Backwash

During the softening period, water flowing down through the zeolite bed leaves suspended matter on the surface of the bed. As the accumulation grows, the pressure drop across the bed increases and so the bed becomes generally more compact. Therefore, when the zeolite bed becomes exhausted, it is necessary not only to regenerate it but to backwash it as well to remove the surface deposits and to insure efficient regeneration and subsequent operation. The backwash operation consists of flushing water upward through the bed at a rate that causes the bed to expand a specified minimum amount. This permits dirt to be separated between zeolite grains (remember, they're pinhead-sized) and floated off at the top to the drain. The operation is something of a cleansing process, so that in the succeeding regeneration step the salt solution can properly contact all the zeolite resin.

Regeneration

As we mentioned, in the course of the softening process, the sodium content of the ion-exchange bed is depleted and requires regeneration by passing a concentrated salt solution through it. The sodium of the salt displaces the calcium and magnesium taken up by the bed during the softening and restores the ion exchange material to its original condition. The salt solution, fed into the softener above the resin bed, flows downward through the bed. The efficiency of this operation is determined by the amount and strength of the salt solution, distribution and rate of flow, and time of contact with the bed. (The solution should be added slowly.)

Rinse

After the regeneration step the ion-exchange bed must be rinsed to flush out the excess salt. This is accomplished by passing water through the bed until all traces of salt have been thoroughly washed away. Manufacturers' instructions vary, depending on size of the unit and capacity. Synthetic resins require longer rinses than natural zeolite.

Service

The softening process is resumed after the regeneration and rinse cycles. Raw water passes down through the fresh bed at the flow rate recommended by the manufacturer of the softener. The softening process continues until the zeolite bed is again exhausted and regeneration is required once more.

Filter Units

Example 7–7 includes several sketches that show how a single-faucet filter can be installed in a water pipe. The filter container pictured in these sketches is AMF's *Aqua-Pure* unit, which can be fitted with various types of replaceable cartridges. In the first sketch, the undersink installation includes a valve that allows the water to be shut off during cartridge changes. If the plumbing line already contains an undersink valve, this need not be added, of course.

The second sketch shows a popular method for filtering drinking water that flows to a special, easily installable faucet adjacent to the regular sink faucets. If this approach is used, a stop valve is a desirable feature but still not necessary if a valve already exists in the main supply line. In the third sketch, two valves are employed, one on each side of the filter unit; this approach is recommended when a filter is installed in a main water line that serves a full branch or several fixture branches.

The small-filter installation is shown in detail in example 7–8. Once

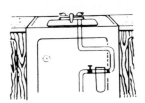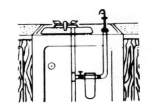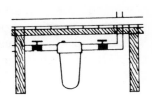

example 7–7. Single-faucet filter installations.

example 7–8. Small-filter installation.

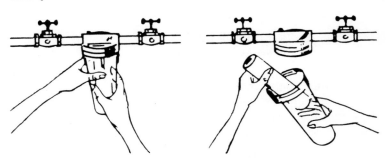

the fittings are in place, installing the filter and replacing it are as simple as unscrewing the container portion, dropping in a new cartridge, and screwing the filter back into position. The valve on the downstream side of the filter prevents gravity draining of the water that might be in those pipes. Without this, all the water in the pipes will, if they are located at a higher elevation than the filter, drain down into a puddle at your feet while you're trying to make a simple cartridge change.

Full-size filters have the same general appearance as a water softener; the main difference between filters is the material contained in the tank. Like a water softener, a filter may be automatic or manual. The difference between the appearance of the manual and the automatic filter is small; the manual has a couple of controls at the top and the automatic contains a small control box. (Compare the two Myers units in example 7–9.)

Manual filters should be backwashed and rinsed once a week. Iron filters should be charged occasionally (usually monthly) with potassium permanganate (usually),

example 7–9. Manual (A) and automatic (B) filters.

A B

which reoxygenates the mineral that does the filtering.

The rate of backwash is very important, but this varies from one maker to another and according to the capacity of the filter. To keep a filter operating properly, it is of utmost importance to maintain the right backwash rate (generally, this will be at a flow of about 8 gallons per minute for each square foot of filter-bed area).

Neutralizing filters require heavy backwash at fairly high backwash rates, though the rate of flow may be slow. Once a year or so, a small amount of neutralizing material will have to be added to the unit.

Disinfection Units

Where chlorine is used as the disinfectant, a chemical feeder is employed. Manual and automatic feeders are available, but local codes often stipulate that a manual feeder is not acceptable because of the attention it requires in order to maintain water potability. Where iodine is used as the disinfectant, the equipment requires a special valving approach and is not used with other chemicals that might be required to condition the water.

Chlorinator

Where big-plant chlorination equipment requires chlorine in gaseous form for injection into a water supply, most home-sized units are fed with a solution mixed with ordinary household bleach. In a typical system, we might mix 2 quarts of laundry bleach with perhaps 10 gallons of water. This reservoir serves as the supply for the chlorinator, which meters it into the water supply ahead of a holding tank. The *chlorinator* is a small boxy piece of equipment (example 7–10) that contains a couple of controls to allow

example 7–10. Chlorinator.

the metering quantities to be established initially (they're maintained automatically).

Usually, a chlorinator is installed after the pump but ahead of the pressure tank, as shown in example 7–11. The chlorinator sucks the chlorine solution up from a capped container (usually of polyethylene) and meters it into the water line. When the pump is engaged, the water is mixed with the chlorine solution as it is pumped into the pressure tank, where it must be allowed to remain for 20 minutes to insure safe water. If the tank is not large enough to allow the water to be detained before use for the required period, a separate holding tank must be used. You can calculate the required tank size to assure the 20-minute retention period: it should be 20 times the gallons-per-minute pump capacity plus an additional 25% (as a safety factor).

Adjustment of the chlorinator isn't difficult. If the chorine residual is less than 0.2 ppm, add more bleach; if it's above 0.5 ppm, dilute the solution with water. Once this mix is established, you only have to check the water for chlorine residual on a regularly scheduled basis.

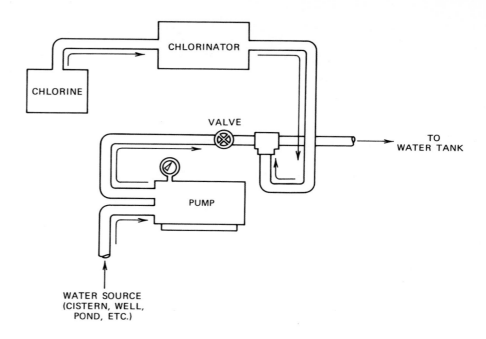

example 7–11. Typical chlorination system.

One of the distinctive advantages of using chlorine to disinfect a water supply is that the feeder can be used to meter in other chemicals as well. The result is one single system of chemical addition that can be used to alter the pH of water, remove small amounts of iron and manganese, and kill harmful bacteria.

Iodinator

The iodinator is not to be used with any other chemical or feeder. It's a stand-alone unit and must be installed between the water source and a holding or pressure tank. As shown in example 7–12, the unit is a simple "fire-extinguisher" sized tank that comes with inlet and outlet tubing and a complement of four valves. The unit is connected into the water supply in such a manner as to allow all water to be diverted to or around the iodinator. The critical element in this system is the initial adjustment of the four valves in the system. Once they have been set properly, no further attention is required except for an occasional check of the iodine residual and periodic replacement of the iodine crystals.

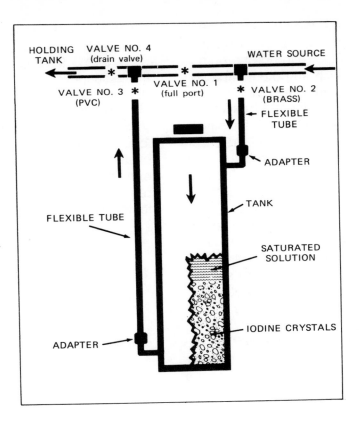

example 7–12. Iodinator with associated piping and valves.

To determine the right tank size for use with the iodinator, multiply the gallons-per-minute rating of the pump by 9. A pump with an output capacity of 6 gallons per minute would require a tank with a capacity of 54 gallons.

Example 7–12 shows the piping and valves for the iodinator; each valve is numbered for ease of reference in the adjustment procedure. Adjustments are made to units in service according to a need established by simple residual tests similar to those for chlorine residual.

Note that not all water is diverted to the iodinator; the valves are set so that only a small quantity of water goes into the unit. The output solution then mixes with the fresh water and enters the pressure or holding tank, where it must be allowed to remain for a 20-minute contact period. By using a gate valve on the main line, slight back pressure is created, forcing a small portion of the water to flow through the iodinator. This flow (equal to about one six-hundredth of the total water flow in the line) percolates up through the bed of crystals, after which the iodinated water is returned to the line. In the tank, bacteria, fungi, viruses, and cysts are completely destroyed.

Valve 1 (example 7–12) adjusts the iodine level; opening the valve decreases the iodine content, and closing it increases the balance. Valves 2 and 3 allow access to the iodinator for checking crystal content or adding new iodine crystals. Valve 4 prevents backflow when adjustments require removal of the iodinator unit temporarily.

Section 8

Hints for House Buyers

Section 8

Are you considering buying a house that's been occupied previously? Especially if the house is an older one, you should inspect the plumbing just as thoroughly as you would inspect the foundation, framing, or electrical wiring. In most communities a building official will inspect the house and indicate code violations. Local building-code requirements must be met, and these vary from one locale to another. What might seem to be a minor addition or repair may require a major revamping of an entire plumbing system for compliance.

Questions You Should Cover

Architect M. S. Timmins recommends that you ask, and answer, these questions before making a final purchasing commitment:

1 Are there water stains in the building, indicating the existence of leaks in the water supply or drainage piping? If so, have the leaks been repaired satisfactorily?

2 Is the flow of water from the faucets good and strong, indicating the absence of corrosion or scaling in the water supply piping? If not, can the deficiency be corrected without great expense or difficulty?

3 Do all fixtures drain quickly and quietly and maintain the water seals in the traps, indicating an adequate vented drainage system? If not, how many drains don't empty quickly? Are they on a single branch or is the entire venting system inadequate? How expensive would the repair be?

4 Are all fixtures and piping anchored and supported well?

5 Does the water closet flush completely, relatively quickly, and shut off in less than a minute? Does it refill quietly?

6 Do faucets and valves operate freely and close completely?

7 Are the fixtures chipped and stained? Is replacement required?

8 Do the stoppers built into lavatories, sinks and tubs hold? If not, how expensive would replacement or repair be?

Quick Checks

Here are some considerations, inspection tips, and remedial actions based on the above question sets.

Water Stains

Examine the wall area adjacent to every fixture for water stains; when the stains appear on side walls, they are probably caused by water spills and careless splashing. But when they are on walls that conceal pipework, they might indicate an existing pipe leak. Some leaks in pipes don't show themselves until the water reaches a specific high temperature—these point to a fractured hot water pipe or a faulty joint. Repair isn't difficult unless the wall itself is inaccessible. When you see a wall stain, ask about it. But don't stop there. Examine the wall to see how difficult it would be to get at the pipe. Does an access panel exist? If the wall serves as a partition between rooms, look at the side opposite; very often you can gain access to the piping in a wall by removing an inconspicuous panel from the other side.

While you're looking for wall stains, examine tiled and plastic-paneled walls in bathrooms for evidence of leakage behind the wall surface. Tiled walls that have endured leaks for long periods will have moist or gritty grout that easily rubs off. Plastic paneled walls concealing leaks will often have a telltale bulge resulting from repeated swelling of the wall material behind the paneling. Ceramic tile that conceals a wet wall area will be loose, but this doesn't necessarily indicate a leak; it might be a sign of an improper tile installation job. But if there are pipes immediately behind such a wall, it would be wise to look for a leak as the culprit.

If the house has a basement, don't fail to spend some inspection time there, too. Acoustical ceilings in basements are dead giveaways to leaks from the floor above; they'll show a multiplicity of stain rings in one general area, and often the ceiling tiles will bulge and exhibit considerable deterioration at the leak site.

Even though such stains don't give positive identification of a plumbing leak, they'll offer enough circumstantial evidence to justify suspicion of one. If the acoustical ceiling material has to be replaced in such spots—when it's obviously deteriorated to the point of sagging or warping beyond what you'd consider an acceptable amount—ask the owner

to let you remove one tile so that you can examine the support structure to which it's fastened. (But don't pull the tile unless you plan to replace more than one; tiles generally are tongue-and-groove connected, which means that you may have to pull a whole line of tiles from the soiled area all the way to the wall. And, of course, replacement will involve inserting the tile at the wall and sliding each piece up to the soiled area individually. This can prove time-consuming—a real pain in the neck for the homeowner.)

If you do remove a soiled tile, look at the two-by-four or two-by-six framing under it. Is it warped or saturated? If there is insulation batting under the tile, make a thin slice in its bottom with a knife. If water is there, you are in for an expensive repair job, because it means the insulation can't be doing its job, for one thing; and for another, it means there is very likely a plumbing leak from the floor above that might be a real bear to pinpoint. It might also mean the floor above is warped.

In one Pennsylvania home the prospective purchaser found five or six heavily stained acoustical tiles in the ceiling of the basement, one or two of

which had deteriorated to the point of shredding. He removed those tiles and found a heavy, sagging vinyl sheet immediately above it. A small knife puncture released a stream of water that appeared sufficient to keep a good-sized garden for a year. A subsequent check of the floor immediately above the basement in the area of the leak revealed a bathroom floor that was warped so badly that the door to the room couldn't be closed all the way. The homeowner had dismissed the door problem earlier as a "bad hinge." To repair this, the owner had to pay to have the entire bathroom floor ripped out and replaced. When the wood is warped there's no solution other than total replacement. Needless to say, this kind of repair is very expensive—certainly enough to justify your having second thoughts about buying the house at the asking price.

Why would water cause the floor to warp so badly as to require replacement? Because the vinyl sheeting that held the insulation batting wouldn't release the water it contained. The floor was wetted by a small initial leak and the water found its way into the insulation, where the material became swollen to the point where it exerted a continuous and heavy force against the floor. Each time the floor became wet the wood softened, and the constant pressure of the swollen, wet insulation managed eventually to change the flat floor structure to a waviness that was so severe the door couldn't be closed.

Water Pressure

Unless the water pressure is obviously adequate, it might be a good idea to measure the flow rate. Remember that a home in use often has water running from more than one faucet at a time. One of the most tiresome irritations of daily living is that of experiencing a drastic pressure reduction from a bath or shower faucet when someone turns on a faucet somewhere else in the house. It can be dangerous too, especially if someone turns on a cold-water faucet and it results in a drastic temperature increase in the bath.

As an absolute minimum, a bathtub faucet should give you a flow of 2 gallons per minute while another faucet in the house is turned full-on. To check this, get a gallon jug and a watch with a sweep second hand. Turn on a kitchen faucet fully, and then take the empty gallon jug into the bathroom. Turn the bath faucet on all the way and start timing the instant you put the container under the tap. If the container is filled before 30 seconds have elapsed, the water delivery may be considered adequate.

(But a flow of 1 gallon in 30 seconds isn't much more than adequate.)

If 30 seconds tick by and the gallon container still isn't filled, it's a real indication of a problem—which might be quite difficult to correct without considerable expense. If the pipes are properly sized ($\frac{1}{2}$ or $\frac{3}{4}$ inch for lavatory, $\frac{3}{4}$ inch for hose bibbs and wall hydrants, $\frac{1}{2}$ inch for showers, tubs, and kitchen sinks), the problem can be traced to incorrect pressure at the water source or to excessive scaling, corrosion, or mineral deposits in the supply piping.

Check the source first. If the house you're considering for purchase is fed from a community reservoir, there won't be much you can do to correct the problem at your end—other than register a complaint with the individuals, group, or company responsible for maintaining water pressure. But if it's pump-fed from an individual well or spring, you should examine the pump and tank carefully. Check the gage on the pressure tank; it should not read less that 20 psi, and ideally between 35 and 50 psi. Section Five in this book includes all the information you need to track down troubles related to private water systems.

Once you've checked out the source and found it to be satisfactory, and yet the water delivery still isn't up to par, you have good reason to suspect that the piping has accumulated enough deposits along its inside surface to seriously restrict water flow.

No book should ever state that nothing can be done about scaling and liming buildup in water pipes. It's a fact that certain kinds of buildup can be dissolved in heated acetic acid solutions; but there is no established procedure for heating vinegar and forcing it through the piping of a house with strictured-conduit problems. And unless some systematic method can be found for eliminating the buildup of years of scaly water, the only recourse is an unfortunate choice: live with the problem or replace the pipes. *Caveat emptor!*

If you consider the alternatives and decide that it might be worthwhile to explore the possibilities of replacing certain pipes, here are a few guidelines to help you.

First, locate the water main—the large service pipe furnishing the water for the entire house. Then look at the pipe routing between the main and the bathroom branch. If there are many fittings, 90-degree elbows, and small pipe sections joined by couplings or unions, see if it might not be possible to simplify the routing. Remember, every fitting, every bend, every separate pipe section contributes to the problem of excessive resistance to good water flow. If you can replace a large number of pipe sections, fittings, and bends with a single straight run or with a reasonably straight run having only a couple of fittings, chances are you won't have to go any further in your updating attempt. Look for joints connecting metal pipes of different materials—copper and steel, for example. When two dissimilar metals are connected directly, the result is a veritable corrosion factory. Often, low-pressure problems in such cases can be corrected by the expedient of replacing the metal-to-metal joint with a fiber or plastic union. (If you use a plastic union, be sure to connect it with a flange fitting.)

Finally, if the piping seems quite old, determine how difficult it would be to replace certain long pipe runs with plastic piping (CPVC or PVC), which permits considerably faster flow than aged galvanized pipe.

All of the above guidelines are dependent, of course, on the piping being accessible from the basement or a crawl space under the floor. If the pipework appears inaccessible, the task of pipe replacement, even partially, might be too expensive to even consider. Certainly it would be wise to get a few outside estimates on replacement, and then follow up with a purchase offer that reflects the expense you'll be incurring in the plumbing fix-up process.

Fixture Drains

The drainage system consists of the sewer lateral (drainage line between the house and the street), the underfloor drain, the drainage pipes above the floor, and the vents. Pipes may have become clogged or broken or they may be of inadequate size. Venting in particular may be inadequate and far below code requirements.

Bathtubs, dishwashers, kitchen sinks, and laundry trays need drains of $1\frac{1}{2}$ inches diameter, and the drain pipes they feed into should be the same diameter or greater. This is easy to check with an actual measurement. But the right pipe diameter is not in itself assurance that all is well with the drain system. Run water into each basin—tub, lavatory, laundry tray, kitchen sink—and allow the water to drain. If a sink drains sluggishly or with a gurgling noise, it could indicate a vent problem.

A sluggish drain isn't *necessarily* a sure sign of a problem; it might simply indicate that foreign material is in the drain, or that the stopper isn't allowing sufficient water clearance space. Here's how to make sure: once you've spotted a sluggish drain, determine which of the other fixtures in the house are using the same branch, then check them individually; if they drain satisfactorily, chances are the venting is all right. If not, then do this:

1 Put at least a quart of water into each fixture on the questionable branch.

2 Keeping all fixtures except one sealed, release the water from the one by pulling the stopper all the way out. If the stopper is an integral part of the drain, lift it manually as high as possible so the water has as much space as needed for swift drainage.

3 If the water drains slower with all other fixture drains on the branch plugged, a blocked or restricted vent is likely. To doublecheck, drain all branch fixtures completely, then remove the plug from the slowest drain.

4 After all water has been drained from the most sluggish fixture, insert a long, slender stick down into the open drain until it hits bottom. Then withdraw it and measure the wet area. It must be not less than 2 inches nor more than 4 inches. If the wet area is more than 4 inches, additional checks are called for; if it's less than 2 inches, siphonage is indicated, suggesting a clogged or partially clogged vent. Remember, insufficient venting can allow the escape of sewer gases into the room—a real health hazard.

5 If more than 4 inches of water is in the trap, it would be wise to remove the trap or unscrew the trap cleanout, if one is available. This shouldn't pose too great a problem; very often a P-trap can be loosened without any tool. But get a bucket to catch the water trapped in the drain.

6 Once the trap has been cleaned out, you can be sure a sluggish drain can't be attributed to wads of hair, curlers, or other debris; almost certainly vent problems exist. Reconsider your house purchase, or ask the seller to make an allowance for necessary plumbing repairs.

Fixture and Piping Stability

The stability of the fixtures and piping will tell you a lot about the plumbing and how much trouble you're going to have with it. Loose fixtures mean problems; so do loose pipes, especially hot-water pipes.

If the lavatory is a pedestal or wall-hung type, grasp it firmly and attempt to move it up and down. If it gives easily, chances are that jarring will weaken the supply- and drain-pipe joints. Even more likely, the owners have already experienced pesky leaks from time to time.

If the supply pipes are accessible from the basement area, check them too. Try to move a supply pipe back and forth along its length axis—that's what sudden hot water surges will do. If the pipes move more than a few inches, examine the piping close to the end of the run to see which pipe joint or bend is absorbing the shock of the movement. That's a potential trouble spot.

Inspect the pipe hangers. They should give good support to the pipes but not prevent some longitudinal freedom. Plan to replace all broken hangers or be faced with the unattractive alternative of having a plumbing system that is noisy and leak-prone.

Are the hot-water pipes well insulated? Are they insulated at all? Metal pipes are extremely good heat radiators; hot-water pipes that are not insulated will double your water-heating bill, or worse.

Examine the drain line if it's accessible. This should be absolutely immobile. Look at the entire waste-disposal pipeline to see if the materials are all the same. Small pipe sections of one material interconnected with a main run of another, point to probable earlier problems. And plumbing problems have a way of recurring unless the cause of the problem is removed or corrected.

Plastic piping deserves special attention. Where hanger iron (pipe strap) is used to support plastic pipes, examine the vinyl pipework at each hanger to see if it's been gouged or scored badly. Remember, plastic pipes meet certain specifications as to pressures and temperatures; when a wall is scored, the pipe must be derated accordingly: it won't meet its original specs. No plastic pipe should ever be suspended from hanger iron without some form of interface protection such as padding, shim stock, or the like.

Water Closet

Place one square of tissue on the surface of the water in the toilet and flush it. Within 10 seconds, the water and the tissue square should be out of the water closet and into the waste line. If it takes longer, you can be pretty sure that the water closet will have emptying problems at least some of the time. Slow flushing is indicative of several toilet ills. A defective or worn out flush valve is one probability, but if you examine the flush valve (see Section Three) and find it to be in fair condition, you can suspect a buildup of lime and scale deposits in the water closet's innards.

As a quick check, remove the top to the water-supply tank for the water closet and lift the flush valve by hand. If the flushing operation isn't faster this time around, inspect the inside surface of the water tank, including the outer surface of the brass or copper tubing in the tank. If there is corrosion evident but plenty of water and a flush valve that works right, the tubing inside the water closet is probably excessively constricted with deposits from the water.

See if the water closet is securely mounted. If it's a one-piece assembly and not wall-mounted, grasp the tank firmly with both hands and attempt to jiggle the unit. Excessive movement indicates that the previous owners have had trouble with the toilet and have pulled it from its floor position in the past. But it also indicates that a bad seal exists between the floor flange and the water closet, which may be attributable to a broken floor-flange holddown bolt or a loose nut on one of these bolts.

If the holddown bolts are capped, lift one of the caps to see if the bolt has come free of the floor-flange slot. You can examine the illustrations in Section Three of this book to see how the mounting arrangement *should* look. Just remember, a loose water closet can introduce a multitude of problems, both with supply piping and with the drain line.

Faucets and Valves

Check the shutoff valves at the service entrance and at various points in the system to determine if they have become frozen with age. If any of them are, check further along that branch to see if the water pressure is lower than at other points in the house. A shutoff valve that is frozen is of no use, and one that won't shut off all the way probably won't have a fully cleared water passage, either. Chances are, valves that are in poor condition are creating one-spot strictures that will affect pressure and flow elsewhere.

Check the faucets generally while you're examining the fixtures and piping. Dripping faucets can be very costly, as we've pointed out previously; but faucet replacement is also a costly business—especially if there are many needing replacement. Generally, a faucet that leaks or drips can be corrected with little effort other than replacing the washers or grinding-in the seats. But if you're planning to move into an older house, there will be a great deal of renovating in store for you anyway, and faucet refurbishing will be one more task that you will have to face.

Fixtures and Stoppers

A brief examination of sinks, lavatories, and tub will tell you a lot about the water in the house you are thinking of buying.

Reddish-brown stains on fixtures indicate excessive acidity in houses where galvanized pipe is used for the water-distribution lines. Bluish-green stains indicate acidic water in houses with copper piping. The stains can be removed with certain chemicals, but no chemical in the world will restore the porcelain eaten away by the acids in the water. Absence of stains is no indication of a good fixture. Feel the porcelain surface with your fingertips; if the water is acid, the surface will be pitted and rough. And if the porcelain is damaged, you can imagine what the pipes must look like on the inside.

Appendix

Plumbers' Tools and Plumbing Diagrams

Appendix

Much of the information in this appendix is going to be old hat to you. You know a great deal about tools already. Odds are you're pretty good with them, too. If you didn't have something of a knack for handling fairly complicated maintenance jobs around the house, you probably wouldn't give a plumbing repair book a second glance. But there are quite a few tools in the plumbing inventory you may not be familiar with—tools like calking irons, drain augers, and strap wrenches. And even the common ones we use every day, like hammers and wrenches and screwdrivers, have uses you may not know. So we'll cover all the tools, common and uncommon.

The information in this appendix is not the kind of material you have to read from beginning to end, although if you do you'll get a good overview of the tools you'll be using from time to time and perhaps a preliminary feeling for the way they should be handled and cared for. Most important, this section provides a reference source for you. As you read the various sections in this book dealing with different phases of the art of plumbing and associated repairs, you'll note

352

references to specific tools you need to make a repair or perform a necessary task. When that happens, and you're in the dark about the tool you need or how to use it, come to this section for a few minutes, read the appropriate passages, and then get right back to the work at hand.

Borrowing from computer-age terminology, we can divide our tool requirements into two categories: hardware and software. The *hardware* are those tools and other special materials we need to physically carry out the installation, modification, or repair. The *software* items are the plans and schematics—with their peculiar symbology that might be considered the plumber's shorthand—that we use to keep a record of how a house is laid out from a piping viewpoint.

Plumbing Hardware: Tools and Equipment

Those very few plumbing repairs that don't require tools of some sort merely involve replacing or altering parts by hand. For the majority of the repairs you'll encounter, you'll need the right tool and the know-how to use it—not only to save time and effort but to avoid injuries. But no matter how abundant your tool collection becomes, its utility will be nullified if you can't get your hands on the right tool when it's needed. It doesn't matter, really, where you keep your tools—of course they shouldn't be exposed to rust-causing moisture—so long as you keep them in one place (and isolate cutting tools to protect their edges); your familiarity with their location will eventually make them as accessible as pens clipped to your pocket. You'll be surprised how efficient you can become by organizing your tools so those used most frequently can be reached easily without having to dig through an entire toolbox. And keeping junk out of your tools, piping, and other supplies will make your work go even quicker.

A Basic Plumbing Tool Kit

Because highly specialized tools can be costly, buy only what you need for *repeated* operations. It is pointless, for example, to buy bending, flaring, and reaming tools to work the only piece of copper tubing you're ever likely to handle. Fortunately, the more frequently used tools are screwdrivers, hammers, wrenches, and pliers; and these are tools you no doubt have already on hand.

Hammers

No tool kit would be complete without at least one hammer. Probably the most common type found in the home is the *carpenter's hammer,* primarily used for driving or extracting nails; its head, attached to a wood, steel, or fiberglass handle, may have a flat or slightly domed face.

Machinists' hammers are designed for work on metal. Unlike carpenters' hammers, they have a peen rather than a nail-pulling claw opposite the *face* side of the head. The ball-peen hammer is probably the most familiar of this group: because it has a ball smaller in diameter than its face, it can reach areas the face can't.

Ball-peen hammers are made in different weights: 4, 6, and 12 ounces and 1, 1½, and 2 pounds. For most work 1½-pound and 12-ounce hammers will do fine. A 4- or 6-ounce hammer will come in handy for light work, such as tapping a punch to cut a pipe-flange gasket.

Machinist hammers may be further classified as hard-faced (made of forged tool steel) or soft-faced (brass, lead, or a tightly pulled strip of rawhide). The most recent addition to the last class is a hammer tipped with plastic or having a solid plastic head that contains a lead core for weight and balance.

When there is danger of damaging the work—as when pounding on a machined surface—use a soft-faced hammer. Most of these have replaceable heads; lead-faced hammers, for example, have to be fitted with new heads when old ones become battered—a small price to pay for their usefulness in striking a solid, heavy, nonrebounding blow (for such jobs as driving shafts through tight holes). If a soft-faced hammer is not among your tools, you can protect the surface to be hammered with a piece of soft metal (copper, for example) or hardwood.

Whatever hammer you use, it won't have much effect if you hold the handle too close to the head. Such "choking" reduces the force of the blow and makes it hard to keep the face parallel to the work, which is necessary for safety—you don't want the head to glance off to one side. Otherwise, you can hold a hammer in whatever way feels most comfortable to you.

It's particularly dangerous to store a broken hammer with the intention of fixing it later; if someone who doesn't know about the problem uses it, a loose head could fly off and cause a serious injury to someone standing nearby. If the wedge that secures the head to the handle (example A–1) starts to come out, drive it in; if the

WEDGE

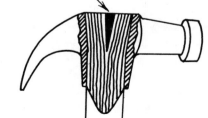

example A–1. Handle expanded in hammerhead with wedge.

wedge must be replaced but another isn't immediately available, you can make a substitute by filing a piece of flat steel or by modifying the tang of a worn file. It is also wise to keep the faces of hammers free from oil or grease; these substances could cause them to give a glancing blow.

A hammer used carelessly can be as dangerous as any other misused tool. The following is a "refresher" summary of safety tips for using hammers.

- Never use a hammer handle as a tool, and especially never as a pry bar; such abuses will cause the handle to split and become a hazard that can cut or pinch.

- Don't try to repair a cracked hammer handle by binding it with tape or wire—*replace it.*

- Never strike a hardened steel surface with a steel hammer; small pieces of steel may break off and injure you or someone nearby. (But it's okay to hit a punch or chisel with a ball-peen hammer because the steel in its head is slightly softer than that of the regular steel hammerhead.)

Mallets and Sledges

Mallets are short-handled tools that are useful for forming or shaping sheetmetal—a hard-faced hammer would often mar the work. Mallet heads are made from soft materials such as wood, rawhide, or rubber, with each head intended for a specific job. Their driving faces are reinforced with iron bands.

Sledges are steel-headed, heavy-duty driving tools, the short-handled versions being suitable for driving bolts and large nails and for striking cold chisels. Long-handled sledges are used to break rock and concrete, to drive bolts or stakes, or to strike rock drills and chisels. Sledge heads are usually made of high-carbon steel and weigh from 6 to 16 pounds; the shape will vary according to the job the sledge is designed for.

The safety rules applicable to using hammers should be followed in using mallets. The only special problem associated with mallets is wood or rawhide heads drying out and cracking, especially if they are left in the sun; they should be conditioned with a light application of oil for moisture retention.

Punches

There are several kinds of punches designed to do a variety of jobs. Most are made of tool steel, with their handles octagonal or knurled for a good grip. The shape of the cutting end depends on the particular job the punch is intended for.

The *center punch* (example A–2) is aptly named; it is made for marking the center of a hole before drilling to prevent the drill bit from wandering. Center punches are also handy for making corresponding marks on two pieces that must match to make an assembly; for greater precision, a *prick punch* can be used. To make the corresponding marks, clean the places you have chosen to mark, then scribe a line across the joint between parts and punch a line on both sides of the joint. (Using single and double marks on opposite sides of the part helps to eliminate alignment errors). In reassembly, refer first to the sets of punch marks to determine the approximate position of the parts, then align the scribed lines to determine the exact position. Because it's difficult to regrind a center punch accurately by hand, avoid using one on extremely hard materials.

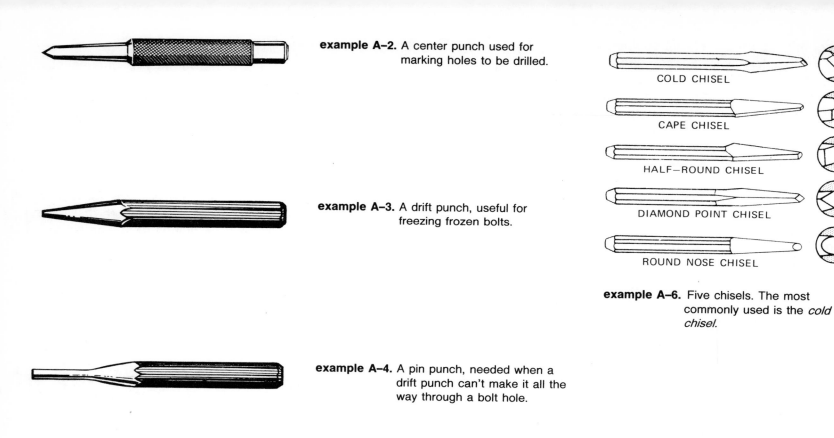

example A–2. A center punch used for marking holes to be drilled.

example A–3. A drift punch, useful for freezing frozen bolts.

COLD CHISEL

CAPE CHISEL

HALF–ROUND CHISEL

DIAMOND POINT CHISEL

ROUND NOSE CHISEL

example A–6. Five chisels. The most commonly used is the *cold chisel.*

example A–4. A pin punch, needed when a drift punch can't make it all the way through a bolt hole.

example A–5. Gasket punch, which has a hollow shank to take in gasket material through which it is being driven.

Drift punches, sometimes called starting punches, have a long taper from the tip to the body to withstand the shock of heavy blows. They can be used for knocking out rivets that have had their heads chiseled off or for freeing frozen bolts.

A drift punch (example A–3) may be only able to loosen a bolt partially because it will only partially enter the bolt hole. In this case, it should be followed by a pin punch (example A–4) that will go through the hole without jamming. Always use the largest punch that will fit the hole (these punches usually come in sets of three to five assorted sizes).

Soft-faced drift punches, made of brass or fiber, are especially designed for removing pump parts. To mate bolt holes in parts being assembled, an aligning punch is helpful.

For cutting holes in gasket materials, there are *hollow-shanked gasket punches* (example A-5). They come in sets of various sizes for standard bolts and studs. The cutting end has a sharp tapered edge that makes a clean hole. To use a gasket punch, place the gasket material on a piece of hardwood or lead (to protect the cutting edge); then strike the punch with a hammer, driving it through the gasket material.

To prevent a punch from slipping sideways when you hit it, and to keep from hitting your fingers with the hammer, hold the punch perpendicular to the work and strike it squarely.

Chisels

Chisels can be used for chipping or cutting any metal that is softer than the chisels themselves. Chisels are made from good grades of tool steel and have hardened cutting edges and beveled heads.

Cold chisels (example A–6) are classified according to the shape of their points, the width of the cutting edge denoting their size. Used more often than any other type, the cold chisel will cut rivets, chip castings, split nuts, and cut thin metal sheets. The other, more specialized chisel types have no application in plumbing repair.

As with other tools, there is a "correct" way of using a chisel. Select a chisel large enough for the job and be sure to use a hammer that matches it: the larger the chisel, the heavier the hammer needed. A heavy chisel will only absorb the blows of a light hammer, and will do virtually no cutting.

As a general rule, hold the chisel steady but not tightly with the thumb and first finger about an inch from the top. Your finger muscles should be as relaxed as possible; this way, if you miss a blow and hit your hand, your fingers will slide down the tool handle, thus lessening the force of the blow. Watch the cutting edge of the chisel while you work—don't watch the head. And, to avoid slippage, swing the hammer straight down on the chisel head, whatever the angle of the chisel body. If you have a lot of chiseling to do, slide a piece of rubber hose over the handle to cut down the shock to your hand.

Screwdrivers

Even though screwdrivers are as familiar to us as hammers are, few realize screwdrivers are not jacks of all trades; using them as chisels, pry bars, punches, or scrapers, for example, abuses them and the fasteners they are used on. To avoid such abuse, screwdrivers come in many sizes and varieties.

Standard screwdrivers are classified by size, the combined length of the shank (shaft) and blade. For each size there can be nearly as many shank-to-blade proportions as there are different human physiques. Screwdriver handles of wood, plastic, or metal may be fitted with round or square shanks that enter the handles partially or go all the way through. Square-shanked screwdrivers are the only ones intended for use with a wrench (*not* pliers).

Although there is an abundant variety of special screw slots, the most common are the *single slot,* seen in screws used almost always in woodwork, and the Phillips configuration, consisting of two crossed slots (example A-7). Three standard Phillips screwdrivers will fit

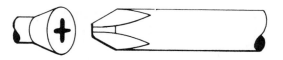

example A-7. Phillips head and screwdriver blade.

almost any Phillips-head screw. (Reed and Prince screws, less common relatives of the Phillips type, require their own special drivers.) However, you must have an assortment of flat-bladed screwdrivers for dealing with the many varied single-slotted screws you are likely to encounter. To prevent damage to a screw slot—and to prevent slippage and possible injury—the screwdriver must match the slot in both length and width.

Pliers

Pliers can take almost as many forms as there are trades that use them, but the most generally useful for plumbing repairs are the ones made for a single purpose: getting a good grip on fairly wide objects.

Adjustable-Jaw Pliers. Slip-joint pliers (example A-8) have straight, grooved jaws pivoting on a screw that holds them together. They are adjustable to two positions, giving a limited jaw opening range for small and large work.

Water-Pump Pliers. Whereas common slip-joint pliers offer two jaw opening positions, water-pump pliers (example A-9) may have as many as seven. The coarse teeth are well suited to gripping cylindrical objects.

The most popular version of water-pump pliers, the *channel locking* type, has extra long handles to give it a very powerful grip. Pliers of this class are adjusted by engaging the groves of one jaw into the *lands* of the other (example A-10). Because of this arrangement, they are less likely to lose their grip on what they are

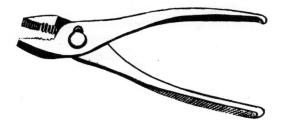

example A–8. Slip-joint pliers, commonly included in automotive emergency tool kits.

example A–9. Water-pump pliers, with seven jaw-opening positions.

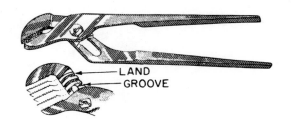

LAND
GROOVE

example A–10. Channel-locking pliers, with extra-long handles for a very powerful grip.

holding than are slip-joint pliers; but they should be used only when the most obviously suitable tool won't work.

Locking Pliers. *Vise-Grip pliers,* more accurately but less commonly known as *wrench pliers,* can be likened to portable, pocketable vises with a handle (example A-11). They can grip and lock in their jaws objects of almost any shape. Vise-Grips can be adjusted to various jaw openings by turning a knurled adjusting screw in the end of the handle. Because their serrated teeth can chew up metal easily, they should not be used on chromed fittings and fasteners.

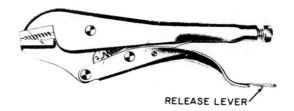

RELEASE LEVER

example A–11. Vise-Grip pliers: like a portable vise with a handle.

Wrenches

A wrench is used mainly to apply a twisting force to pipes and the bolts, nuts, and studs used to hold machined parts together. All these fasteners must be installed and removed with the right wrench, one that has the proper shape and size to do the job. No matter what the fastener, if your grip on it is unsure, the wrench will slip and damage the fastener, the work surface, or vulnerable parts of your body. To improve your efficiency, use the right wrench and so better your chances of completing a repair unscathed.

The special wrenches designed to do certain jobs are, in most cases, variations of the basic wrenches just described. The best wrenches are made of chrome vanadium steel, which is expensive but light and close to unbreakable. More common wrenches are made of forged tool or carbon steel or a molybdenum steel alloy. These materials make good wrenches that are generally a little heavier and bulkier to match the strength of chrome vanadium steel.

Wrench size is determined by the size of the opening between the jaws, which is slightly larger than the bolt or nut head designated on the tool. (Hex, or six-sided nuts and other types of nut and bolt heads are measured across opposite *flats,* as shown in example A–12.) We are all familiar with the result of using a wrench that's too large; the points of the fastener's head become round, making necessary a tool with teeth next time around.

Open-End Wrenches. Open-end wrenches (example A–13) are solid, nonadjustable tools with openings in one or both ends. Usually they come in sets of 6 to 10 wrenches, with sizes ranging from $\frac{5}{16}$ to 1 inch. Wrenches with small openings are usually shorter than those with large openings; this proportions the leverage of the wrench to the size of the fastener and helps prevent damage to the wrench and fastener.

In close quarters, it may be impossible to swing an ordinary large wrench because of its length. The *Bonney wrenches* shown in example A–14 are open-end wrenches for large fasteners and have the advantage of being thick, yet short.

example A–12. Wrench size, which is given as the distance across *flats*.

DISTANCE ACROSS FLATS

example A–13. Open-end wrench. Some have, unlike the one shown, only one opening for nuts.

OPEN–END WRENCH

example A–14. A Bonney wrench. It has a short handle, allowing for use in confined areas.

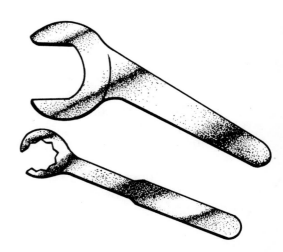

Open-end wrenches can have jaws parallel to the handle or turned away from the handle to any angle up to 90 degrees, the average being 15 degrees. These angular variations permit the selection of a wrench for places where there is room to turn a nut or bolt only partially; turning the wrench over after the first swing makes it fit the same flats so the fastener can be turned farther. After two swings of the wrench, the nut is turned far enough to present a new set of flats to the wrench.

Handles are usually straight but may be curved; S-wrenches have curved handles. Other open-end wrenches have offset handles to allow the head to reach nuts sunk below the surface of the work.

Box Wrenches. Box wrenches (example A–15) are always safer to use than open-end wrenches: They are less likely to slip off the fastener because they completely surround its head.

The most frequently used box wrench has 12 points or notches around the head and can be used with a minimum swing angle of 30 degrees. Six- and eight-point

POINT

example A–15. A 12-point box wrench. Because its end fully encloses a fastener, this wrench is safer to use than an open-end wrench.

wrenches are used for *heavy* duty, 12 for *medium,* and 16 for *light.*

One advantage of 12-point construction is the thin wall, which is suitable for turning nuts difficult to reach with open-end wrenches, and the wrench will operate between obstructions where space for the handle to swing is limited. A very short swing of the handle will turn the nut far enough to allow lifting the wrench so the next set of points can be fitted to the nut corners. The disadvantage of this wrench is the time lost when lifting and repositioning it on the nut when clearance is insufficient for spinning the wrench in a full circle.

Combination Wrenches. A loosened nut can be unscrewed much more quickly with an open-end wrench than with a box wrench. This is where a *combination box- and open-end wrench* (example A–16) comes in handy. You can use the box end for loosening nuts or for snugging them down and the open end for faster turning. The box portion of the wrench is often designed with an offset; in the illustration, the 15° offset allows clearance over nearby parts.

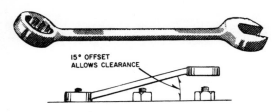

example A–16. A combination wrench with box and open ends, one to follow the other in tightening and loosening nuts.

The correct use of open-end and box-end wrenches can be summed up in one simple rule: be sure that the wrench fits the nut or bolt head properly. When you have to pull hard on the wrench, say, to loosen a tight nut, make sure the wrench is seated squarely on the nut's flats.

Pull on the wrench—*don't* push it; if you push a wrench, it can slip or the nut can break loose, while you skin your knuckles. If you have to push on a wrench, do it with the open palm of your hand.

Only actual practice will give you the feel for using the right amount of force on a wrench. The best way to tighten a nut is to turn it until the wrench has a firm, solid feel. This will tighten the nut properly without stripping its threads.

Socket Wrenches. The socket wrench is one of the most versatile tools you can have in your toolbox. The *Spintite wrench* (example A–17), also known as a *nut runner,* is a special type with a hollow shaft to accommodate a bolt protruding through a nut; this wrench has a hex head and is used like a screwdriver.

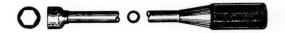

example A–17. Spintite wrench. As it is turned to tighten a nut, its hollow shaft swallows the protruding bolt.

A complete socket wrench set includes several different handles: an extension, an adapter, and a variety of sockets. The four types of handles used with sockets, each with its special advantages, are shown in example A–18. Experience will tell you which is best suited for the repair at hand. The square driving lug on a socket wrench handle has a spring-loaded ball that fits a recess in the socket receptacle. This mating ball recess feature keeps the socket on the drive lug during use. Yet, a slight pull on the socket removes it.

A typical socket (example A–19) has a square opening on one end that fits a square drive lug on a detachable handle. The other end has a 6- or 12-point opening very much like the opening of a box wrench; for square nuts there are 8-point sockets. Because a 12-point socket has to be swung only half as far as a 6-point socket before it has to be repositioned for a new grip, it can be used where there is less room to move the handle. (A ratchet handle keeps you from having to repeatedly lift the socket and refit it to the nut.)

Sockets are classified according to (1) drive size, which is the size of

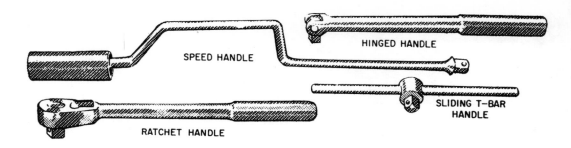

example A–18. Handles used with sockets.

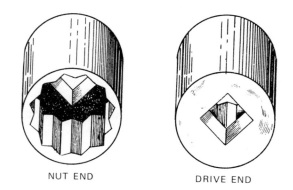

example A–19. A 12-point socket, which doesn't have to be repositioned on the nut as often as a 6-point socket.

NUT END   DRIVE END

the square opening for the drive lug of the handle, the most useful being $\frac{1}{2}$ and $\frac{1}{4}$ inch and (2) the size of the opening for the fastener, which is usually graduated in $\frac{1}{16}$-inch increments. Deep sockets are available to fit over long bolt ends.

A socket wrench is only as good as its handle. And you have a variety of these to choose from. The *ratchet handle* has a reversing lever that operates a pawl in the tool head. Pulling the handle in one direction makes the pawl engage the ratchet teeth, turning the socket. Moving the handle in the opposite direction makes the pawl slide over the teeth, permitting the handle to be backed up without moving the socket. This lets you turn the nut or bolt fast with many rapid, short turns of the handle. With the reversing lever in one position, the handle can be used for tightening, and, in the other position, for loosening.

A *hinged handle* is very convenient. To loosen tight nuts, swing the handle at a right angle to the socket; this gives the greatest possible leverage. After loosening the nut to the point where it turns easily, move the

handle into the vertical position and turn it with your fingers.

A *speed handle* works like a wood-working brace. After nuts are loosened with a sliding bar handle or a ratchet handle, a speed handle can be used to remove the nuts quickly. But in many instances a speed handle won't be strong enough to break loose or tighten nuts. A speed socket wrench should be used carefully to avoid damaging nut threads.

To complete a socket wrench set, there are several accessory items you can add (example A–20). Extension

bars (available in several lengths) extend the distance from the socket to the handle for greater torque. A universal joint allows turning nuts with the wrench handle at an angle; universal sockets are also available. The use of universal joints, bar extensions, and universal sockets with the appropriate handles makes it possible to form a variety of tools that will reach otherwise inaccessible nuts and bolts.

Another adapter allows you to use a handle having one drive size with a socket having a different drive size.

example A–20. Accessories, which can be combined with various sockets and handles to form a versatile variety of tools.

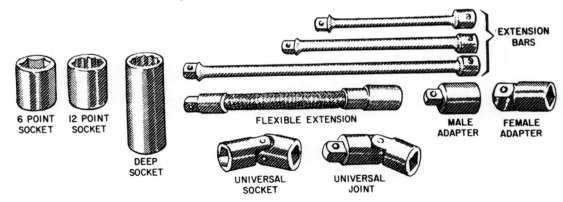

6 POINT SOCKET   12 POINT SOCKET   DEEP SOCKET   FLEXIBLE EXTENSION   EXTENSION BARS   MALE ADAPTER   FEMALE ADAPTER   UNIVERSAL SOCKET   UNIVERSAL JOINT

For example, a ⅜- by ¼-inch adapter makes it possible to turn any ¼-inch (square) drive socket with any ⅜-inch (square) drive handle.

Adjustable Wrenches. An adjustable open-end wrench (everybody calls them *Crescents,* but since that's a tradename we'll use the formal nomenclature) is a handy all-around tool that should be in your toolbox. This wrench (example A–21) is not intended to take the place of the regular open-end wrench and can't be used on extremely hard-to-turn fasteners. But it's hard to beat for odd-sized nuts.

Adjustable wrenches are available in lengths ranging from 4 to 24 inches, but if you need one that would be over a foot long, a *hex-nut wrench* would be a more practical alternative. The size of the wrench you select for a particular repair will depend on the size of nut or bolt head you need to turn; as the maximum jaw opening increases, so does the length of the wrench.

Adjustable wrenches are often called "knuckle-busters," and for good reason; mechanics frequently

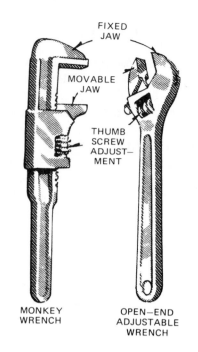

example A–21. Adjustable wrenches. The wrong size can be a "knuckle buster."

suffer such consequences as a result of using these tools improperly. To avoid accidents, choose the correct wrench size; that is, don't pick a 12-inch wrench and adjust its jaw for turning a ⅜-inch nut. Make sure the jaws are adjusted to fit the nut snugly. Position the wrench so the nut is completely within the jaws. Pull the handle toward the wrench side having the adjustable jaw (example A–22) to prevent the adjustable jaw from swinging open and slipping off the nut. If the location of the work won't allow you to take these safety measures, select another type of wrench.

Pipe Wrenches. Pipe wrenches are highly specialized in that they are intended for gripping smooth round metal stock (pipe) rather than the flats of fasteners. Handle extensions cannot be used with pipe wrenches because they will supply more torque than the wrenches are designed for—*something* would give!

For rotating or holding round work an adjustable pipe wrench, or *Stillson,* (example A–23) is the tool to have. Its movable jaw pivots to permit a gripping action on the work. However,

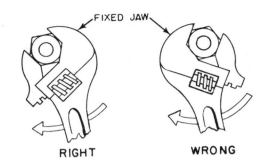

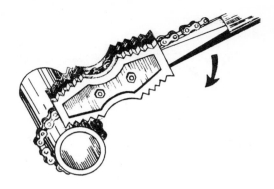

example A–22. To prevent an adjustable wrench from slipping off the nut, pull the handle toward the movable jaw.

example A–24. Chain pipe wrench, needed for dealing with large sections of pipe.

the serrated jaws will mark the work unless it is protected. The jaws should be adjusted so they bite on the work near its center.

Used mostly on large pipes, the *chain wrench* shown in example A–24 works in only one direction, but it can be backed partially around the work so another hold can be taken without freeing the chain. To reverse the operation, the grip is taken on the opposite side of the head, which is double-ended and can be reversed when the teeth on one end are worn.

The *strap wrench* (example A–25), similar to the chain wrench, uses a heavy web strap. This wrench is used for turning the pipe or tubing when you want to avoid marring the work surface. To use this wrench, place the strap around the pipe and pass it through the slot in the wrench body, then pull the strap tight. As you turn the wrench, the strap will tighten further, turning the pipe with it; if the strap still slips, rosin will help it grip.

Spanners. Some faucet packing nuts have notches in their outer edges and require a *hook spanner* (example A–26) for turning them. The wrench

example A–23. Adjustable pipe wrench. Although the ticket for handling round stock, it can chew up plating badly.

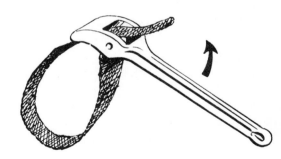

example A–25. A strap wrench which can be used safely on exposed, plated piping—it won't leave marks.

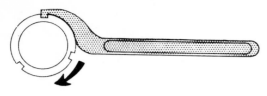

HOOK SPANNER

ADJUSTABLE HOOK SPANNER

example A–26. Hook spanners, used to turn nuts with notches in their outer edge.

has a curved arm with a lug or hook at the end that fits one of the notches. Spanners either fit one size or have a hinged arm for adjustment to a range of nut sizes.

Pin spanners (example A–27) have a pin instead of a hook; the pin fits a hole in the outer part of the nut. *Face pin spanners* have pins that fit into holes in the face of the nut.

Allens and Bristols. Setscrews and capscrews with recesses in their heads require Allen or Bristol wrenches for their installation or removal, depending on the form the recess takes. Allen wrenches (top, example A–28), made from hex L-shaped bars of tool steel, fit recesses with six flat sides; Bristols fit recesses that resemble a Maltese cross. Allen wrenches range in sizes up to $\frac{3}{4}$ inch. It is important that you use the correct size of Allen wrench and make sure the fastener's recesses are clear to prevent rounding or spreading the screw heads; a snug fit within the recess means you have the right size.

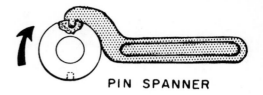

PIN SPANNER

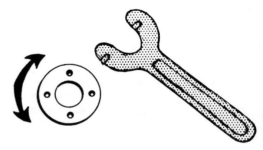

FACE PIN SPANNER

example A–27. Pin spanners, which engage holes in the edge or face of a nut.

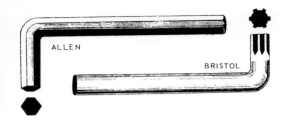

example A–28. Allen and Bristol wrenches. Fasteners with recesses in their heads require these wrenches for their removal.

Torque Wrenches. There are times when, for structural reasons, a definite force has to be applied to another bolt with a torque wrench. A common application for this tool in making plumbing repairs is torquing down bolts around a flanged connection. The *deflecting beam* torque wrench indicates the amount of torque on a scale mounted on the tool handle.

Here are some torque wrench tips to bear in mind:

- Make sure the wrench has been calibrated (zeroed) before you use it.

- Make sure the threads of the fastener you are going to tighten are clean; dirty bolt threads lead to inaccurate torque readings.

- Hold the torque at the desired value until the dial needle steadies.

- Never use a torque wrench to loosen fasteners; if you feel the need to retorque a fastener, loosen it with a conventional wrench, then apply a torque wrench.

Files

Your complement of tools should include a few files. You'll need at least one triangular file to sharpen saw teeth and a rough flat file for shaping up items like battered mallet heads. There are a number of different file types in common use, each ranging in length from 3 to 18 inches. Files are graded according to their fineness—the spacing and size of their teeth—and whether they have single- or double-cut teeth (example A–29).

Single-cut files have rows of parallel teeth set at an angle of about 65 degrees to the long axis of the tool. They are useful for sharpening tools and finish filing, but they really stand out as instruments for smoothing the edges of sheet metal.

Double-cut files have crisscrossed rows of teeth that form a raised diamond pattern for fast cutting. They are especially good for removing metal quickly and for rough work, like enlarging a pipe hole in sheet metal.

In addition to the three file grades shown in example A–30, there are *dead-smooth files,* with very fine teeth, and *rough files,* with very coarse teeth. The fineness or coarseness of file

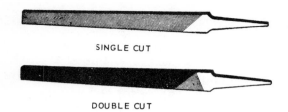

SINGLE CUT

DOUBLE CUT

example A–29. File cuts. All files have either single- or double-cut teeth.

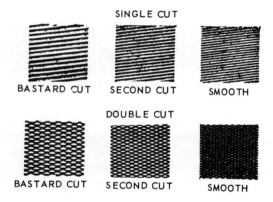

SINGLE CUT

BASTARD CUT   SECOND CUT   SMOOTH

DOUBLE CUT

BASTARD CUT   SECOND CUT   SMOOTH

example A–30. File grades. Files finer than smooth are *dead-smooth;* those coarser than bastard cut are *rough.*

teeth is also influenced by the length of the file, the distance from the tip of the heel (unscored portion), excluding the tang. If you compare the size of the teeth on a 6-inch single-cut smooth file with those on a 12-inch file of the same configuration, you will notice that the shorter file has more teeth per inch.

Even though several file shapes are available, *double-cut half-round files* are the most useful for deburring newly cut pipe; these files are tapered in width and thickness.

Hard files, somewhat thicker than flat files, taper slightly in thickness, but their edges are parallel. *Round files* serve to enlarge round openings. Small round files that taper to nearly a point are called *rattail files;* these are very handy for reaming small-diameter tubing when a standard reamer isn't available.

Probably the only filing techniques you'll need to employ—other than reaming—are crossfiling and the filing of round metal stock. *Crossfiling* involves moving the file across the surface of the work in a crosswise direction. For best results, spread your feet apart to steady yourself as you file with slow, full-length strokes. Because

a file cuts as you push it, ease up on the return stroke to keep from dulling the teeth.

As a file is passed over the surface of round work (example A–31), its angle with the work changes; the file produces a rocking motion that permits all the teeth to make contact and cut as they pass over the work's surface, thus tending to keep the file relatively clean.

As you use a file, its teeth may clog up with some of the filings and scratch your work, or they may lose some of their effectiveness. This condition, called *pinning,* can be prevented by keeping the teeth clean. Rubbing chalk between the teeth will help prevent pinning too, but the best method is to clean the file frequently with a file card (example A–32) and brush. To use a file card, brush with a pulling motion, holding the card parallel to the rows of teeth.

Never use a file for prying or pounding; the tang is soft and will bend easily, and the body is very brittle. Even a slight bend or a fall to the floor can snap a file in half. Don't strike a file against anything to clean it—if you don't have a file card, use an old brush for cleaning suede or an old

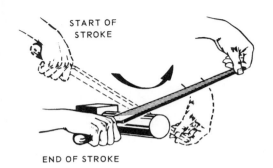

START OF STROKE

END OF STROKE

example A–31. Filing round stock. The rocking motion produced keeps even wear on the teeth.

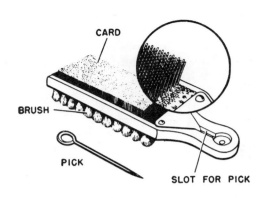

CARD

BRUSH

PICK

SLOT FOR PICK

example A–32. A file card, the best file-maintenance tool.

toothbrush (but don't wet it). If the tang becomes loose in a wood handle, tap the handle end on a hard surface.

Drills and Bits

The portable electric drill is more common these days than manual drills and braces. It is designed for drilling holes, but by adding accessories you can use it for such varied jobs as sanding, sawing, polishing, driving screws, and grinding. Electric drills are classified according to the largest-diameter bit shank that will fit their chucks. For heavy work, such as drilling masonry or steel, use a drill with a $\frac{3}{8}$- or $\frac{1}{2}$-inch chuck capacity (a $\frac{1}{4}$-inch drill is more suitable for drilling small holes in wood or sheet metal). If you buy just one drill, make it a $\frac{3}{8}$-inch size—this represents the best general compromise.

Just as a pair of roller skates is useless without a skate key, so is a drill without its chuck key—the tool used to tighten the chuck jaws against the drill bit. Most home-shop workers who have experienced the exasperation of losing a chuck key keep it taped to the drill's power cord about halfway between the drill and the plug.

The *carpenter's hand brace* (example A–33), viewed as outmoded by some, is just as useful for its specialty as is its most recent descendant, the rechargeable electric drill. In the long run, it is unquestionably the most economical tool for drilling holes for pipe runs through beams and planking.

Like an electric drill, the carpenter's brace has a chuck that takes bits—more precisely, *auger bits.* There are six parts to an auger bit's anatomy (example A–34), four of which do the work. The *screw,* the first part to touch the wood, centers the bit and pulls it into the wood; the *spurs* touch the wood next, scoring a circle; the *cutters* follow, doing the actual work; and the *twist* takes shavings out of the hole.

Auger bits are sized in $\frac{1}{16}$-inch increments, the number stamped on their tangs being the numerator in the fractional inch measurement across the holes they bore; for example, 10 indicates a bit for a hole $\frac{10}{16}$ ($\frac{5}{8}$) inch in diameter. However, this size range ends at 1 inch. An expansive bit (example A–35) is really the most practical to have for the holes needed in piping installations; of the two sizes

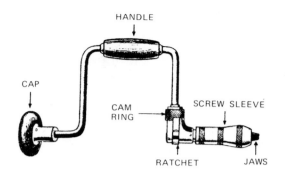

example A–33. Carpenter's brace.

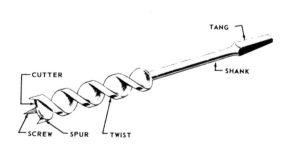

example A–34. The anatomy of an auger bit.

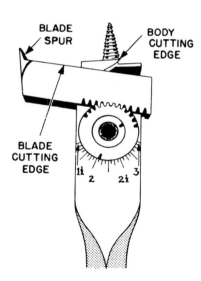

example A–35. An expansive bit, which lets you dial the hole size you need.

usually available, the largest has three cutters that bore holes up to 4 inches across.

There's one very nice thing about using a carpenter's brace: there's no chuck key to lose. To open the chuck of a carpenter's brace, merely grasp it and rotate the brace handle. Some brace chucks have a notched driving socket that must be mated to the corners of the bit's square tang.

To bore horizontal holes through wood, it is best to support the head (knob) against your body whenever possible. (No matter what direction the hole takes, always keep the bit perpendicular to the surface being drilled.)

To bore a hole in wood without splitting out the opposite surface, back the bit out of the hole as soon as the screw sees daylight; center the bit on the screw hole in the opposite surface and complete the hole by drilling in the opposite direction.

Levels

Any plumbing fixture that holds water has to be installed level, with its top perpendicular to a floor or its sides perpendicular to a wall. The device you'll need for this "squaring up" is a *spirit level,* a simple instrument that consists of at least one curved tube filled with liquid and mounted on a simple but square (true) frame. Almost every hardware store carries a few sizes, and some grocery stores and discount outlets have them planted among other tools in bargain bins for impulse buyers.

To check the trueness of any object with right-angled sides against the vertical or horizontal, place the level lengthwise on top of or against the side of the object: if the bubble centers in the curved tube, the surface is level. If it's a water closet you're concerned with, lay the level across its top. If the level is too short to straddle the bowl or tank, use a flat board, then place the level on the board.

A single reading may be misleading; the part may be level in only one axis. For accuracy, rotate the level into several positions across the bowl. If the bubble is displaced from the tube center at any point in the level's rotation, the object is off-level in the axis parallel with the level.

Plumber's Friends

Stoppages in fixtures, invariably caused by materials lodging in the drain, trap, or waste line (materials that shouldn't have been let in), can be dealt with by using mechanical devices, chemicals, or both, depending on the severity and location of the problem.

A *force cup,* commonly known as a *plumber's friend,* is the aid to try first for clearing clogged drains in lavatories, sinks, and tubs. The best-known version is a heavy rubber suction cup fitted to a simple wood handle. To use it properly, place the cup over the drain of the constricted fixture (which should be partially filled with water) and work it up and down in alternate suction and compression strokes. Either of the two forces should do the trick.

For clearing stoppages in water closets, there is a specially designed force cup shaped to fit that fixture's drain.

If the obstruction is in a trap and cannot be cleared by the plunging action of the force cup, clear the trap by inserting a *trap auger,* or *snake,* through the cleanout plug at the bottom of the trap. (Not every trap has a cleanout plug, of course; if none is available, start by inserting the auger through the regular drain opening. Then, if that doesn't do the job, remove the trap and push the auger into the pipe beyond that point.)

Trap augers can also be used to clear waste pipes. They are made of coiled tempered wire in various lengths and diameters and can easily follow bends.

The *closet auger* is a cane-shaped tube enclosing a coiled-spring snake and equipped with a handle. The coiled spring is retracted into the curve of the tube, which is then inserted into the water closet or trap. Turning the handle rotates the spring as it is pushed into the trap.

Always use drain-purging chemicals as a last resort. Not only are they inherently dangerous to work with, but if you use one first and it fails, what is left of it could be brought up later, possibly splattering into your face by the action of a force cup or a snake.

A Plumbing Workbench

The tools discussed thus far are usually toted about in a toolbox, being carried to piping and fixture sites where on-the-spot repairs are needed. But others, with which this section deals, are seldom transported for such visits: these are the tools used for specialized operations on the bench and those used so infrequently that they are best left protected in the workshop.

Vises

Pipe working is usually a two-handed operation. Vises provide the safest and most efficient way of holding work when it is being sawed, drilled, or shaped.

Bench and pipe vises (example A–36) have integral jaws for holding pipe $\frac{3}{4}$ to 3 inches in diameter. The maximum main-jaw opening is usually 5 inches; the jaw itself is 4 to 5 inches wide. The base can be swiveled to any position and locked. These vises, equipped with an anvil, are made to bolt onto a workbench.

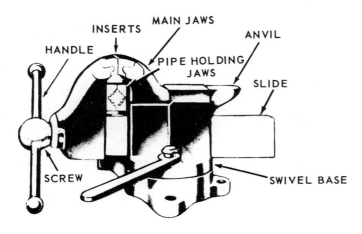

HANDLE

INSERTS

MAIN JAWS

ANVIL

PIPE HOLDING JAWS

SLIDE

SCREW

SWIVEL BASE

example A–36. The bench-and-pipe vise, the most versatile holding tool for working almost any shape of metal stock.

The bench-mounted pipe vise (example A–37), designed to hold round stock or pipe, has a capacity of 1 to 3 inches. The pipe is positioned under a hinged jaw, which is brought down and locked. Some pipe vises use a section of chain to hold the work. A small chain pipe vise can accept pipe diameters of from $\frac{1}{8}$ to $2\frac{1}{2}$ inches or so.

When holding heavy objects in a vise, placing a block of wood under the work as a prop will prevent it from sliding down and possibly falling on your foot.

Vises should be cleaned and wiped with light oil after they are used. The screws and slide should be oiled frequently, but the swivel base of a swivel-jaw joint should be kept dry; otherwise it will lose its holding power. When your vise is not in use, bring the jaws lightly together or leave a very small gap; the movable jaw of a tightly closed vise may break if the metal gets hot and expands. It's also a good idea to leave the handle positioned vertically, for obvious reasons.

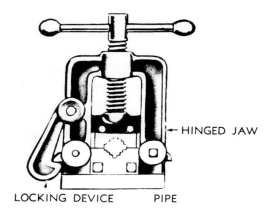

example A-37. The pipe vise, designed for one purpose: holding round metal stock.

Measuring Pipe

Before you cut a piece of piping for a replacement section you should have some idea of how pipe measurements are taken. Three methods of measuring threaded pipe are commonly used: end to end, end to center, and center to center (*center* is the central point of an associated fitting, as shown in example A–38). How much thread enters the fitting must be taken into account when determining the total length of the pipe without fittings. Fortunately, thread depth relative to pipe diameters is standardized: $\frac{1}{2}$- and $\frac{3}{4}$-inch iron and steel pipes have a thread length of $\frac{3}{4}$ inch; on 1- to $1\frac{1}{2}$-inch pipes, it's 1 inch; on 2-inch pipe, $1\frac{1}{8}$ inches; and on $2\frac{1}{2}$-inch pipe, $1\frac{1}{2}$ inches. With this knowledge, determining the length of the new pipe section is a matter of simple arithmetic: the installation measurement, minus the total fitting size, plus twice the thread length (for the given pipe diameter) equals the new section length (when using fittings with identical center-to-face measurement).

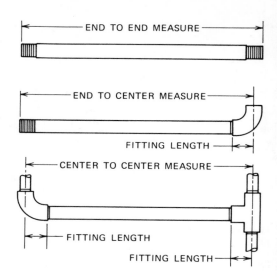

example A-38. Three methods of measuring sections of pipe for replacement. Variations can be derived from the ones shown.

example A–39. A telescoping gage, which can be used to find the right stock size for a pipe replacement easily.

Gages and Rules

Telescoping gages provide a quick, easy way of measuring the inside diameter of pipes and tubing. Available in sets of various size ranges, they consist of a handle with perpendicular arms at the end (example A–39) that telescope into each other and are held apart by an integral spring. Using a telescoping gage entails compressing the arms, inserting the gage into the pipe and allowing the arm to expand, then locking the gage using a nut in the handle end. The gage is withdrawn and the distance across the arms measured. You probably won't need a set that includes a gage smaller than that for $\frac{1}{2}$-inch pipe or tubing.

Of course, your needs may not justify the expense of a telescoping gage. Instead, a rule may work well enough, although it can't be counted on for extreme accuracy.

There is a right way to use a rule for measuring the inside diameter of a pipe. Butt one end of the rule slightly within one side of the pipe end and swing the other end in a narrow arc across the other side, taking the largest inch marking as the measurement (example A–40).

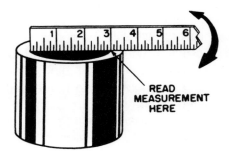

example A–40. Measuring the inside diameter of pipe with a rule.

Pipe-Working Tools

Pipe and Tubing Cutters. Pipe cutters (example A–41) are used to cut pipe of steel, brass, copper, or plastic. Tubing cutters (example A–42) are used for cutting thin-walled iron, steel, brass, copper, aluminum, or plastic tubing. These cutters are not interchangeable.

Two sizes of hand pipe cutters are generally used: the No. 1 pipe cutter, with a cutting capacity of $\frac{1}{8}$ to $1\frac{1}{4}$ inches, and the No. 2, with a capacity up to 2 inches. Pipe cutters have a cutting wheel and two pressure rollers that are adjusted and tightened by turning a handle.

Most tubing cutters look like pipe cutters, but they are lighter in construction. A screw-feed tubing cutter with $\frac{1}{8}$- to $1\frac{1}{4}$-inch capacity has two rollers with off-center cutouts so that cracked flares may be held and cut off without wasting tubing. It also has a retractable cutting blade that is adjusted by turning a knob. The tube cutter with a triangular portion shown in example A–42 is designed to cut tubing as large as an inch across the outside. Rotating the triangular portion within the conduit eliminates burrs.

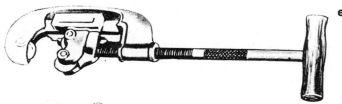

example A–41. A No. 1 pipe cutter, which will accept $\frac{1}{8}$-inch to $\frac{1}{4}$-inch pipe.

example A–42. Cutters are designed to be used on tubing, which has much thinner walls than pipe.

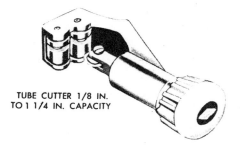

TUBE CUTTER 1/8 IN.
TO 1 1/4 IN. CAPACITY

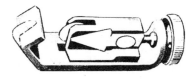

TUBE CUTTER 1/8 IN.
TO 3/4 IN. CAPACITY

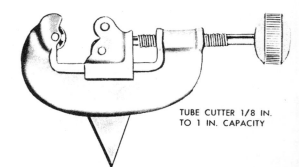

TUBE CUTTER 1/8 IN.
TO 1 IN. CAPACITY

Before cutting, make sure there are no nicks or burrs on the cutter wheel and make sure you have the right wheel for the material. If the cutter seems to be in good shape, open its jaws and position it exactly on the cutting marks, close its jaws, rotate the cutter to make sure it is seated, and tighten the handle an extra quarter turn to get a bite on the pipe. Rotate the cutter around the pipe in the direction shown in example A–43. The pipe has to be cut in a series of complete cutter revolutions, each beginning with the tightening of the handle by a quarter turn. The cutter should be kept perpendicular to the pipe throughout the operation. The new section is ready for installation only after the shoulder on the outside has been filed away and the burrs have been removed with a reamer.

Copper tubing is especially easy to cut. The operation (example A–44) is the same as that for ordinary pipe, except that a vise is unnecessary; just keep the cutting wheel as tight as possible against the tubing without flattening it.

Some tubing cutters have a backup roller (example A–45) to facilitate cutting off a flare at its base.

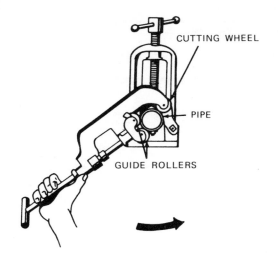

example A–43. A pipe cutter is turned toward the back of its frame.

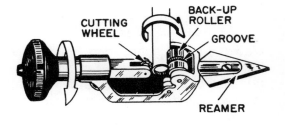

example A–45. Specialized tubing cutter, useful for removing flares.

example A–44. Rotating the cutter: keep a slight pressure against the cutting wheel with the screw adjustment.

The flare is placed in the backup roller's groove so the base of the flare contacts the cutting wheel. The method of operation is then the same as for cutting regular tubing.

Hacksaws. When you want to cut pipe that is too heavy, too hard, or too thin for pipe and tubing cutters, you have to use a hacksaw. Hacksaws have two parts: a frame and a blade. Adjustable frames (example A–46) can hold blades 8 to 16 inches long; solid frames take only one size of blade (the size being the distance between the two pins that hold the blade in position).

Hacksaw blades are made of tempered high-grade tool steel, and may be either *all-hard* or *flexible*. All-hard blades, as the name implies, are hardened throughout; on flexible blades, only the teeth are hardened. Hacksaw blades are about $\frac{1}{2}$ inch deep and have from 14 to 32 teeth per inch of length, the toothier versions being the most suitable for plumbing work.

Hacksaw teeth are pushed away from the blade in opposite directions, the amount being known as the *set*. There are four kinds of set: alternate, double alternate, raker, and wave; the

example A–46. Adjustable (top) and solid hacksaw frames.

latter is the most common. Example A–47 shows three types of set.

One of the most common mistakes with a hacksaw is using it with one hand, while the other holds the work. The material to be cut must be placed in a vise for utmost efficiency and safety. A minimum amount of overhang from the vise will reduce vibration, give a better cut, and lengthen the life of the blade. The blade should be installed in the hacksaw frame with its teeth pointing away from the handle (a hacksaw should cut on the *push* stroke) and tightened so it remains under constant tension; this is absolutely necessary for making cuts that are consistently straight, and for blade longevity. The blade should be selected according to the type of work to be done. Coarse blades (with relatively few teeth per inch) cut faster and are less liable to become clogged with chips than fine blades. But cutting hard and thin metals requires finer blades. A general rule is to select a blade so that each tooth starts its cut while the tooth ahead is still cutting.

The proper method of cutting with a hacksaw involves your hands as a sawing "team." Notice in example

A–48 how the forward pointing index finger helps guide the frame. Let your body sway forward and back with each stroke of the blade, applying pressure only on the forward stroke. Using long, steady strokes will keep your muscles from getting so sore that you need frequent breaks.

Hacksaws are great for removing frozen nuts. Start the blade close to the bolt threads parallel to one face of the nut (example A–49A). Saw parallel to the bolt, stopping when the teeth of the blade almost reach the lockwasher (lockwashers can be hard enough to ruin hacksaw blades). Example A–49B shows when to stop sawing. Remove one side of the nut completely by opening the saw cut (kerf) with a cold chisel and hammer. Put an adjustable wrench across the new flat and the one opposite, and again try to remove the nut. Since very little original metal remains on one side of the nut, it should either give or break through, permitting its removal.

The main danger in using a hacksaw is cutting the hand with a broken blade. When too much pressure is applied, when the saw is twisted, when the cutting speed is too fast, or when the blade loosens in the

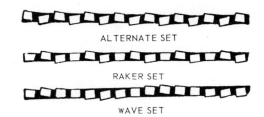

ALTERNATE SET

RAKER SET

WAVE SET

example A–47. Three hacksaw sets. The wave set is the most common.

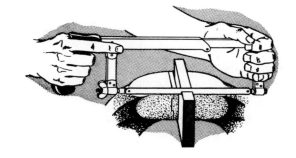

example A–48. A hacksaw worked by two hands as a team for maximum safety and efficiency.

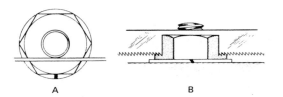

A   B

example A–49. To cut a frozen nut, start sawing near the bolt and parallel to a flat (A), and stop close to the lockwasher (B).

frame, the blade will dull quickly and very likely break. And if the work is not tight in the vise, it could slip, twisting the hacksaw's blade enough to break it.

Pipe Reamers. If the ends of a section of cut pipe aren't deburred, the flow of water through the pipe will be restricted. The tool for the job is a pipe reamer such as the one shown in example A–50. Pipe reaming, being a two-handed operation, also requires the use of a vise. Once you have secured the pipe in a vise, insert the reamer in the pipe and turn the handle clockwise in short, even strokes until the pipe is devoid of burrs. To remove the reamer, reduce pressure on it while rotating it in the same direction.

Tubing cutters often have reamers attached, like the unit shown in example A–51. As is necessary following the cutting of any water conduit, the burrs should be removed from tubing before it is used in a plumbing installation.

Dies and Taps. If you have needed a section of replacement pipe, you probably have made a trip to a

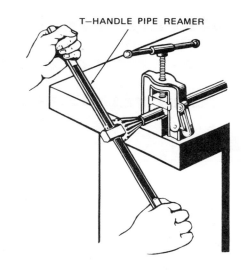

T–HANDLE PIPE REAMER

example A–50. Reamers are necessary for deburring pipe for maximum water flow.

example A–51. A reamer attached to a cutter.

hardware store or a plumbing supplier, only to find that threaded pipe is available only in specific lengths or that raw stock can be threaded only at extra cost. With a set of taps and dies to complement your cutters, you can be your own plumbing supplier.

Dies and taps are used to cut threads in metal, plastics, or hard rubber, dies being used for external threads and taps for internal threads.

Dies, made in several different shapes, are either solid or adjustable. The square version, for cutting external threads in pipe (example A–52), will cut only American standard pipe threads (ASPT). It comes in sizes for cutting threads on pipe from $\frac{1}{8}$ to 2 inches in diameter.

Two-piece rectangular pipe dies (example A–53) for cutting ASPT are held in ordinary or ratchet diestocks such as those pictured in example A–54. The jaws of the dies are adjusted with *setscrews*. An adjustable guide keeps the pipe aligned with the dies. The smooth jaws of the guide are adjusted by means of a cam plate; a thumbscrew locks the jaws firmly in the desired position.

Threading sets are available in many combinations of taps and dies,

example A–52. A square die for cutting external threads into pipe.

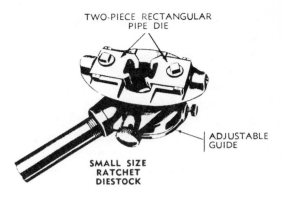

TWO-PIECE RECTANGULAR PIPE DIE

ADJUSTABLE GUIDE

SMALL SIZE RATCHET DIESTOCK

example A–54. A ratchet diestock with an adjustable guide for holding pipe dies.

example A–53. Two-piece rectangular pipe die set.

together with diestocks, guides, and tap wrenches. Example A–55 illustrates a typical threading set for pipe.

To cut external threads on galvanized and plastic pipe, first determine the nominal size. For 1-inch or larger pipe the nominal size is very close to the inside diameter. To measure these pipes, use a rule across the inside diameter, taking the nearest $\frac{1}{32}$-inch marking as the nominal size. For sizes below 1 inch, measure the outside diameter to the nearest $\frac{1}{32}$ inch and read the corresponding nominal size in example A–56. The chart in this figure can also be used for the measurement of smaller pipe across its outside diameter.

Adjustable pipe dies have a reference mark that is aligned with a similar mark on the diestock to give a standard-size thread. You adjust the dies one way or the other from the reference mark to cut a thread with the fit you want. Expanding the die will make it cut shallower threads, which will produce a tighter but weaker fit.

To begin cutting, pass the pipe through the guide in the diestock and into the tapered face of the die. Turn the diestock clockwise for right-hand

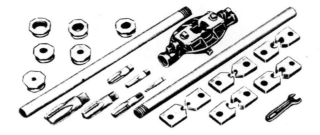

example A–55. Pipe threading set with rectangular adjustable dies, diestock, wrench, guides, and taps.

example A–56. Outside- to inside-diameter conversion chart for standard iron pipe.

SIZE OF PIPE, INCHES	DIMENSION A, INCHES	SIZE OF PIPE, INCHES	DIMENSION A, INCHES	SIZE OF PIPE, INCHES	DIMENSION A, INCHES
$\frac{1}{8}$	$\frac{1}{4}$	$1\frac{1}{2}$	$\frac{11}{16}$	5	$1\frac{1}{4}$
$\frac{1}{4}$	$\frac{3}{8}$	2	$\frac{3}{4}$	6	$1\frac{5}{16}$
$\frac{3}{8}$	$\frac{3}{8}$	$2\frac{1}{2}$	$\frac{15}{16}$	7	$1\frac{3}{8}$
$\frac{1}{2}$	$\frac{1}{2}$	3	1	8	$1\frac{7}{16}$
$\frac{3}{4}$	$\frac{9}{16}$	$3\frac{1}{2}$	$1\frac{1}{16}$	9	$1\frac{1}{2}$
1	$\frac{11}{16}$	4	$1\frac{1}{8}$	10	$1\frac{5}{8}$
$1\frac{1}{4}$	$\frac{11}{16}$	$4\frac{1}{2}$	$1\frac{3}{16}$	12	$1\frac{3}{4}$

threads; apply pressure only when starting. (After the die has taken hold, it will feed itself.) Continue turning, applying thread-cutting oil during the operation, until the end of the pipe has gone through the die flush with the near face. The thread produced should have the length called for in example A–57 to insure a tight fit. Notice the several *unseated* pipe threads in the fitting for the assembled joint shown; they permit further tightening in the event of a leak.

Freshly cut threads should be protected because their thin edges make them especially vulnerable to damage from impact. One good way to protect pipe threads is to install a cap or fitting once the threads have been cut; the fitting can be removed before the pipe is used.

Thread Restorers. Damaged threads can render a pipe section unusable. There are two tools available for restoring pipe threads. The die shown in example A–58, used conventionally in a diestock, removes metal. The restored thread that remains will not be perfect—the cross section of the restored thread will be flat rather than

example A–58. Die designed to restore pipe threads by removing metal.

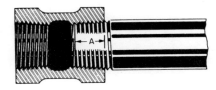

example A–57. Recommended pipe-thread lengths for various pipe sizes.

pointed—but for practical purposes it can be considered as strong as the original.

The thread restorer shown in example A-59 doesn't cut—it re-forms threads by forcing them back into their original shapes. Four sizes are necessary to cover thread diameters of $\frac{1}{4}$ to 6 inches. Each size will fit any pitch thread within its capacity without having to change blades or dies.

To use an adjustable restorer in right-hand threads, slip it over the threads with the directional arrow facing you and tighten the jaws snugly into the damaged threads close to the undamaged ones. Then turn the tool counterclockwise, as indicated by the arrow in example A-59.

Flaring Tools. These are "musts" if you plan to do much work with tubing. Flaring tools (example A-60) are used to flare soft copper, brass, or aluminum tubing. The end of the tubing is forced into a funnel shape so it can be held by a flare nut to a threaded fitting, as shown in example A-61. The tube flaring tool illustrated consists of die blocks that join to form holes for $\frac{3}{16}$- to $\frac{1}{2}$-inch outside diameter tubing (struck flaring tools are

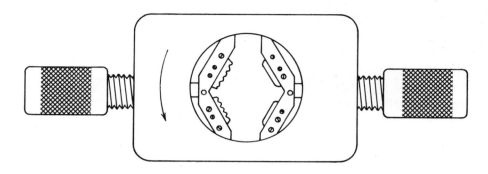

example A–59. Thread restorer, which, rather than cutting over threads, forces them into their original shape.

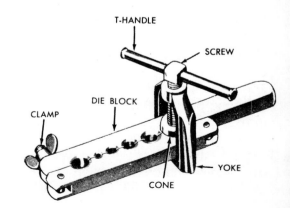

example A–60. A flaring tool used to make funnel-shaped ends on tubing so the tubing can be joined with a nut to a threaded fitting.

available for larger sizes), a clamp that locks the tubing in the die blocks, and a yoke that slips over the die blocks. Turning a compressor screw in the yoke moves a cone with 45° sides into the tubing, which is held in the die blocks, thus forming a flare.

Before you flare the end of a piece of tubing, make sure it has been cut off squarely and has no burrs. Slip a flare nut on a piece of tubing and open the die blocks at the appropriate hole (example A-62A). Insert the tubing into the die blocks so it protrudes slightly above them, then tighten the wing nut (example A-62B). The amount of tubing extending above the block determines the finished diameter of the flares; this is a matter of trial and error. The flare must be large enough to seat properly against the fitting, yet small enough to allow the flare nut threads to slide over it. Position the yoke over the tubing and tighten the handle, forcing the cone into the end of the tubing (example A-62C). The completed flare should be slightly visible above the face of the die blocks. For your first few attempts, use tubing that has not been cut to size to give you some margin of error.

FLARED COPPER TUBING

example A–61. Coupling using flared tubing and flare nuts.

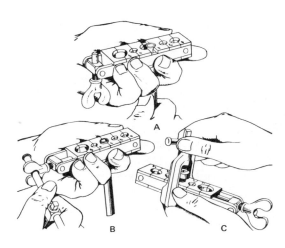

example A–62. A flaring tool, which entails (A) opening the right set of die blocks for the tubing size used, (B) tightening the die blocks, and (C) screwing the cone down into the end of the tubing.

Tubing Benders. You can put a gradual bend in annealed or untempered copper tubing by hand, possibly filling it first with sand to prevent its total collapse; but for right-angle bends, a tube bender is an absolute necessity. Available in sizes to match standard tubing stock, benders give a truly professional touch to plumbing installations. Not only do they insure geometric accuracy, but they leave tubing unmarred and free from pinches that restrict water flow. The only difficulties in using them are placing them correctly at the point of bend and controlling the bend angle.

There are two types of spring benders: the *external* type for unflared tubing and the *internal* type for flared tubing. The external type, shown in example A-63A, selected for the particular tubing size, is slipped over the tubing, grasped near the bending point with both hands, and bent to the desired angle as shown in example A-63B; it is pulled off by the bell-shaped end. The spring contains the tubing, preventing it from pinching at the bend. Internal spring benders are used similarly, but are inserted into the tubing; they look like external types without the flared ends.

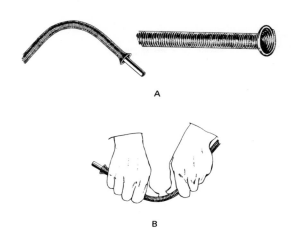

A

B

example A-63. Spring tubing benders, which, rather than offering any advantage in leverage, merely contain the tubing as it is bent (B) to keep it from crimping.

Hand tube benders consist of a handle, radius block or *mandrel,* clip, and slide bar. This type of bender is available in $\frac{3}{16}$-. $\frac{1}{4}$-, $\frac{5}{16}$-, $\frac{3}{8}$-, and $\frac{1}{2}$-inch sizes. For larger sizes of tubing, there are similar mandrel benders geared for greater leverage.

To use a hand bender, raise the slide bar and place the tubing on the radius block close to the slide bar. Set the slide bar zero mark (example A-64A) opposite the one on the radius block. After dropping the handle clip over the tubing (example A-64B), move the handle in the direction of the bend, stopping when the mark for the desired degree of bend coincides with the zero mark on the slide bar (example A-64C).

Blowtorches

Your most common use of the blowtorch will be in making sweated joints. Besides the flared joints made with copper tubing, you may also have to make sweated joints between sections of tubing or pipe. A sweated joint is usually made with solid (coreless) $\frac{50}{50}$ solder—half tin, half lead—in the soft soldering process. The process depends on molten metal

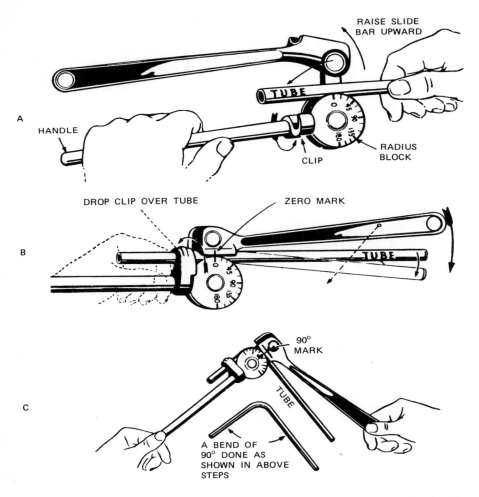

A

RAISE SLIDE
BAR UPWARD

TUBE

0 45 90 135 180

HANDLE

CLIP

RADIUS
BLOCK

B

DROP CLIP OVER TUBE

ZERO MARK

TUBE

0 45 90 135 180

C

90°
MARK

TUBE

A BEND OF
90° DONE AS
SHOWN IN ABOVE
STEPS

example A–64. Using a hand tubing bender (see text for descriptions of the steps).

flowing into the joint by capillary action, the same action that moves water from the soil through a plant.

The type of blowtorch you will need for a sweated joint will depend on how much metal is involved in the area of the joint. More metal means greater heat conduction away from the joint. For pipe 2 inches or less in diameter, a small hand propane, acetylene, or hotter-burning MAPP gas torch fitted with the proper tip will do. However, if the pipe is being joined to something rather massive, like a valve fitting, it must be compensated for by using a tip larger than that used for pipe alone. For larger pipe, larger equipment and bigger torch tips are needed.

Before you make a soft solder joint, be prepared for the possibility of fire; the temperatures involved are extremely hazardous. Expose at least an inch of new metal at the end of the tubing using steel wool or very fine sandpaper. Treat the inside of the fitting similarly. If it is a previously soldered joint, clean and reflux the parts. Next, spread a thin film of paste flux on the tubing end with a clean brush or other applicator. It is important that no oil contaminates the joint—don't use your fingers to spread the flux. Carefully insert the tubing into the fitting, making as close a fit as possible; no capillary action will result if the fit is loose. If the fit is loose, you'll have to tin the end of the tube by applying a little solder to it before it is inserted into the fitting. Adjust the torch flame so it is about two pipe diameters long and heat the fitting until it reaches the melting temperature of the solder; the heated fitting, not the flame, should melt the solder. If you are working with copper tubing, don't heat it after it begins to tarnish.

Feed the solder at the edge of the fitting. When a continuous ring of solder forms at the end of the fitting, the joint is complete; applying more solder beyond this point might cause excess solder to obstruct the pipe. If you are using a hand torch, keep it vertical; holding it horizontal will make liquefied gas flow to the valve, causing the flame to flare. Some of these units can be obtained with hosing connecting the tank and torch, permitting the tank to remain upright no matter what position you have to work in.

Sweated joints should be cleaned with a wire brush, with soap and water, or with an emery cloth. Any flux left on the joint will cause corrosion.

Never use acetylene without reducing the pressure through a suitable regulator. Acetylene pressure above 15 pounds per square inch (psi) must be avoided. Hose connections can be checked for tightness by coating them with a soap solution and looking for telltale bubbles that indicate a leak.

Calking Equipment

Calking is a process that uses molten lead to join sections of cast-iron soil pipe. The lead is melted over a melting furnace in a melting pot, spooned out with a plumber's ladle, and packed in against oakum with a calking iron. Pots and ladles for this job are usually made of cast iron (see example A-65), but some ladles are steel.

Calking irons (example A-66) are classified as either *inside or outside irons,* the former being used to pack lead close to the hub side of the joint, and the latter being used for packing lead close to the spigot side of the joint.

The mishandling of lead can be very dangerous. Therefore, safety precautions regarding its use should be explained. Most of the injuries that occur during a calking operation are caused by hot lead splashing out of the pot, not so much because of clumsiness but because of water somehow getting into the molten lead. When moisture contacts the lead, it produces steam that expands rapidly,

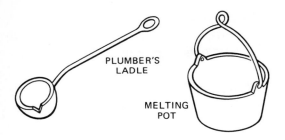

PLUMBER'S LADLE

MELTING POT

example A-65. Plumber's ladle and melting pot.

example A-66. Calking irons, used to pack lead into a soft-soldered piping joint.

OUTSIDE CALKING IRON

INSIDE CALKING IRON

making the lead splash. Even perspiration falling into the pot can be extremely dangerous. It is essential to wear goggles, gloves, and protective clothing when working with molten metal.

Melting Furnaces

Several types of melting furnace are available, the most common being kerosene or propane. Some tank-mounted propane units, however, are so top-heavy as to be dangerous. Whatever type you have (or decide to get), follow the manufacturer's recommendations regarding its care and use.

When lead is melted, products of oxidation, or *slag,* form on its surface. Slag must be removed from lead before it is used to make a joint; do this by scooping it out with the ladle. Because a cool ladle can solidify molten lead, ladles should be preheated before they are used. A ladle can be kept hot while in use by hanging it over the edge of the pot.

Before loading the ladle, use its bottom to push back the scum or *dross* on the surface of the molten lead, but otherwise don't disturb the lead any more than is necessary. Dross and lead oxide can be removed from the pot with a sharp chisel followed by a wire brush.

Poured-Lead Joints

Before making a poured-lead joint in cast-iron soil pipe, wipe the mating ends to remove any foreign material and inspect them for cracks. Be especially fastidious about removing moisture. It is possible to pour lead into a wet hub by first sprinkling powdered rosin or oil in the joint to reduce the chance of molten lead flying out of it, but it's safest to dry the hub completely.

You will need a pound of lead per inch of pipe diameter for each joint and an amount of oakum equal to 10% of the lead required.

Center one pipe end within the other—this is necessary for making a uniform joint—and make sure they are aligned. Using a *yarning iron* (example A-67), pack in oakum that is slightly tacky to within an inch of the top of the hub. Ladle lead into the joint, filling it even with the hub. Try to fill the joint with a single ladle full of lead. When heating one end of a tee or ell, plug

example A–67. The yarning iron, somewhat related to the calking iron, is needed for pushing oakum into a joint.

the other with a piece of tubing; this will prevent dry flux and oxidation from contaminating the piece being heated.

Horizontal Lead Joints

If the joint to be poured is horizontal, you will have have to use a joint runner (pouring rope), which will retain the lead within the hub. One type, made of asbestos, is fastened in place with a spring clamp as shown in example A-68. Another type, grooved on the upper side to direct the molten lead, is made of fiber and hard rubber and fits snugly around the pipe against the hub.

The procedure that requires a joint runner is practically the same as for a poured-lead joint. Pack the joint with oakum, then place the joint runner around the spigot end of the pipe and push it flush against the hub end of the other pipe, sealing the joint except for a small hole at the top. Tap the joint runner lightly against the hub of the joining bell to prevent lead from leaking out. Place a small piece of oakum between the clamp and pipe to seal the gap and prevent hot lead from running out of the joint. From this

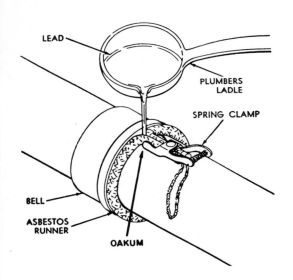

LEAD

PLUMBERS LADLE

SPRING CLAMP

BELL

ASBESTOS RUNNER

OAKUM

example A–68. Using a joint runner to retain molten lead against the hub of a joint.

state, follow the procedure given for a poured-lead joint.

Lead-Wool Joints

A lead-wool joint, a cold calked joint that doesn't require molten lead, is needed where a line is under water or in a wet place, where molten lead is unusable. The tools needed are a yarning iron and calking irons.

Connect the pipes, making sure they are centered and aligned, then block them with braces so they won't shift.

Pack oakum in the joint with a yarning iron within an inch of the top of the hub, using a ball-peen hammer to tamp the oakum tightly. Pack two $\frac{1}{2}$-inch layers of lead wool over the oakum, tamping both lightly into the joint with calking irons and a ball-peen hammer (example A-69).

example A–69. Tamping lead wool into a cold-calked joint.

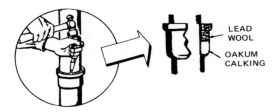

LEAD WOOL

OAKUM CALKING

Hanger Iron

The most popular support for piping from basement ceiling beams and other anchor points is hanger iron, sometimes called *pipe strap* or *plumber's tape.* Commonly available in packages of ¾-inch wide 10-foot lengths, the ''iron'' is actually a rolled strip of galvanized iron (typically 22-gage) perforated with holes to accept screws, bolts, or nails. Where local codes don't prohibit it, a strip of hanger iron can be slung around piping and nailed to the ends of either side of a beam, or one end can be nailed to the beam with the other wrapped around the pipe and bolted to itself; the former application does, of course, provide greater security.

Plumbing Software: Plans and Symbols

Some of the most useful plumbing might not be considered as tools at all by some. But as surely as a TV technician is aided by the schematic of a color set he's called upon to fix, so is the plumbing repairman helped by the schematic of a home's piping—the hot water lines, the vents, the cold water distribution system, and the fixtures and fittings. And, as in electronics, there are specialized symbols to represent the various elements of a complete plumbing system. First, you have to understand the language of plumbing plans.

Plumbing Plans

The drawing is the universal communications medium of engineers, technicians, and skilled craftsmen. But seldom are the originals—usually made on the translucent paper or cloth with ink or pencil—of this kind of document seen by those outside the drafting room. Rather, most people are familiar with reproductions of these drawings in the form of plans that consist of white lines on a blue background (the original blueprint), black or brown lines on a white background, and other forms, depending on the reproduction technique.

The Parts of a Plan

Besides the body of the plan—the drawing of an object or system layout—most plans include pertinent information located on the drawing's periphery. Generally, this gives the origin of the drawing, communicates what can't be said graphically, and contains notes helpful in the plan's interpretation.

Title Block. The title block, located in the lower right-hand corner of all plans prepared by architectural firms, contains a drawing number (including sheet numbers), the name of the object or system, the firm's name, the scale to which the drawing was done, and other special bits of data; sometimes there is a space with a diagonal line through it, indicating that the information "crossed out" is unnecessary or can be found elsewhere on the drawing.

Scale. In drawings for plans, very small parts are enlarged for clarity and large objects are reduced to fit the paper. So that the person using the drawing has an idea of the object's real size, a scale is given. It tells how much the object has been reduced or enlarged. The scale, indicated within one of the title block spaces, may be shown as 1 inch = 2 inches, 1 inch = 1 foot, and so on; it may also be indicated as full-size, half-size, or quarter-size.

Graphic scales are often found on plot plans. They indicate the number of feet represented by an inch, but are not accompanied by the words *inch* or *inches*. Sometimes a fraction is used; $\frac{1}{100}$ means that 1 unit on the plan is equal to 100 like units in real space.

There are several reasons why you should use the dimensions on a plan along with the scale, rather than taking measurements from the drawing: the plan may be a reduction of the original, you might not account for the drawing's scale, or the paper may have changed in size due to humidity.

Revision Block. Usually located in the upper right-hand corner of a plan, the revision block gives drawing updates, consecutively numbered with a description of the revision. A revised drawing can be identified by a letter given in addition to the drawing number. For example, the letter "B" would indicate a second revision.

HOT AND COLD WATER
CONNECTIONS 1" I.P.S.

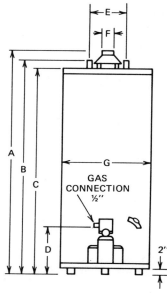

GAS
CONNECTION
½"

2"

MODELS: GL50–75–5
GL75–75
GL100–75

HOT AND COLD WATER
CONNECTIONS 1" I.P.S.

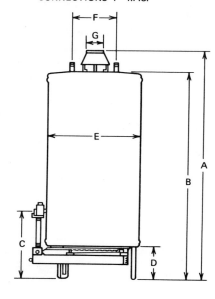

MODELS:
GL50–96–5
GL67–120–5

example A–70. Tagging the dimensions of the water heater in the drawing allows the manufacturer to use it to describe a number of models with different capacities. (Courtesy Ruud Manufacturing Company.)

DIMENSIONS							
MODEL NO.	A	B	C	D	E	F	G
GL50–96–5	$60\frac{11}{16}$	56	$19\frac{13}{16}$	$8\frac{3}{4}$	$24\frac{1}{4}$	13	5
GL67–120–5	$64\frac{11}{16}$	60	$19\frac{13}{16}$	$8\frac{3}{4}$	$26\frac{1}{4}$	13	5
GL50–75–5	$63\frac{7}{16}$	$60\frac{13}{16}$	$59\frac{15}{16}$	14	11	5	$20\frac{1}{4}$
GL75–75	$63\frac{3}{8}$	61	$59\frac{7}{8}$	14	11	5	$24\frac{1}{4}$
GL100–75	$69\frac{1}{8}$	$66\frac{3}{8}$	$65\frac{5}{8}$	14	11	5	$26\frac{1}{4}$

Notes and Specifications. Although plans contain all the graphic information pertaining to an object or system, not all details can be stated in terms of lines and symbols, details needed by contractors, manufacturers, and service workers.

Notes on a drawing give additional clarification of the subject; leader lines indicate the particular part being described. Specifications describe or enumerate particulars or details of an object not shown. Notice how certain dimensions of the water heater shown in example A-70, a simple specification sheet, have been tagged with letters so the same drawing can be used to describe more than one model.

Legends and Symbology. Legends are used to explain or define special symbols or marks used on a plan. Generally, the legend is placed in the upper right-hand corner of the plan.

Interpreting the Lines. If you study a plan for any time, you will notice that many different kinds of lines are used: some heavy, some light, some broken in different ways, some ending in arrowheads. Each has a special application in a plan. For example, heavy unbroken lines are always used to define the visible edges of an object. The variety of lines available to the draftsman are shown in example A-71; an example of their application can be seen in example A-72.

LINE STANDARDS			
NAME	CONVENTION	DESCRIPTION AND APPLICATION	EXAMPLE
VISIBLE LINES		HEAVY UNBROKEN LINES USED TO INDICATE VISIBLE EDGES OF AN OBJECT	
HIDDEN LINES		MEDIUM LINES WITH SHORT EVENLY SPACED DASHES USED TO INDICATE CONCEALED EDGES	
CENTER LINES		THIN LINES MADE UP OF LONG AND SHORT DASHES ALTERNATELY SPACED AND CONSISTENT IN LENGTH USED TO INDICATE SYMMETRY ABOUT AN AXIS AND LOCATION OF CENTERS	
DIMENSION LINES		THIN LINES TERMINATED WITH ARROW HEADS AT EACH END USED TO INDICATE DISTANCE MEASURED	
EXTENSION LINES		THIN UNBROKEN LINES USED TO INDICATE EXTENT OF DIMENSIONS	

example A–71. The draftsman's vocabulary of lines. Each has its own application in a drawing.

NAME	CONVENTION	DESCRIPTION AND APPLICATION	EXAMPLE
LEADER		THIN LINE TERMINATED WITH ARROW-HEAD OR DOT AT ONE END USED TO INDICATE A PART, DIMENSION OR OTHER REFERENCE	¼ X 20 THD.
PHANTOM OR DATUM LINE		MEDIUM SERIES OF ONE LONG DASH AND TWO SHORT DASHES EVENLY SPACED ENDING WITH LONG DASH USED TO INDICATE ALTERNATE POSITION OF PARTS, REPEATED DETAIL OR TO INDICATE A DATUM PLANE	
STITCH LINE		MEDIUM LINE OF SHORT DASHES EVENLY SPACED AND LABELED USED TO INDICATE STITCHING OR SEWING	STITCH
BREAK (LONG)		THIN SOLID RULED LINES WITH FREEHAND ZIG-ZAGS USED TO REDUCE SIZE OF DRAWING REQUIRED TO DELINEATE OBJECT AND REDUCE DETAIL	
BREAK (SHORT)		THICK SOLID FREE HAND LINES USED TO INDICATE A SHORT BREAK	
CUTTING OR VIEWING PLANE VIEWING PLANE OPTIONAL		THICK SOLID LINES WITH ARROWHEAD TO INDICATE DIRECTION IN WHICH SECTION OR PLANE IS VIEWED OR TAKEN	
CUTTING PLANE FOR COMPLEX OR OFFSET VIEWS		THICK SHORT DASHES USED TO SHOW OFFSET WITH ARROW-HEADS TO SHOW DIRECTION VIEWED	

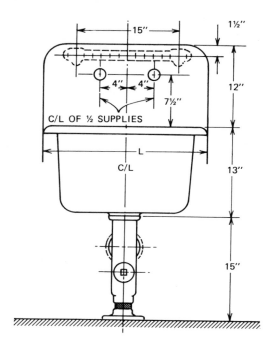

example A–72. Sink measurements. Six of the lines in the draftsman's illustrative lexicon are used: visible, hidden, center, dimension, extension, and leader lines.

Reading Views

One of the great difficulties in reading a plan is being able to visualize an object when only a few views of it are shown. Of course, any method of representation can only encompass a few views—a flat, two-dimensional drawing can't "walk you around" the object in every direction. In interpreting a *multiview drawing,* you can get a general idea of the object's shape by looking at all views, then pick one for more careful study. By referring to an adjacent view, you can determine what each line represents. Given only the top view in example A-73, the concentric circles could represent anything from an end view of a tube to two holes, one inside the other; the front view (below) removes all ambiguity from the top view. When three views are needed to describe an object, all must be consulted for an idea of the correct shape.

Usually no more than three views of an object are furnished. A three-view *orthographic projection* drawing generally shows the object's front (selected as front because it shows the most characteristic feature—in example A-74, the notch),

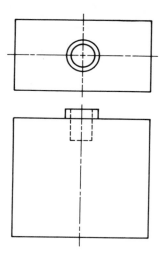

example A–73. Two views of an object. The bottom view explains the concentric circles in the top view.

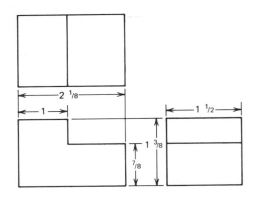

example A–74. An orthographic drawing. The top and right views are extensions of the front view, which shows the object's most distinguishing feature, the notch.

its top, and a side or its bottom, the last two views being derived from the first. After studying each view of an object, you should be able to visualize it in the manner illustrated in example A–75.

Phantom Views. Phantom views indicate an alternate position for a part (example A–76), repeated detail, or the relative position of an absent part.

Sectional Views. Sectional views show interior or hidden features of an object that cannot be seen clearly in conventional exterior views. These features of an object are shown by drawing the object as if a part had been cut away. Included in these views are *cutting-plane* lines that show how the "cut" is made; for example, the line terminating in bent arrows in example A-77A; example A-77B shows the cutting plane (the arrows show the viewing direction). If the object were actually cut in half, as indicated by the cutting-plane line, it would look like the drawing in example A-77C; when sectional views are drawn, the part cut is shaded with diagonal *section* lines (example

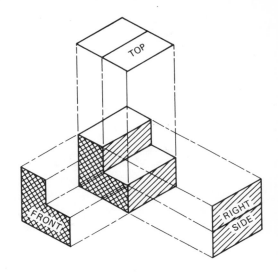

example A–75. Visualization of the object depicted in Fig. A–74.

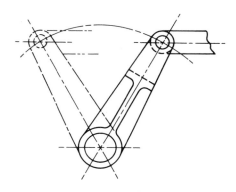

example A–76. Phantom view (broken lines) showing an alternate position of a part.

example A–77. The cutting plane line, terminating in bent arrows (A), shows the placement of the cutting plane (B); if the object were actually cut, it would look like C, but view D would be the most probable representation of it.

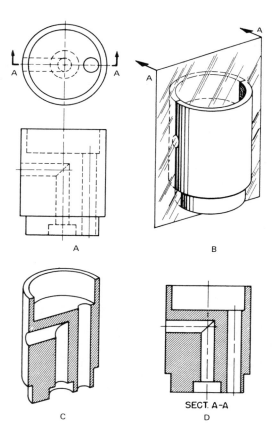

A

B

C

SECT. A-A

D

A-77D). When more than one part is shown in the same view, each is sectioned or *crosshatched* with lines in different patterns, as seen in example A-78.

Offset Sectional Views. There are instances when a sectional view with a single, straight cutting plane just doesn't show enough—when it would miss showing the interior of a special feature, for example. In cases like this, a broken plane is needed. The broken offset cutting plane in example A-79 makes it possible to show the second hole in the right-hand part of the object (again, the arrows show the viewing direction).

Exploded Views. An exploded view is probably the easiest kind of drawing to understand. It shows the relative location of parts—how they align and fit together—and is particulary useful as a guide to assembling a complex object, such as a pump. The parts, drawn pictorially, are usually aligned in the order of assembly around a broken line. If you are having trouble putting together a manufacturer's product, ask if an exploded view is available.

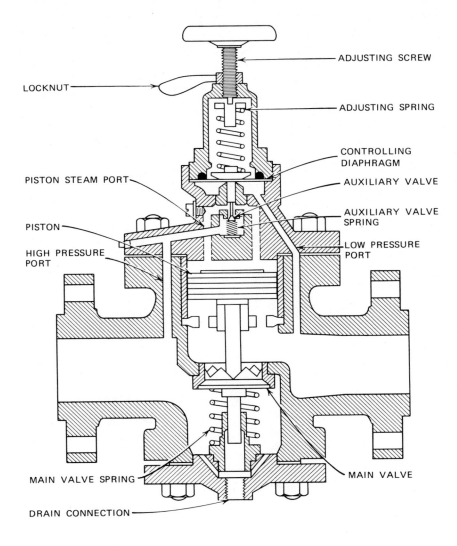

ADJUSTING SCREW

LOCKNUT

ADJUSTING SPRING

CONTROLLING DIAPHRAGM

PISTON STEAM PORT

AUXILIARY VALVE

AUXILIARY VALVE SPRING

PISTON

LOW PRESSURE PORT

HIGH PRESSURE PORT

MAIN VALVE SPRING

MAIN VALVE

DRAIN CONNECTION

example A–78. Cross section of valve. Because so many parts of this valve had to be shown in cross section, two dot patterns besides crosshatching in two slants were used to differentiate them.

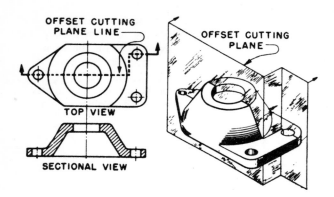

OFFSET CUTTING PLANE LINE

OFFSET CUTTING PLANE

TOP VIEW

SECTIONAL VIEW

example A–79. An offset (zigzag) cutting plane line is used when a straight line would miss showing the cross section of an important detail.

Plumbing Symbols

What makes plumbing diagrams plumbing diagrams are the symbols used; otherwise, they are quite similar to any other plan. Fortunately, most of the symbols resemble closely the things they represent; but they are usually accompanied by graphic shorthand that may be difficult to interpret without some background information.

Symbols Showing Joints

Joints between pipes and fittings may involve ends welded together, hub-and-spigot unions, ends screwed together, flanges mated by bolts, and other connections; for each union there is a symbol (example A-80). On piping diagrams where symbols for fittings are absent, two intersecting (connected) pipe runs are represented by the symbols in example A-81A; where piping runs cross over each other, the symbols in example A-81B are used. An example of such a plan is example A-82, an *isometric schematic* of a pictorial illustration in Section Four. To interpret this diagram, start at the sink indicated by

example A–80. Types of pipe-fitting connections shown symbolically.

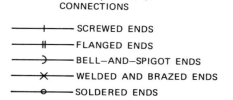

PIPE FITTINGS, TYPES OF CONNECTIONS

SCREWED ENDS

FLANGED ENDS

BELL–AND–SPIGOT ENDS

WELDED AND BRAZED ENDS

SOLDERED ENDS

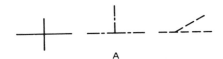

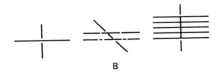

A

B

example A–81. Pipeline intersections (A) and crossovers (B).

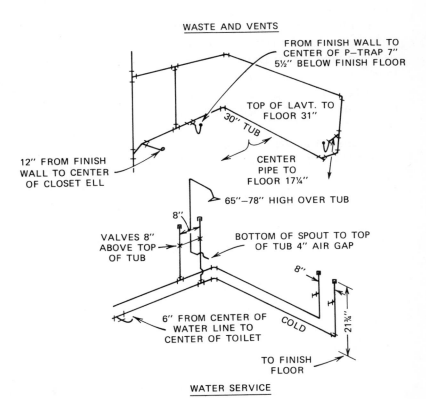

WASTE AND VENTS

FROM FINISH WALL TO CENTER OF P–TRAP 7″ 5½″ BELOW FINISH FLOOR

TOP OF LAVT. TO FLOOR 31″

30″ TUB

CENTER PIPE TO FLOOR 17¼″

12″ FROM FINISH WALL TO CENTER OF CLOSET ELL

65″–78″ HIGH OVER TUB

8″

VALVES 8″ ABOVE TOP OF TUB

BOTTOM OF SPOUT TO TOP OF TUB 4″ AIR GAP

8″

COLD

6″ FROM CENTER OF WATER LINE TO CENTER OF TOILET

21¼″

TO FINISH FLOOR

WATER SERVICE

example A–82. Isometric diagram of the bathroom shown in Fig. 4–22.

the P-trap symbol in the upper part of the diagram labeled WASTE & VENTS. The portion of the tee section leading upward is the vent section, and the portion leading downward is the drain portion of the plumbing. Follow the drain pipe along the wall and you will see that it reaches the corner where a 90-degree elbow brings the drain around the corner. Another section of pipe is connected between the elbow and the next tee, a branch of which leads to the P-trap below the bathtub. The other branch goes to the tee necessary for the vent (leading upward between the tub and toilet), and continues on to the wye bend with a heel that leads to a 4-inch main house drain (the vent pipe runs parallel to the floor drain, slightly above the sink).

Symbols for Fittings

The schematic symbols for fittings are given in example A-83. In addition, cleanouts are represented by a small circle accompanied by the abbreviation CO; a small square labeled FD shows the location of a floor drain. At this point, you should be able to recognize immediately that the piping and fittings in example A-84 are screwed together. But this illustration introduces a few new symbols, most notably those representing valves. Not specified in the drawing is how these valves are connected to the system; the symbols for the valves commonly used in plumbing installations are given in example A-85, along with the shorthand denoting how they are connected to piping.

example A–83. Pipe-fitting symbols.

ELBOWS	
FITTING	SYMBOL
ELBOW, 90 DEGREES	
ELBOW, 45 DEGREES	
ELBOW, OTHER THAN 90 OR 45 DEGREES, SPECIFY ANGLE	30°
ELBOW, LONG RADIUS	LR
ELBOW, REDUCING	
ELBOW, SIDE OUTLET, OUTLET DOWN	
ELBOW, SIDE OUTLET, OUTLET UP	
ELBOW, TURNED DOWN	
ELBOW, TURNED UP	
ELBOW, UNION	

TEES	
FITTING	SYMBOL
TEE	
TEE, DOUBLE SWEEP	
TEE, OUTLET DOWN	
TEE, OUTLET UP	
TEE, SINGLE SWEEP, OR PLAIN T–Y	

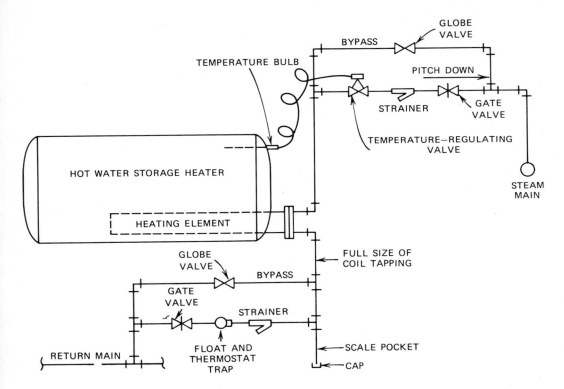

example A–84. Piping for a hot-water storage heater.

GLOBE VALVE

BYPASS

TEMPERATURE BULB

PITCH DOWN

STRAINER

GATE VALVE

HOT WATER STORAGE HEATER

TEMPERATURE–REGULATING VALVE

STEAM MAIN

HEATING ELEMENT

GLOBE VALVE

FULL SIZE OF COIL TAPPING

BYPASS

GATE VALVE

STRAINER

RETURN MAIN

FLOAT AND THERMOSTAT TRAP

SCALE POCKET

CAP

example A–85. Valve symbols.

GATE VALVE	⊢▷◁⊣	—▷◁—	✕▷◁✕	✳▷◁✳	—◦▷◁◦—
GLOBE VALVE	⊢▷◁⊣	—▷◁—	✕▷◁	✳▷◁✳	—◦▷◁◦—
SAFETY VALVE	⊢▷✕◁⊣	—▷✕◁—	✕▷✕◁	✳▷✕◁✳	—◦▷✕◁◦—

Fixtures and Traps

Fixture symbols need hardly any explanation because they look so much like profiles or top views of the fixtures they represent. If you imagine that you are looking down on the bathroom in a floor plan, you could easily identify the fixtures without needing labels. Example A-86 gives some of the symbols a draftsman might select from his or her graphic repertoire in drawing a floor plan.

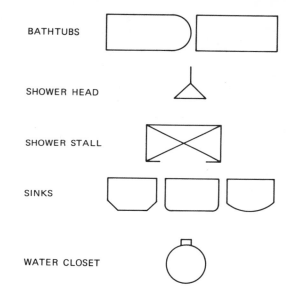

BATHTUBS

SHOWER HEAD

SHOWER STALL

SINKS

WATER CLOSET

example A–86. A few of the many fixture and trap symbols available to the draftsman.

Index